PRINCIPLES OF COMMAND AND CONTROL

Edited by
Jon L. Boyes and Stephen J. Andriole
Foreword by Russell E. Dougherty

AFCEA/SIGNAL MAGAZINE C^{3}I SERIES
AFCEA International Press

Washington, D.C.

- Brussels
- Seoul
- Manila
- Ottawa

Library of Congress Cataloging-in-Publication Data

Principles of command and control.

(AFCEA/SIGNAL magazine C^3I series; v.6)
Includes bibliographies.
1. Command and control systems. 2. Command and control systems—United States. I. Boyes, Jon L., 1921– . II. Andriole, Stephen J. III. Series.
UB212.P75 1987 355.6 87-3574
ISBN 0-916159-12-4

Contents

Foreword

In 1980, the Armed Forces Communications and Electronics Association (AFCEA) generously awarded me its David Sarnoff Award for noteworthy achievement in the area of military communications—all due to my good fortune in having been the Commander in Chief of Strategic Air Command, and its thousands of talented communications/electronics personnel, throughout the mid-1970s. On the night of the black-tie award dinner, I secreted a vintage, military bugle underneath the speaker's podium. When Dr. John McLucas, then AFCEA's president, read the citation and presented the award I held off accepting it, produced the bugle and blew the military calls for "Attention" and "Officers' Call." Then I said to the AFCEA audience, "Now that I have established my own credentials as a military communicator, I proudly accept this award on behalf of all the communicators serving with SAC."

Obviously, this little bugle vignette smacks of schmaltz, but I had a serious purpose with this demonstration of my duty as a military person in 1935, as the bugler for Kentucky's National Guard Cavalry Regiment. I wanted to illustrate that, right up to *my own military lifetime,* as it had in all of the 2,000, or so, lifetimes of man preceding mine, some form of direct dissemination of disciplined sounds (animal horn, conch shell, jungle drums, etc.) had been used as the system of issuing military commands and exerting tactical control.

In just my own lifetime, however, all of this has changed. Nothing short of an explosion in scientific discoveries and technological achievements has changed dramatically the capacity and reach of our command and control (C^2) systems; but, with equal drama, these advances also have expanded the complexity and scope of the military situations that our C^2 systems must address.

This vital, constantly evolving C^2 milieu is what this book is all about. The editors have assembled an expert group of contributors to illuminate the capabilities, constraints and challenges from diverse, but important, perspectives. While change seems to be the constant in all of these contributing developments and technologies, the fundamental tasks to be served by our C^2 systems remain constant. As I see it, they must serve *the process of making, disseminating and implementing informed command decisions in order to obtain optimum effectiveness of the nation's military forces and resources in peacetime, crisis, hostile actions and war.* The success of our C^2 systems is measured by the degree to which they support this process reliably and effectively—at all levels and under all circumstances.

When AFCEA began its book series on command, control, communications and intelligence (C^3I), it intended to provide a focal point for research and offer a collection of responsible thought and expert opinion on an increasingly complex and fastpaced facet of military activity. Previous volumes from the AFCEA/*SIGNAL* C^3I Series have provided the building blocks and assessed the issues from such viewpoints as technological opportunities, operational requirements and policy constraints.

With this new volume, *Principles of Command and Control,* the series offers to both military and industrial readers a chance to look at the subject from strategic, tactical and philosophical points of view. It offers a more comprehensive framework and a longer view with valuable insights from experts who have labored to keep C^2 capabilities more closely tuned to evolving requirements. This book traces the origins of C^2 from its use in ancient times through the present day and into the next century; it discusses systems design procedures, program management and policy; it tackles tactical, strategic, NATO and Warsaw Pact C^2 strategies. It is not always positive and complimentary; some sections are critical and harsh in faulting methods, systems and results. This, too, is important for the subject demands our best and most innovative efforts; the mere presence of modern electronics does not guarantee the success of our C^2 systems.

The true state of our C^2 capability is a matter of increasing concern because many see improved command and control as the primary means to redress the significant shift in the balance of conventional and nuclear force levels between the United States and the Soviet Union. U.S. forces now must prepare to confront military formations of imposing numerical superiority and growing technical equivalency. To do so, we have adopted policies and strategies—now embedded in doctrine and force structure—that put a premium on precise control of our forces through timely acquisition and exploitation of information. This much is readily acknowledged in the literature. What are not so clearly recognized are the inordinate difficulties continually encountered in acquiring sensor, decision aids and information systems that can underwrite such demanding strategies in the face of determined enemy countermeasures. These critical paths are dealt with in this book. The reader will learn why it is so difficult to transfer state-of-the-art technology into useful operational systems.

We are justifiably proud of our scientific base and technological achievements; a degree of rhetoric seems a necessary part of the democratic process for solving problems. This can be harmless so long as those responsible for national security understand what our C^2 systems can and cannot do. We go to extraordinary lengths to discover the strengths, weaknesses and vulnerabilities of hostile C^2 capabilities and we ought to be no less rigorous and forthright in self-assessment—for the same reasons. We think we see in Soviet actions a recognition of the role played by C^2 systems and those actions signify a full appreciation of both strengths and weaknesses, as manifested in robust systems and rigorous, realistic evaluation in their exercises. Can we do otherwise and still be credible? This book will provide the reader with invaluable insight into what questions to ask and what answers to accept.

There are many facets to military C_2, politicians, theoreticians, bureaucrats and engineers necessarily will emphasize certain points over others. As a former commander of a U.S. combatant command, these are the major points that I would hope the reader would consider carefully in reading this book:

- At the center of each C^2 system is the commander; bearing enormous responsibility, uniquely accountable for the the mission, desperately in need of support from modern electronic technology, and all but bypassed by the ongoing information revolution. The commander seems to benefit least from all the technology brought to bear on C^2. Input and data that support his staff specialists are important but, by themselves, insufficient. Improvements to the other elements of the system—sensors, information, communications, data processing, etc.—can be of little utility if we fail to satisfy the basic and urgent needs of the commander. We must first view the process from his perspective and help him satisfy his primary needs without inundating him with secondary data.
- The systems bought to support the C^2 function of a command seldom fulfill the need without extensive tailoring and adaptation. The authors of this book will give you their expert views on the many causes for failures and false starts in the acquisition process of C^2 systems. I would highlight two particularly troublesome areas: the establishing of requirements and the procedures used in testing. The process by which DOD buys practically all material more or less assumes that the customer knows exactly what is required and that the product can be tested to insure that he gets exactly what he wants. Both of these seemingly simple assumptions have proved to be major obstacles for C^2. Innovative ways must be found to decouple our C^2 systems from the classic requirements process and to develop better ways to test such systems for effectiveness without a pro forma comparison to some prematurely contrived standard. The only valid standard is that of insuring that we can accomplish what *needs* to be done and avoiding things that are counterproductive—but it is very difficult to know which is which at the outset of the process. Fortunately, AFCEA has played a leading role in defining changes to the acquisition process to accommodate the special needs of C^2, but implementing these desirable changes is proving to be difficult indeed.
- How can military C^2 best capitalize on the astonishing advances we are experiencing in civil information systems? Our exploding information based economy will invest the most brilliant talent and literally billions of dollars in systems to solve civil management problems that differ only marginally from many of those of military command. Achieving a competitive edge or military tactical superiority are similar objectives and both depend largely on the ability to turn information into accurate decisions and effective actions more quickly than the opposition. However, we need to understand better how to meld commercial achievements into useful decision systems for military purposes. This, too, will not be easy, for commercial markets take a shorter and less circumspect view of serviceable life and maintainability, and owners' manuals cannot

supplant military technical data. Also, the competitive marketplace favors uniqueness in feature and design, and things bought separately for military purposes usually will not work collectively; joint warfare cannot be fought effectively with information systems that do not interoperate. Notwithstanding the difficulties, we need to recognize the opportunities and stay abreast of the astounding progress in the civil sector.

In volume two of this AFCEA series*, while serving as Commander in Chief of Strategic Air Command, I wrote that ". . . the stability of force balance and the reflected deterrent strength of our force capability are dependent on the C^3 systems we acquire for the command and control of our weapons systems." Now, more than a decade later, I find that comment still valid and ever more current. Credible deterrence and effective systems for command and control are inseparable.

General Russell E. Dougherty, USAF (Ret.)

**Air and Satellite Communications*, Part IV.

Introduction

In many respects this book represents a "first." Several years ago when AFCEA International Press was born, it was noted that there was a serious shortage of material on military command and control (C^2). Part of the problem of course was rooted in the classification of C^2 data and processes, but a larger part of the problem was—and to an alarming extent still is—attributable to our unwillingness to codify our C^2 experiences.

This book represents one of the first attempts to assemble a set of materials that describes, explains, prescribes and predicts C^2. An enormous amount of material was canvassed to pull the collection together. A large number of original essays was commissioned, and an equally large number of issues is addressed in *Principles of Command and Control.*

The book reflects the way we "see" the field. While some might quarrel with the way we have partitioned C^2, we hope that many more will see the treatment as comprehensive.

"What is command and control?" This question haunts every inquiry into the field. While the Organization of the Joint Chiefs of Staff has a "definition" of C^2, it is significant that just about all C^2, C^3, and C^3I professionals have their own definitions, and that precious few share much similarity. As it turns out, C^2 is indeed many things to many people. Context plays an important part in definition building in C^2. "Why do you want to know?" and "What do you have to do with it?" are two questions one hears when C^2 definitions are required. The articles in Part I of *Principles of Command and Control* attest to the definitional latitude that we willingly grant our C^2 professionals. It is significant that some of the authors see C^2 as hardware, some as software, some as integrated systems, some as "systems of force" and some as conceptual compasses. Some adopt a behavioral perspective defining C^2 as "decision making." Some see leadership as essential to C^2, while others see C^2 as a strategic or tactical function.

Eclecticism is endemic to the study of C^2. The articles in Part I of the book also suggest how interdisciplinary the study of C^2 must be. By implication, they also suggest how those who build C^2 systems must staff their projects.

In fact, there may be hundreds of commands and controls, each identifiable by a set of situational and operational variables. At the same time, there must be constants since aspects of C^2 are generic: commanders must decide, execute and lead; systems must sense and communicate. This generic/problem-specific paradox explains why C^2 is so many things to so many people—and why it has been difficult to develop theories of command and control.

This book acknowledges these subtleties. The articles in Part I all see C^2 differently. The healthy differing perspectives on C^2 in the first part of the book set the tone for the rest of the book and challenge the reader to prepare for an interdisciplinary journey that is sure to raise as many questions as it will answer.

Part II of the book suggests we look closely at the steps that we take to design and develop our C^2 systems. Requirements analysis techniques figure prominently in this part of the book. In our judgment, many C^2 systems have failed because requirements analyses were either not performed or not performed adequately. It is impossible to design a system from indirect inferences about how or why a system should support a commander; requirements data must be collected and validated on-site. But "on-site" has its own connotations. While it is important to understand the battlefield on which the commander must maneuver, it is equally important to understand the human information processing principles that determine his behavior on the battlefield. It is clear now that we have largely ignored the human component in C^2, and that as C^2 systems become more sophisticated they must remain cognitively compatible with their users and the human systems they are intended to serve.

The significant change in C^2 over the past few years may be traced to the role that we expect our C^2 systems to play. As computers become more distributed and intelligent, we also expect them to become more aggressive, assuming decision making roles that a few years ago would have been described as preposterous. Battle management C^2 systems are on the drawing boards that will assist commanders in ways that will challenge our current understanding of C^2. We expect future systems to generate and evaluate options for commanders, review and critique commander judgment and plan and replan battles in real time.

The design and development of C^2 systems intended to serve commanders much like intelligent staff members serve them is fundamentally different from the design and development of "passive" C^2 systems. New methods that recognize the importance of *analytical* requirements must be perfected if we are to field C^2 systems capable of intelligently supporting their commanders.

We also need to manage the processes by which we build and deploy C^2 systems, as Part III on C^2 Program Management suggests. The relationship between doctrine and C^2 systems development must be examined continuously. We have learned over the years that doctrine can affect the way we design systems and that systems design can affect doctrine. Evolutionary development is a concept that has its roots in high technology, doctrine, cost-benefit calculations and the common sense that dictates a "try-before-buy" strategy. The recent emphasis on prototyping is consistent with evolutionary development requirements and is likely to become an essential step in the C^2 systems design process.

The question of evaluation is also inescapable. How do you know when you have designed an effective system? Clearly evaluation must be considered at the outset, and tests must be conducted as our C^2 systems evolve, not after they are in production. Part of the problem, of course, is our reliance on simulated C^2 environments to test our systems, environments that may or may not capture enough reality to make our evaluations meaningful. While we endeavor to develop measures of system effectiveness (MOEs), it is interesting that we have failed to develop measures of human effectiveness (MOHEs) to help determine if the human component in the overall system is performing adequately. The role of stress in particular must be determined before we field systems that may have inherent stress related vulnerabilities.

The subject of Part IV is Tactical C^2 and the perspectives are many and varied. Some of the articles are directed toward specific systems that now exist while some others describe systems that someday might be. Part V is equally substantive with profiles of some strategic C^2 systems and issues. Today it is impossible to discuss strategic C^2 without reference to the Strategic Defense Initiative (SDI). Suffice it to say here that SDI presents the C^2 community with perhaps its greatest challenge: at no time in the past has the reliance on technology been so great, while the requirements so elusive.

Part VI of the book turns its attention to some adversary and allied principles of C^2. It is, in our view, impossible to assess our C^2 systems without reference to how they might intersect with the systems of our allies and how they might defeat the systems of our adversaries. This section of the book is essential to our C^2 design, development, policy, doctrine, and research and development.

Do we understand allied and adversary C^2? It is important that we never stop describing, explaining and predicting the behavior of those systems that directly or indirectly might control our destiny. The articles in Part VI are essential to our understanding of our own C^2 as well as how our C^2 might be multiplied or crippled by allied or adversary C^2.

Part VII examines C^2 High Technology. It also causes us to examine the dependency we have engineered directly into our C^2 systems. While high technology is necessarily the foundation of our next generation C^2 systems architectures, it is also at times untested and can betray us. Technology push must be balanced against requirements pull.

Part VIII keeps us in the real world of constraints and opportunities that have their origin not in technology or "the threat," but in the realities of security, interoperability and other policies that affect the way we build and deploy C^2 systems.

The final section of this book, Part IX, looks squarely at C^2 in the Year 2000. The contributions here are nearly all original. They stress change and the need for adaptation. They also stress the changes that the world will present to commanders in the year 2000 and beyond. The forecasts here are important since they determine how we design system concepts for the future, and how we plan to allocate research and development resources.

The book's nine parts hopefully comprise a whole greater than their sum. The cumulative impact of the chapters, we believe, will exert an impact on those who currently design, develop, test and fund C^2 systems. It was our intention to assemble a set of materials that, together, would help define a broad field, help organize a set of perspectives and present a set of principles that determine the way we build and deploy C^2 systems. At the same time, we are not so naive to believe that the journey stops here. Indeed, the steps taken in *Principles of Command and Control* are small. As always, the real work lies ahead.

Perspectives on Command and Control

Roger A. Beaumont

Patterns of Evolution

Perspectives on command and control (C^2) vary widely among people, relative to their experience, their immediate role and systems and their views of the future. Within the limitations of space, this article ranges across the spectrum of perspectives and identifies the many paradoxes and anomalies arising from a less than reasonably cohered mass of data and from the momentum of a complex technological evolutionary process.

The mind-set of many who deal with C^2 is necessarily practical, and their primary focus is on what is and what might be rather than what has been. As a result, a general sense of modernity and futurism among those involved in command, control and communications (C^3) tends to overlook how long the human impulse to use C^2 at a distance has been evident in history. The sense of that history, however weak it may be, is a fundamental perspective, not only because it is interesting, but also because the ultimate development of a profession or discipline will depend on a greater sense of that continuum, especially in developing a body of shared principles and terms.

C^2 technologies have, in fact, been around for a long time; in essence, since the Pharaoh's umbilical cord on a pole served as a rallying standard in early Egyptian battles. The Pharos lighthouse, the postal service of Turn ünd Taxis, the Medici and Fugger bank correspondents, the extensive system of semaphore stations that answered Napoleon's incessant query on campaign "What is Paris saying?" all predated the use of electronics in communications. Royal

This article was written especially for **Principles of Command and Control.*

Navy flag signals in the days of Lord Nelson could send messages at about 250 miles an hour in clear weather.

C^2 in land battle was managed through flags, military bands, couriers, lights, sound and smoke signals and the gaudy uniforms of various regiments. At the same time, in business, statecraft and intelligence, intertwined from the Renaissance at least, codes and ciphers came into use. The pattern of evolution in respect to C^2, then, was again much like other fields of technology, with "eo-," "paleo-" and "neo-" phases. "Eo-C^2," being simple line-of-sight systems with minimum informational content in the ancient and medieval world, persists in many forms from the arm-and-signals of traffic police and infantry battle drill to the thumbs-up signal of fighter pilots. "Paleo-C^2" encompasses the array of more sophisticated techniques, such as sophisticated codes and cyphers and flag signaling, that can be traced back to at least the late medieval period. "Neo-C^3" begins with the introduction of the electrical telegraph and continues through telephones, radios, radar, electronic analog and digital computers to the burgeoning of refinements and the coming of C^3. C^3 usually is construed to include the many systems that relate to strategic warning and battle management networks, for example, nuclear weapons-related systems.

Sources of Differences in Perspectives

Like many lines of technology, C^2 has grown faster than the ability of even those closely involved in it to comprehend, let alone to conceptualize it or to maximize its potentials. Before wading any further in this virtual morass of paradoxes, it is useful to invoke the law of location from physics: It says that no matter where you go, there you are. (That is, each point of view is in a sense wholly unique because the place from where it is viewed is different from all others.) While it would be easy to beg off on the grounds that human perceptions vary from person-to-person, it also is obvious that categories and reductions are needed in grappling with the many points of view that dot the complex landscape of $C^{2/3}$ and with the many paradoxes.

Perspectives, of course, constantly are being altered by both a rising curve of technical evolution and by the fragmentation driven both by technological momentum and by task specificity. One result is that many attempts to wrestle C^2 into coherence—in the linked realms of doctrine, policy and operations alike—have been hampered by the disparity of perspectives of the discussants or of the terms used. The range of experience now shaping perception includes that of users in military-naval-air peacetime roles, such as air traffic control, military police nets, maneuver control and links between elements of various major and subsystems; of routine administrative communication; of strategic monitoring and exercises; of simulation; of stress loaded, critical functions in peacetime, including intelligence, crisis management, low intensity and/or covert operations; of component and system design; and of policy makers and analysts in hierarchies.

Qualitative Aspects of Perspective

Beyond the steady changes of perspective wrought by technology are other forces directly and subtly altering points of view. The number of people in the United States military with war experience is declining rapidly, a matter discussed later in more detail. Another experience differential also generating varying perspectives on $C^{2/3}$ systems is a person's level in the hierarchy from platoon/aircraft/ship's division up to the Oval Office. Users at lower levels often have substantially different views about utility, reliability trade-offs between weapons and $C^{2/3}$ systems. The sense of the "them-us" directed up the chain of command is strong and has been well charted.

Adjacent, parallel, overlapping and constantly shaping perspectives regarding $C^{2/3}$ systems and doctrine are the National Security Council (NSC), Office of the Secretary of Defense (OSD), the Assistant Secretary of Defense for Command, Control, Communications, and Intelligence (C^3I), the service secretaries' respective C^3-responsible deputies, major commanders at several tiers, civil and military intelligence agencies and tangential bureaucracies in the executive branch. Outside are vendors, consultants, academics, journalists and politicians. The resultant spectrum of perspectives ranges from abstract idealism to straightforward self-interest. Fractionalization of perspective also can be found within the $C^{2/3}$ community. For example, varying views are found among electrical engineers, computer scientists, ergonomists, psychologists, general systems theoreticians, operations researchers, systems analysts and military analysts.

Sharp and Fuzzy Sets

Perspectives also can be arrayed along axes with optimist-pessimist/believer-heretic polarities. As in other domains of technology, trade-offs and paradoxes abound. A fundamental tension exists between the relatively pristine and quantitatively reducible electromechanical and physical aspects versus the "squishy" or "fuzzy" ones. The latter includes idiosyncrasies of commanders and staffs, unforeseen problems and side-effects; human error; procedural differences; political forces; rivalry and friction among factions, branches and services; and personality clashes. The extent to which the practice of command and intelligence remains—like medicine—art more than science, stands in tension with rationalist perspectives at each level. For those who hunger for surety and a world of rectilinear precision, the blurred, jellolike phenomenology of combat always has been a source of frustration, and the $C^{2/3}$ revolution/evolution has been a generator of paradox and uncertainty for much of the modern era. For example, an advanced C^2 system may attract an enemy to commit resources to block, spoof or turn it around, while a more primitive system might not appear so attractive or pliable is one paradox. Another is that a signal battalion or radar might have more effect on a given battle environment than equal or larger, and more expensive, weapon-wielding units or systems.

The Problem of Echelons

An especially wide gap in perspective is visible regarding the projection of authority down the chain of command, that is, the extent to which $C^{2/3}$ systems allow high echelon commanders and staffs to look over the shoulder of commanders at micro/tactical/operational levels. Resentment toward senior commanders hovering in helicopters and overriding platoon leaders' and company commanders' orders in combat in Vietnam ("the squad leader in the sky" syndrome) persisted long after that war, even though those hovering believed that their perspective allowed them a better "big picture" and a leverage point to aid the forces engaged, as a variant on the force-multiplier argument that "x" amount of $C^{2/3}$ resources can amplify "y" amount of combat units' impact, in some instances synergistically, that is the whole may well exceed the sum of its parts.

Skipping Echelons

A central contradiction here is that refinements in C^2 technology have enhanced the ability to bring high commands closer to real time contact with operations. This profoundly changes many assumptions and logics behind military organization and procedure that have persisted in essence since the 17th century in ground warfare and deep rooted naval forms as well. Although air forces have grown up with radio, strong skepticism about control from on high can be found among pilots, some of whom argue that if war came, it would be time to "light the fires, kick the tires and let the first one to the end of the runway be squadron leader." Certainly the sense of "them-us" regarding higher echelons is no less hardy in air forces than in the other services. Balancing the progress toward refinement and real time high echelon battle management made possible by such systems as the E3-A airborne warning and control system (AWACS) is the fear that the principle of garbage in-garbage out has a corollary in $C^{2/3}$. Another inconsistency is that sophisticated $C^{2/3}$ systems might amplify mediocrity or error as much as competency, for example, the tampering with field operations by Corporal Adolph Hitler made possible by the C^2 systems of the 1940s.

German military theorists were concerned about another aspect of this problem in the late 19th century. Von Bernhardt was concerned that the ability to transmit actual voices would convey panic in battle up the chain of command. In a parallel, if slightly more abstract, realm are differences of perspective at various levels as to how much and what kind of command information is required from operational forces.

Logic Variants

It is not just a difference in available resources that has produced different U.S. and British perspectives in C^2 design. The former tends more to hierarchicalism and aims at providing real time interaction and highly graphic battle display. The British build command networks around elements that will survive after battle begins with less concern for the pyramids existing at the outset, like some reports of Soviet systems. They also move information in verbal and numerical formats, reducing problems of ambiguity, garbled language and too much emotional content. In this respect, it is slightly ironic to recall U.S. experiences in the naval campaigns in the Pacific during World War II, especially the confusion and abuse of the talk-between-ships short-range communication systems in the Guadalcanal campaign and Admiral Chester Nimitz' once-famous "all the world wonders" message to Admiral William Halsey.

Degraded States: Entropy Out of Negentropy

It seems obvious enough that maintaining a capacity to operate at less than a maximum level of effectiveness is prudent, since crisis or war—sophisticated enemy strategems aside—will reveal a host of problems that have not been anticipated in peacetime. The ability to operate in degraded states has been experimented with both in the West and in the Soviet block, perhaps more rigorously in the latter. Here, moreover, a critical shaper of perspective is evident—habituation. Analogically, in hospitals, increasing reliance on a vast nexus of biomedical engineering devices has seen a steady reduction of those able to deliver health care under the old and more technically primitive conditions. Furthermore, the skill level of those who might once have been proficient has faded from lack of use, while materials and equipment have been discarded, many items no longer found outside of antique stores.

However inevitable such processes may have been in systems affected by technological transition, perspectives have changed, not only as a matter of deliberation or in reaction to certain sets of experiences, but also as a product of often subtle and subliminal processes of adaptation. In the parlance of Marshall McLuhan, the military has, along with its parent society, been "switched on," and thus altered over a much longer time and in more ways than are generally appreciated. Nonelectric typewriters, for example, are no longer found in lower echelon U.S. ground combat units.

There are, of course, palliatives and prevention that could offset effects of damage to, or failure of, sophisticated C^2 systems, for example, standing operating procedures (SOPs), and backup systems based on atavistic technology. While such strategems may seem even less than half-a-loaf, their cost is much higher in that special microeconomy of military training where there scarcely is enough time to train in normal procedures and where safety rules and dependency on state-of-the-art technology make even simulated degradation difficult. Whether allowed for or not, degradation has occurred and can and will again.

The Disruption of Command Perspectives: Two Special Syndromes

Since the mid-19th century, various soldiers' battles have occurred in which the breakdown of command and control has led to escalation into large-scale battles—blind, bloody slugging matches, with minimal gains and at least momentary exhaustion on both sides. Another syndrome that developed from the extension of battles in time and space through refinements in communication and transportation technology has been the ever-increasing frequency of surprise, tactical and strategic, as noted by Barton Whaley and Richard Betts. In both instances, the phenomena have devolved from a pattern of diffusion and fragmentation that varies, in essence, in degrees from limited and guerrilla war, but not wholly. Confusion, dispersion and uncertainty had come to mark the normal state of the battlefields that S.L.A. Marshall referred to after World War II as the "eddy currents" of battle.

In the latter part of the 19th century, concern about lag in message flow between higher headquarters and the "man behind the gun" led German military theorists to advocate and, in stages, after 1918, to refine the doctrine of *auftragsbefehlgebung*—general mission type orders—to allow lower echelons more latitude to shape their own destiny in battle. As a *Bundeswehr* doctrine, it both stymies the Warsaw Pact forces' ability to anticipate German actions in advance and politically raises the threat of German conventional deterrence by suggesting the possibility of offense as well as defense. In the U.S. Army, however, substantial enthusiasm for *auftragstaktik* since Vietnam has had overtones of contempt for the "squad leader in the sky" and for political constraints in Korea and Vietnam. The development of a warrior spirit in the U.S. Air Force and Army has paralleled these doctrinal impulses, which are coming from a substantially different perspective than those C^2 logics shaped by concern for constitutionally mandated civil primacy and the concern about nuclear war. It also stems from, at best, an ideal based on a selective use of history, of an age in which soldiers operated in the field without serious constraints imposed by politicians or the media. Here, the rising curves of technology and the flat and regressive curves of tradition's role as a mask to anxiety and doubt stand in sharpest tension in generating sharp polarities in perspectives.

The Profusion of Sub-loops

The proliferation of $C^{2/3}$ loops generates further perspectival variants among designers, vendors and users. More than 40 separate major systems reportedly comprise the World Wide Military Command and Control System (WWMCCS), and many such systems have been left unused upon delivery or diverted to other functions. Due to the short time-in-role of U.S. military and civilian defense leaders, a working cliché is "the buyers don't use what they buy." Perspectives vary on how much $C^{2/3}$ systems may be products of bureaucratic game-playing; or the complex dynamics of defense market competition; or part of a maze used to fend off monitoring and political control; or products of essentially a chaotic process well out of control. The sub-loops in the realm of C^2, for example, NTIDS, Joint Tactical Information Distribution System (JTIDS), TACFIRE, AEGIS, approach the myriad.

Quantity and Quality

Analysis here is hampered by the position and the knowledge base of the observer and is filtered by motive, across a range from skepticism and anxiety to optimism and greed. The quantitatively oriented designers and users are not much more rewarded for taking long strategic views than are politicians. Nor do subtle factors or glitches count for much in a system that views action as superior to reflection and hard numbers and self-confidence as superior to fuzzy sets and caution. Beyond that lies the predisposition of people to want to do what they do best, what they really like to do or what they think they do best.

Stylistic Variants

A penchant for piling on, "can-doism" and firepower long have been hallmarks of U.S. military style. Attempts to form massive linear arrays have marked the majority of U.S. military operations since the Wilderness Campaigns. And not much sensitivity exists in U.S. military circles for the spirit behind Rudyard Kipling's "sweet leaving well enough alone." Restructuring has churned the system since the founding of the Republic, although other nations have shown similar tendencies, including the British who are often (wrongly) seen at a distance as predisposed traditionally to organizational stability. Busyness generally is seen as virtue, and unbroken tends to be fixed anyway. The churning effect on the mechanical aspects of the system, on proficiency in operating it and in doctrine has been considerable and also has been a contributor to the fragmentation in perspective that has stymied those who have attempted to rationalize it. What, then, of the perspective of the United States' great adversary? The less subtlety of approach and rigorous thinking-out of contingencies, the more a system is readable/penetrable/manipulable to a capable foe. At various stages in the Cold War, the Soviets and their clients have sought to "turn on" or "read out" Western $C^{2/3}$ capacity, often with considerable skill. Their systems, logics and basic perspectives are quite different and not always less sophisticated or rational because of technical lag and the cybernetics gap in the late 1950s and early 1960s.

$C^{2/3}$ Semiotics

Aspects of $C^{2/3}$ that shape perspectives profoundly but that have received uneven attention from practitioners are the symbols used to depict system states. These situation reports, measle maps, etc., are the ultimate graphic and functional products of even the most primitive $C^{2/3}$ system. Unavoidably, images of operations must abstract reality to some degree. At higher echelons, such distortion has tended to increase as a function of space and time, leaving the higher echelons with approximation (or worse) to deal with rather than realism. Refinements in technology have closed the space-time gap, but increase in systems' capacity also has generated myriad perspectives and data that must be blended to paint a coherent picture.

Systems Refinement and Command Prerogative

Advances such as packet switching and increased channel capacity and bandwidth not only offer protection against jamming and reduce degradation, but also allow a richer interaction between echelons. At the same time, there is no easy way to anticipate what the combined imagery and data presented to a high echelon commander might look like. Even running a vast number of permutations and combinations with various subsystems up, down and/or degraded at various levels of functioning would not guarantee much more than sensitivity to the range of possible new configurations. Again, the ability for those "up top" to read more quickly and in more detail might add to their sense of caution and concern or might lead to some degree of paralysis from flooding the system, a matter the Soviets have paid some attention to. At this point you encounter polarized perspectives regarding the importance of humans *versus* electromechanical systems. While some, like Rear Admiral Grace Hopper, USN (Ret.), are optimistic that removing people from the Strategic Defense Initiative (SDI) loop will reduce the danger of "crazies" running amok, others see commanders on the spot as the key decision points in the tactical loop. Again paradoxically, substantial historical evidence supports both perspectives. Unfortunately, there is little respect or reward for those who suggest a tolerance for multipolarity, or a perspective based on a range of perspectives. It may be that a corollary to some biological law or locational reflex exist that makes such clinicalism repugnant in itself.

The Man-Machine Gap

The basic dilemma of human capacity falling behind that of sophisticated systems has not, however, led to a great deal of agreement or sensitivity to the importance of selection, training and conditioning, and monitoring and management of those involved in C^3. Much of the recruiting and socialization has remained a process of screening for physical and mental traits of various kinds. Such traits predispose leaders to certain habits, reflexes and attitudes, which are more appropriate to some environments than to others. One can consider therefore how analogically appropriate to $C^{2/3}$ is Christopher Shaw's provocative view that "perhaps the quality of aircraft is of no more importance than the experience and training of its pilots."[1]

To step outside the boundary of your own systemic loop toward that of the adversary in the quasi-peace that has been the operational norm since 1945, you encounter a yawning chasm between East and West perspectives regarding the reliability of systems as they are subjected to counter C^3 (CC^3), electronic countermeasures (ECM) and counter countermeasures (ECCM). Here, a special contradiction arises. As more peacetime systems and procedures resemble those of wartime, in practice, a clever foe is allowed easier read-out and entrainment. More $C^{2/3}$ can be an avenue to surprise, intentional and self-induced, as well as a source of signatures, from the patterns of electromagnetic radiation read by enemy electronic intelligence (ELINT) specialists to actual content deciphered or decoded by signals intelligence (SIGINT) analysts. Separate from enemy intrusion or monitoring is the wrenching and disorienting transition from peacetime routine to the uncertainties, frictions and chances in war, which General von Clausewitz saw as far from rationalistic planning and academic analysis.

History is not wholly reassuring in respect to the extent to which the idiosyncracies of commanders and staffs can get in the way, for example, MacArthur's insistence during the Leyte operation that messages between Kincaid's 7th Fleet and Halsey's 3rd Fleet pass through his headquarters. Beyond the information glut that policy produced, even in that third year of war, battle experienced systems were wracked by delays and widely engaged in overgrading message urgency and wild guessing by operators of urgency and sequence. Kincaid's staff finally resorted to decoding all traffic, while many messages were "obscurely worded and misinterpreted. . . ."[2] All of this suggests the need for maintaining a perspective on $C^{2/3}$ similar to that evident in the growing awareness of the problem of what Charles Shrader has called "amicicide"—casualties induced by your own forces or by friendly fire.[3]

At the same time, romanticist visions of the warrior spirit and *Wehrmacht* prowess are somewhat ironic in view of their overreliance on *Enigma* encryption which made *Wehrmacht* high echelon signals open to Allied monitoring through much of the war. In any case, skipping echelons was seen at the beginning of the Eo-$C^{2/3}$ age, as major political leaders began to query and prod field commanders soon after telegraphy began, from Napolean III's bedevilling his marshals in the Crimean War in the 1850s, through Abraham Lincoln, David Lloyd George, and Georges Clemenceau. Telegraphy also gave the media and vast reading publics a more immediate view of campaigns and battles that was to hamper command in the field and governments at war from that point on. A century later, nuclear weapons, long-range aircraft, computers, television and satellite links further undercut command prerogative from various angles.

In any case, an ideal of freewheeling warfare, unfettered by political or administrative constraints, is hardy and thriving in many quarters in the U.S. defense nexus, although it rarely has been seen in modern times. Some of the skepticism regarding $C^{2/3}$ has arisen from that vision, and can be expected to generate a schism in perspective until perhaps a full generation has grown up within the new complex of technologies. The dilemma lies in determining how much to keep and how much to throw away, a variant of an age-old "dialogue between harmony and invention."

Command Dynamics and $C^{2/3}$

A slightly more abstract but major difference in perspective can be seen in the strong assertions by opposing schools of thought of the extent that $C^{2/3}$ technology has altered the nature of the command staff process. Some perspectives view some basic essence and principles as unalterable and some $C^{2/3}$ technologies as relatively peripheral. In the province of C^3, reliability of systems and human controllers has come under ever tighter scrutiny in respect to monitoring, crisis management, alert, battle management, damage limitation, post-attack recovery, war fighting and war termination, for example the analyses of Desmond Ball, Paul Bracken, Ashton Carter, John Cushman, Daniel Ford, Don Gordon and John Steinbrunner.

Chains of Command

The question of how much modern war commanders really are in touch with operations has been dramatized in fiction, for example Norman Mailer's *The Naked and the Dead,* and in military history, for example, the spontaneous pursuit of the Spanish fleet at Santiago. To a great extent, the fighting and crucial decisions have fallen to ever lower levels of the chain of command, for example, LTCmdr. McCluskey's key observation and decision at Midway. Much of the command process can be simulated, but the reality of stress and personal danger cannot. The problem of higher echelon commanders leaving the center of information flow at their headquarters and moving forward was a major dilemma in World War I. This produced what the British deemed a fighting commanders' syndrome, in which those who did well at lower levels failed when promoted up, out of the zone of action and out of close contact with their troops and the situation.

Leader-Led Perspectives

The still mysterious chemistry of leader-led relationships is a very fuzzy set indeed and, although a matter of mass emotion and unquantifiable, it is no less a major vital dimension of battle dynamics because of its intractability. Whatever the glue that has bound humans in battle, ignoring that it is a potential major element—and a widely ranging and squishy variable—in plotting perspective points in $C^{2/3}$ architecture, analysis and operations will be done at a high risk. The work of "anatomists" of battle, for example S.L.A. Marshall, Glenn Gray, John Keegan, John Ellis, Richard Holmes, and Norman Dixon, has paralleled substantial man-machine systems and social science research on human performance and behavior across a range of parameters.

The SDI Paradox

The human issue comes to a sharper point in the realm of SDI where the search for various weapons technologies is paralleled by a strong awareness that the C^3I system, given all of the complexities, will decide to go to war and orchestrate the opening battle with little or no direct human involvement. Ironically SDI's basic C^3 logic is another major step away from the warrior spirit and indeed from human involvement in combat. The trend of more and more people building, maintaining and training and fewer and fewer fighting has been underway for some time and was taken for granted in respect to aviation two generations ago.

Here, varying perspectives regarding "teeth" and "tail" often have generated debate and factionalization far from the rational or functional. One of the most fascinating reversals in polarity of perspectives has been that of military professionals recruited—or self-recruited—into systems, attracted by microwarrior roles, and reversing toward bureaucratic roles during career progression. The powerful nostalgia for early career roles, and the increased changes that C^2 technology offers senior officers to regain some contact, however abstract or remote, should not be forgotten in $C^{2/3}$ design, analysis, operations or policy or doctrine formulation. At the same time, recognition of that at lower levels must also be weighed into the balance.

Any closing of the widening gap between human and nonhuman performance in $C^{2/3}$ will require substantial consideration across a range of disciplines from psychology and pharmacology to neurophysiology and computer science. It also must be accepted that the gap may be intractable beyond a certain point, and that the gap will grow, as levered further by future and unforeseeable technologies. "Tribal" groups now are wired together in the complex fabric of modern military forces, including, most critically, strategic systems. And, as suggested earlier, perspectives looking up, down and across the resultant complex landscape are very different. The resultant strain has been visible since the late 18th century at least, a tension resembling the "them-us" dynamic seen in many preliterate cultures but also reflecting the "scalar versus functional" polarity seen in commercial, industrial and bureaucratic organizations. In many cases, structuring of authority is based on criteria not directly correlated with the ability of those having rank to do the job, as much as it can be measured or is deemed worthy of measurement, and even then, not wholly rationalized or defined.

As with the introduction of other forms of technology in navies and armies in the modern era, much anxiety has stemmed from the threat of new criteria of competency and new arcane skills replacing old ones, co-optation and hostility, which has from time-to-time blocked ignored or rejected advantages implicit in the new lines of development. The frequency with which hostility and co-optation have blocked and blunted the full potential of new systems is proverbial, and the angle of apposition of perspectives has most dramatically approached 180 degrees.

Coefficients of Error

Designers or analysts of $C^{2/3}$ phenomenology also do not share a common perspective. Disparities of perspective can be seen well beyond the lack of a glossary/lexicon/epistemology. Some see the human element in the loop as a fairly tractable range of givens, while others have been concerned about the perturbing affects of human foibles, both in direct controller and staff roles and indirectly. Irreducibly the electromechanical $C^{2/3}$ systems and procedures are extensions of humanity and thus are tainted with irreducible coefficients of error, as General William D. Thurman, USAF, suggested in his *defensa* of the B-1, in invoking the truism that perfection is unattainable in complex systems. It being human to err, the question of machine preeminence, already a dimension of $C^{2/3}$ in battle management, has and will continue to move to the networks of logistics and planning. How much information exists in a machine versus human world is not now quantifiable, but the world is a long way from library drawers, carbon paper, and McBee cards. The lag in perspectival adjustment to that great revolution is often found to be as great among those in its vanguard as anywhere else.

Perspectives on Transition

Just as battleship admirals minimized the potential of aircraft carriers, and cavalry generals, of tanks, $C^{2/3}$ debates have been flavored by concerns generated by the increased ability of higher command echelons—civil and military—to meddle in operations on the one hand and to meddle in needs for new kinds of human ability to operate new systems on the other. The macho image of the military aviator that appeared in several nations after World War I stands forth as an example of tribalism/scalarity exerting dominance over technology/functionalism. Given the especially strong feelings evident in this area, it seems likely that a carefully struck balance will remain out-of-reach for some time to come and that the perspective gap may yawn even more widely if left to the tides of fate. It often has been asked, how can organizations, against all of the counter currents of conformity and orthodoxy, generate visions of contingency and exigency? How much does emphasis on one set of selection-promotion templates purge out traits of value to the organization overall? How realistic, for example, is it today to make invidious comparisons between staff and line and "teeth" and "tail" when teamwork is growingly at a premium and when in major conflict no clear front line or safe zones would exist?

One quandary that has emerged from such discussions is the extent to which a too sophisticated C^3 system can be penetrated, spoofed or nullified. A derivative paradox also stems from the converse—too good a C^3 system undercuts deterrence by making war fighting (planning to fight a nuclear war through to some level of survival and victory) more possible and therefore more likely. A greater paradox derives from the variance in perspective here: whether or not to target or tamper with enemy $C^{2/3}$ systems and thus impair their ability to communicate and/to terminate or limit a nuclear conflict.

Beyond the question of stress-layering is the extent to which perspectives and reflexes are set under conditions of stress and how they may be read by a foe and used to trigger reactions according to a menu or agenda, a defined goal of the Soviets' doctrines of *upravleniye refleksnoye*—reflex control. Few specifics regarding Soviet practices in the nuclear-strategic realm are to be found in open literature. As noted earlier, there is a substantial linking of psychology, communications technology, computers and command method in ways both different and more sophisticated than in the West. It is not clear how they have managed the warrior spirit, which chafes at the need for compliance and restraint deriving from the nuclear standoff.

Perspectives on Authority

The behavior of military and civilian commanders under stress is a fundamental problem of C^2 phenomenology. Handing over battle management and decision authority to nonhuman systems, that is eliminating the "man in the loop," touches on an array of paradoxes. Under what conditions does it make sense? How would humans regrasp the tiller? Should they? When? What advantages are gained in time and sophistication in handing over control to $C^{2/3}$ systems? Are they worth the loss of "common sense," "intuition," "insight" and the "seat of the pants" sensibilities of people on the spot? On the one hand are the many anecdotes about $C^{2/3}$ system false alerts and glitches; on the other is the increase in strategic and tactical surprise in modern warfare, as noted by Barton Whaley, Richard Betts et al., in which C^2 systems often have been used against the users. Beyond that is the fact that, as J. F. C. Fuller suggested in "Magic and War," and John Keegan in *The Face of Battle,* the main goal of combat is breaking the will and disorienting the mind of enemy commanders and troops.

Penetrability

The mind of human battle managers, then, could be influenced more easily when a skillfully manipulative foe is "wired in" to a complex network and when its mind-set is relatively evident. U.S. military professionals, analysts and policy makers generally have been uneasy with such nuances as "psychological operations," except during the brief renaissance in that area driven by desperation in World War II. Contempt for hearts-and-minds and preference for the application of military force in Vietnam were proverbial. Nevertheless, the Soviets long have folded substantial amounts of academic psychology into their military doctrine and practice on a pragmatic and well-thought out basis, for example, Marshal Boris Shaposhnikov's *The Brain of an Army;* Boris Teplov's work on *maskirovka* and his essay on "The Mind of the General"; and the recent work of Kontorov and Druzhinin, for example, the U.S. Air Force translation of *Concept, Algorithm, Decision*.

The practice of command at high levels in the U.S. military system has been left to individual whim to some extent, and, as noted earlier, U.S. C^2 systems and exegeses alike reflect preoccupation with hierarchy, command prerogative and autonomy. The "fuzzy" factors of group dynamics and monitoring and maintenance of the human components in $C^{2/3}$ are addressed unevenly across the spectrum and up and down the chain of command. Relatively weak understanding of the command-staff process, and many "fuzzy sets" that swirl around leadership, morale, charisma and cohesion derive in some part from the substantial differences among the services, within them, and among vendors, as well as the politics of defense procurement. While such variance in perspectives can yield creativity, at least as a byproduct if not side-effect, and can confound adversaries, they also perplex many within the system.

Reaching for Coherence

The great suspicion of power that underlay the framing of the U.S. Constitution in the late 18th century is reflected in the structure and values of U.S. government and in the military realm. Waste, corruption and confusion may be irreducible in any system, and dwelling at length on such matters may be naive and a waste of time in a world incapable of perfection. When there is a clear-cut choice, few would prefer greater efficiency at the expense of lost liberties. Perhaps there is a not fully conscious anxiety abroad about the threat to personal freedom that could be posed by ever more sophisticated $C^{2/3}$ being turned on society, and in the 15 years since Judith Merkle raised the issue, it has not been a matter of much evident concern.

The Essentiality of Coherence

Nevertheless, the extent to which coherence of view and rationalization are attained in respect to $C^{2/3}$ is not merely a goal that would be "sort of nice" to attain. Whatever advantage that resources and personnel gave the winners, victory in the U.S. Civil War and in World War I and World War II went to the side that most effectively orchestrated the full spectrum of C^2 technologies and methods that existed at the time. Disturbingly, the Soviet view of the spectrum of *sistemotekhnika sybernetika* encompassing the world is systemic, while the U.S. world-view is much more stimulus-response and short-focus. The image of stamping out brushfires is a standard perception in U.S. business, government and military, as is the constant turnover of senior military and civilian officials. While the much longer times Soviet leaders spend in senior jobs might make them more stale and dull, it also allows them to learn their jobs and to focus on problems. Crucial in respect to $C^{2/3}$ is the question of how much the rigor and lucidness evident in extant Soviet material on C^2 reflects a substantially greater mastery of the alignment between technology and performance.

Objectivity

Beyond that lies the special problem of monitoring crisis and nuclear war related functions. C^3 systems are not gross conduits, and beyond their mechanical dimensions, they link human perspectives, filtering, blocking, queuing, etc. They can aid commanders in making war and policy makers in preventing it. The question of objectivity in data gathering and graphic presentation is reflected in such terms as "war room." How might those whose perspective is based on war prevention, damage limitation, or crisis minimization design a system differently from those concerned with alert, surprise and response? Whether or not there is one best way in $C^{2/3}$ architectures is more often addressed from the standpoint of cost-effectiveness, engineering or the immediate task, rather than from such concerns. It is crucial enough in the tactical and battle management realms, but, ironically, the stakes could be as great or greater in the alert, damage limitation, post-attack reconnaissance and war termination functions.

Paths to Coherence

Can the fragmented perspectives be brought to some level of order where realism effectively displaces subjectivity, impressionism or surrealism? The development of a glossary, or lexicon of clearinghouse journals was called for in the Department of Defense (DOD) sponsored workshops of 1979 in the wake of NIFTY NUGGET. A journal is in hand; an unclassified one is in the works; and many in-house efforts have been made. C^2 like much other important pansystemic phenomenology, lies athwart many disciplines and is far from being an academic discipline or even a subdiscipline. Most research and reporting tends to be hardware dominated, festooned with jargon, acronyms and enthusiasm for "neat stuff," not unlike the mystique of the air age between the world wars. Wrestling a doctrine on top of that riptide of technology was not easy, and it often was frustrating and bloody. Those who have labored in the vineyard in pursuit of joint tactical fusion would have something to say about frustration, if not blood. You hope, of course, that the latter does not come, but if it does, the amount will be influenced by how much has been done in the vineyards in regard to attaining conceptual coherence.

In pursuing functional commonalities of language, logic and phenomenology, the limits of human mental capacity often are thrown into dramatic relief. So is the power of situation, location, acculturation, experience and training and mind-sets within mind-sets. The U.S. approach to $C^{2/3}$ design and analysis has tended to be mechanistic, and psychology has remained an adjunct rather than the mainstay in Soviet doctrine. The need for turning our analytical spotlight into new crannies was stressed recently by RAdm. Grace Hopper:

> . . . we need to turn our attention much more to communications and . . . to the information itself and the information flow . . . (which) is going to change the organization of some institutions and companies[4]

The abstruse, complex and often academic overtones of $C^{2/3}$ design, development and analysis should not be allowed to blur the reality that the extent to which $C^{2/3}$ is left to existentialism, impression and reaction may be far more frustrating and more costly. In conclusion, the cliché that ends so many publications seems no less apt for its being a cliché: More research and a great deal more hard thinking need to be done. At this point, a healthy uncertainty based on the image of Plato's cave is far from being the least realistic perspective.

Footnotes

[1]Christopher B. Shaw, "The Brewster Buffalo" in Charles W. Cain, ed., *Aircraft In Profile* (Garden City: Doubleday, 1971) p. 156.

[2]E. B. Potter, *Bull Halsey* (Annapolis: Naval Institute Press, 1985), p. 187–88.

[3]Charles R. Shrader, *Amicicide: The Problem of Friendly Fire in Modern War* (Fort Leavenworth: Combat Studies Institute, 1982) [Research Survey #1].

[4]Kathleen Myler, "Amazing Grace," *Houston Chronicle*, November 16, 1986, sec. 7, p. 3).

Command and Control and Decision Making

Frank M. Snyder

The purpose of this paper is to explore command and control from the particular perspective of decision making. Although the term "command and control" is used so frequently that we might assume that it means the same thing to everyone, it seems prudent to begin by defining it and then by identifying the types of decisions that commanders make.

The DOD Dictionary definition of "command and control" is not entirely helpful. The first part of that definition—"The exercise of authority and direction by a properly designated commander over assigned forces in the accomplishment of the mission"—is really the classic definition of the command function, while the second part of that definition—"Command and control functions are performed through an arrangement of personnel, equipment, communications, facilities and procedures employed by a commander in planning, directing, coordinating and controlling forces and operations in the accomplishment of the mission"—is fairly close to a good working definition of a "command and control and communications" (C^3) system rather than of the command and control process itself. For the purposes of this article at least, the term "command" will be used to mean the *function* to be performed, the term "C^3" will stand for the supporting *system,* while the term "command and control" will denote the *process* that commanders follow (in planning, directing, coordinating, and controlling) as they utilize C^3 systems in order to discharge the functions of command.

Before examining the process of command and control from the perspective of decision making, it is necessary to identify the types of decisions that commanders actually make. Operational commanders assume that their key

This article was written especially for* ***Principles of Command and Control.*

decisions are operational in the sense that they concern the way an operation is to be conducted, decisions that result in orders to subordinate commanders to achieve specified operational objectives. These operational decisions are easily seen as the basic stuff of the commander's function of command. But consider what actually happens. Commanders at every level find themselves worrying about three things: whether or not they will be informed about significant events that affect their operations; whether or not they will be able to transform the information they receive into sensible and timely decisions; and whether or not they can get their decisions executed in time to affect the outcomes of their operations. Note that prior to making each operational decision, a commander will have made decisions of two other types—organizational and informational. Organizational decisions are necessary, not only to establish a chain of command for the execution of operational decisions, but also to establish the flow of information and to identify the intermediate processing that will be necessary to support decisions about what is actually happening and what orders to issue. Information decisions are those necessary, but often unstated, decisions about what actually is happening, about the state of nature, that need to be resolved in the commanders' minds before they are able to decide what actions need to be taken.

For each commander, then, there is a command and control process for making *operational* decisions and getting them executed, decisions that in turn are based on prior *information* decisions, both of which are made within a structure established by earlier *organizational* decisions.

The organizational decisions that create a command and control structure establish "who decides what." The Unified Command Plan is an example of such a structure, as is a directive from an Officer in Tactical Command of a battle group, identifying and tasking warfare commanders in accordance with the composite welfare commander concept. In each example, these organizational decisions establish command and reporting relationships that, at a minimum, create requirements for communications, not only for the physical links themselves, but also for staffs with shared vocabularies, dedicated to performing communications functions in the larger sense. Because organizational decisions shape the C^3 system and commit C^3 resources, commanders need to take into account the capabilities and limitations of individual subordinate commanders, their staffs and the C^3 facilities to make the necessary information and operational decisions implied in the organizational decisions.

A great portion of existing C^3 systems is devoted to supporting the information decisions to be made during the command and control process. The uncertainties that surround an assessment of the situation are many: The information available is usually ambiguous, incomplete, and conflicting; and often arrives late, after having been transmitted imperfectly or received with errors. While uncertainties are usually reduced over time, as amplifying and clarifying reports are received and understood, operational decisions usually have to be made long before the situational uncertainties can be completely resolved. Furthermore, the speed with which real time reports can be understood is dependent on the amount of previous information exchange. While we customarily view a report as somehow transporting information, it may be more useful to think of a report as announcing which situation (among all the possible situations shared by both sender and receiver) is now operative. The more understanding that both sender and receiver share about possible situations, the more efficient and reliable is real time reporting. Therefore, we should expect that when a truly unexpected situation arises, the reporting system has to work much harder if it is to convey effectively information that is truly unanticipated.

Information does not always move in its raw state directly from sensor to decision maker, so we need to consider not only the "links" in a reporting system but also its "nodes," where intermediate decisions are made during filtering, correlation and analysis. Because the observations gathered by a sensor may be numerous, filtering is used to remove reports that are redundant or that fall outside some reporting threshold. While the establishment of reporting thresholds is important as a control mechanism, they need to be reevaluated periodically so that unexpected but necessary information will not be filtered out.

Correlation is the process of relating several reports on the same object: Reports by different sensors at the same time, by the same sensors at different times; or by different sensors at different times. Correlation attempts to establish positions, courses, speeds, identities and perhaps readiness. Uncertainties exist here, too, and some correlations may be tentative or may have been made with less than a full level of confidence. Commanders who feel it urgent to find out what it happening also may find it easy to discount or completely disregard such uncertainties. One of the challenges of C^3 systems design is to be able to portray graphically the degree of uncertainty that may exist in any correlation.

A third process that takes place at nodes somewhere between sensor and commander is that of analysis, which is the drawing of inferences from correlated data. Such inferences might concern an enemy's intentions, tactics or rules of engagement. Uncertainties exist here as well, either because the information is insufficient, because of the possibility of deception or because the C^3 system is not sufficiently capable.

It is important to recognize that the organizations that make intermediate information decisions at the nodes of information flow are not necessarily the same organizations that will carry out operational tasks; organizations that provide filtered, correlated and analyzed information to a commander are increasingly likely to be working for some other commander. To this extent, part of each commander's C^3 system is beyond his direct control. Yet, it is important that the organizations who filter, correlate and analyze information not be "out of control" and that commanders who are served by nodes outside of their control be able to see that these intermediate decisions are made in accordance with their sense of priorities. While the same raw data may be useful to several echelons of command, there is no assurance that the data will be analyzed for significance in ways most appropriate to support each echelon.

Operational decisions, the decisions about what action to take, have to be made in the face of uncertainties about future events, and they have to be communicated in a way that resolves whatever uncertainties may exist in the minds of executing commanders about what actions their operational commander intends them to take.

For commanders about to choose from among several courses of action, first, there is the uncertainty about what courses of action their opponent will adopt, and then there are the uncertainties about the outcomes that would result from the interactions between alternative own courses of action and the alternative courses open to an opponent. Increasing attention is being paid to providing decision aids that would assist in making such predictions, but at present, decision aids are more often used to make rapidly the many calculations necessary to optimize the use of own forces. The prediction of engagement outcomes is not unlike the assessment of battle damage that takes place in war gaming, except that instead of converting a number of probabilities (of detection, of correct classification, of component reliability) into discrete outcomes so that the game can continue, the commander making an operational decision needs to convert some of the same basic probabilities into probabilities of a higher order, into probabilities of mission success. Operational decision making does not require that uncertainties about outcomes be resolved, but that such uncertainties be expressed in terms of probabilities of engagement outcomes that commanders can use as the basis for operational decisions.

The uncertainties that confront the executing commanders are usually overlooked. Commanders are likely to presume that decisions clear in their own minds surely will be understood by subordinate commanders. This presumes that the decision is communicated to the subordinate in a timely manner, that the decision as communicated will be clear to the subordinate, and that the conditions on which the decision is based will continue to prevail. But a decision made is not necessarily a decision faithfully communicated or clearly understood unless great care has been taken to create the shared understanding that makes communications effective. This aspect of the command and control process usually is taken for granted, but it should not be.

While commanders at every echelon understandably see the purpose of the command and control process to be the support of their own particular decision making needs, it turns out that the decision making needs of the *executing* commanders are the most important of all. If the executing commander does the right thing, the command and control process has been successful; if he does the wrong thing, it has failed. Executing commanders benefit from organizational decisions that create workable command relationships, that insure information flow that supports effectively information decision making at the scene of action and that provide the staff and other C^3 facilities needed to predict outcomes in support of operational decision making at the scene of action. Executing commanders also benefit from operational decisions at higher echelons that provide clear operational direction and from information decisions that provide them with insights about the existing situation. Because command (the exercise of authority) starts at the top, and with modern technology much of the information comes from above as well, it is not surprising that a great effort has been made to facilitate the command and control process at higher echelons. However, it is easy to forget, that for executing commanders, decision making demands are similar and the uncertainties and urgency may be even greater. Decisions made at the scene of action are the true test of the command and control process at all levels of command.

A Perspective on the Command and Control Agenda

William J. Crowe, Jr.

Introduction

Drowning in data, starving for information; it's an old saw, but in many ways an instructive one. It reminds us that even as the age of technology sweeps us forward, the human factor remains central to all our operational capabilities. Though supercomputers can crunch billions of bits of data, and telecommunications networks can speed them all from one place to another instantly, the capacity for reflective judgment, intuition, sensing, caring and learning which humans possess cannot be mechanized. Decisionmaking aids have been multiplied vastly and have been made more powerful by the march of technology, but it is the power of the mind that ultimately matters most, and its ability to communicate with and energize other minds.

But as everyone who has tried to lead or manage knows, there is a limit to how much information we can effectively use. It is one thing to have all the information in the world available, but quite another to be able to assimilate and apply it wisely in peacetime, crisis and war. Ladling vast quantities of data in nano seconds to decisionmakers, whether at strategic, theater or tactical levels, will only overload them, and especially in time-compressed and emergency situations, they do not need that extra burden.

Americans have an insatiable appetite for information, and an enthusiasm for finding more and more advanced ways to assemble and to move it. But how to separate the wheat from the chaff has become a major problem. Even in the national security business, we are still in our infancy in being able to sift and sort large volumes of data dependably. We can take some small

This article was written especially for **Principles of Command and Control.*

comfort in the knowledge that we are farther along in this than almost everyone else, and certainly we are more capable than our major adversaries. But there is no room for complacency in this effort to apply technological advances in ways that amplify the capabilities of the men and women in Washington and the field who must direct, manage, and lead our military forces.

Chairman's Position and Role

As the Chairman of the Joint Chiefs of Staff, I do not command, but I assist those who do. I am the principal military advisor to the President and Secretary of Defense, and the operational chain of command runs from the President to the Secretary of Defense through me to the Commander in Chief (CINC) of each unified and specified command. In effect, with the Joint Chiefs I provide a channel of communication upward to the National Command Authority (NCA) and of direction outward to the CINCs. I sit at the center of the processes that construct our forces and decide on and direct their employment.

It is my job, furthermore, to help translate national objectives into a military strategy that puts actual capabilities, tailored to operational needs, in the field and programs for their enhancement. Much of our effort here goes, of course, to developing forces, weapons, and equipment—and to building defensive coalitions with our friends and allies abroad—that can accomplish wartime missions and deter our enemies from attack. For forty years we have founded this strategy on three main principles: our commitment to deterrence, our defensive orientation, and our determination—should deterrence fail—to fight to restore peace on favorable terms. I believe strongly that the story of this postwar defense posture is, on balance, a story of success. There have been difficulties, of course, and there always will be. But in the end the abiding reality is this: we have had no nuclear holocaust involving the superpowers or anyone else; and no great war has swept across Europe or Asia. American and Allied military strength has had a great deal to do with keeping the preconditions for that peace firmly in place.

Key to that success story has been the ability to command and control our strength sensibly. This nerve center of our capability always must keep pace with the evolving imperatives of deterrence and warfighting, across the whole spectrum of conflict. The credibility of the U.S. strategic deterrent depends on systems that provide positive control for the effective employment of the Triad. Such systems require security, speed, flexibility, reliability, survivability, and endurability to assure performance before, during and after a nuclear attack. At theater and tactical levels, we need secure, interoperable, and enduring systems for joint and combined operations. At all echelons, command and control systems must provide leaders with the information they need to assess the size and scope of any attack, determine an appropriate response, and issue orders. They also must enable field subordinates to receive those orders and orchestrate battle actions needed to carry them out. We must place continuing emphasis on ensuring that these systems are as survivable, endurable, and effective as the forces they support. That is easy to say, but the practical problems are many and complex, involving the integration of diverse facilities, equipment, procedures, technologies, and—of course—people.

As we grapple with the task, I see several broad issue areas that are likely to dominate our agenda in the future.

Identifying Requirements

Perhaps the overarching challenge is best understood as one of adequately identifying requirements. And this places special burdens on leaders—the individuals who know best what they and their people need to get operational missions accomplished.

The C^2 function is one that cannot be delegated. Commanders at all levels have to be personally involved in defining their command and control systems needs. They do not have to become technical experts, but they need to understand the capabilities, limitations, and cost drivers of our complex systems. Exercises have proven to be a great help in that regard—both suggesting what new systems or enhancements might be needed and testing C^2 systems on hand in simulated combat conditions. But though this experience can help commanders define their requirements, the very nature of the modern command and control challenge greatly complicates the effort.

Technology today makes it possible to collect and transmit vast amounts of information far surpassing the mind's ability to absorb it all. When we are at peace, we have time to massage this mass of data into something more manageable for decisionmakers. The real challenge is to create systems that can smoothly transition to the pressures and tempo of combat.

The nature of the threat, the diversity of our global commitments, and the imperatives of coalition strategy further complicate the commander's problem in specifying his needs. The United States participates in a collective defense structure that combines the strength of our allies with that of the United States. To anchor that defensive aggregation, we keep forces forward-based, on land and afloat, and we also provide for rapid response to contingencies in any region where our interests must be defended. As a result, our forces must be prepared to deploy in widely varying climatic environments, to operate with forces of other nations against diverse adversaries, to establish C^2 even in areas having little or no supportive infrastructure, and be ready to accomplish all this on very short notice. The possibilities are not endless, but clearly the array of scenarios we must prepare for is impressive—in other words, expected combats are many, and the characteristics and requirements of each can differ markedly from others.

Though the specific characteristics of needed future command and control systems are hard to define precisely, certain parameters *are* discernible. We know that we have to integrate the capabilities of all U.S. armed services *and* those of allies, surmounting differences in equipment and procedures and constantly folding emerging technologies into the calculation. We know also that the modern battlefield will place severe demands on the systems we develop. Our forces must have the responsiveness and flexibility they need to defeat a numerically superior opponent. American superiority in technology, if maintained and effectively applied across the board, can be the force multiplier that provides commanders with the critical edge to overcome this disadvantage. But it will not be enough merely to provide our forces with modern weapons; we must be able to direct their operations efficiently in the heat of battle. To that end, we must have a variety of systems that will enable all levels of command to gather information about enemy forces; assist them in analyzing that information; provide jam-resistant, secure communications links; and through the use of electronic warfare, to disrupt enemy communications and intelligence gathering. These systems must have the capability of being rapidly reconfigured, extended, expanded and reconstituted through all levels of crisis and war.

In sum, our military forces are heavily committed worldwide and I see no prospect for these demands to subside. Fundamental interests will not change, and the threat will not diminish. Indeed, American interests, and those of our friends abroad, are now and will continue to be threatened on a number of fronts: burgeoning Soviet military power; Moscow's continuing proclivity to use that strength directly and indirectly; terrorism that can occur at any time and anywhere on the globe; the activities of Moscow's surrogates who seek to promote instability wherever they can; and proliferating local conflicts in the Third World.

Faced with these realities, we are going to need forces and supporting command and control systems capable of responding to a full spectrum of challenges, all of which will evolve over time. In Washington and in other headquarters, much of our attention must inevitably be devoted to the difficult, often perplexing task of determining what we need, and steering command and control capabilities into the future.

Issues

Resources. Another fact of life we face is that the future into which we are headed will be a resource-constrained one. Identifying requirements in deliberate fashion will help us identify the *right* systems, but then the task will be to get them the priority they deserve in a competitive budgeting process, and to avoid debilitating budget competition with other (e.g., weapons) programs. We also will have to strain to maintain a long haul, Big Picture focus as we fight the smaller battles of the budgeteers and accountants.

How much is enough? It is a rhetorical question in a way, but it does remind us that there *are* limits, and that we need *affordable* systems which support our national objectives and contribute to our war fighting capabilities.

We also will inevitably have to face up to the need to balance support for already-fielded systems with the requirement to modernize and exploit technological opportunities.

The realities of the planning, programming, and budgeting process are that even if requirements are well conceived, well established, supportable and supported, tough decisions and trade-offs will have to be made to stay within budget ceilings. I want to emphasize that in evaluating the affordability of any program, the monetary cost must be balanced against the expected reduction in risk provided by the proposed program. Although there is growing emphasis on improving all types of C^2 capabilities, their relationship to the threat, and to risk reduction, is in many cases not as directly obvious as it is for tanks, submarines, and combat aircraft. In presenting and defending our command and control programs, leaders must be sure to make this relationship as plain as possible, and to outline clear priorities that emphasize them.

Technology. Rapid advances in microelectronics, computers, and materials science have put us where we are today; and while we are constantly looking ahead to the enhancements that further advances are likely to make possible, to date we haven't exactly been at the controls of this fast-moving train.

Post World War II developments in platforms, weapons, sensors, tactics, strategy, and organization have been impressive, to say the least. They made necessary the development of C^2 systems to match. But this explosive growth did not follow any particular plan.

At times, we have tended to be mesmerized by promises of what a new system will do. If technology was at hand, that seemed to be reason enough to apply it. The better way—indeed the *only* way in the future we face—is to ensure that the technology we adopt contributes to expected missions. We need to be selective and realistic in fitting the advances we pursue into a threat- and mission-oriented design.

To do this, we who are charged to support commanders and operators need to pay close attention to the needs they present. I said earlier that commanders carry a heavy burden in understanding and articulating their needs. It is equally true that those of us on staffs that serve leaders must listen to what they say and help those requirements drive the research, development and acquisition process. Too often in the past we have turned over analysts' wish lists to the engineers, and they in turn have built systems that have not functioned the way commanders desired. We no longer can afford to do business that way. The leader's operational needs must be at the core of future systems as they evolve and mature. The planning process for this can never be elegantly rational, to be sure. But it is not too ambitious to aim to integrate those needs into an embracing concept or master plan for command and control.

Saying you need master plans conformed to operational needs is one thing; preparing them is another. Anyone who has served in the Washington budgeting milieu can attest to that. But there is no denying that that is the direction in which planners and programmers must strain. The result, no doubt, will be imperfect in some ways, and the effort agonizing. But the process itself will move our operational capabilities forward, and that is the name of the game.

People. Discussions about C^2 tend to emphasize—perhaps inescapably—technology and "systems." But the fact remains that *people* are the most important element of any military capability. If I have given you even the glimmer of an idea that command and control or management functions can be reduced to a computer or a computer network, I'll correct that impression now. Even the most advanced C^2 systems of the future must consider man's integration into the system. We—you and I—are the ultimate users of information and the primary designer and implementer of operations. The mechanics and electronics we adopt must be congruent with our very human needs and capabilities.

We also must emphasize those programs that give us quality people to operate our C^2 systems—people who understand those systems and can use them well. We have some excellent ones in place now, notably the command and control curriculum at the Naval Postgraduate School and the Joint Command, Control and Communications Staff and Operations course at the Armed Forces Staff College. In addition to these resident courses, several short courses are taught at CINC and component commands headquarters. Unfortunately, only a small number of personnel can attend. Widening this experience should be a major effort in the years ahead.

Survivable and Enduring Command and Control

The President in 1981 recognized that deployment of more survivable and effective command and control systems could appreciably reduce the risks of nuclear war. They were accordingly given a priority in strategic force modernization equal to the new Trident submarine, the B-1 bomber, MX missiles and other key strategic high pay-off weapon systems. Indeed, the command and control program, since 1981, has become the fastest growing part of the DOD budget.

Communications join the various elements of our command and control systems, connecting warning sensors with command centers; enabling those centers to exchange vital information; and providing links to operating forces. Although our first priority is to ensure that those forces can receive retaliatory orders, we also are working toward more survivable two-way communications from commanders to the field.

The level of survivability required is, naturally, dependent on our appreciation of the threat—and that assessment is not an easy task. But as long as our arms reduction efforts fall short of producing a nuclear-free world, we will be faced with the reality that we cannot put the atomic genie back into the bottle; nor can we prepare credible deterrent forces without considering how to manage forces in the event of the worst case conflict scenario. For the foreseeable future, C^2 planners will be compelled to think through the proverbial unthinkable, as a matter of the highest priority.

Interoperability

During World War II General Eisenhower observed that the day of the single-Service or even single-nation conflict was over; wars of the future would be fought jointly by all Services and in combined operations with allies. The crises and conflicts of the past four decades attest to the continuing validity of that observation.

U.S. military strategy recognizes that if we go to war, we will fight in coalitions—in multi-service operations and alongside allies. Within our own forces we still have a good deal of work to do in the "jointness" arena, though we have come a long way, especially in the last few years. With our allies, the problem of ensuring interoperability becomes even more complex. Attaining it—or working to attain it—under all conflict conditions and in any location around the globe will be difficult. But it is, and will be, an absolute imperative for command and control; unlike most weapons platforms, these systems are not autonomous and all have some interoperability requirements.

It is easier, however, to describe the interoperability imperative than it is to respond to it in a cost-effective manner. It contemplates full integration of planning, training, doctrine, procedures, force structure, experience, hardware, common standards and, most fundamentally, the incorporation of all this into an overarching design. This places a great burden on planners in joint assignments; increasingly they must ensure *global* systems support not only for our own forces but also for key allies.

Acquisition

Systems are acquired over periods of time in which the whole world of electronics and telecommunications undergoes great changes. Such rapid advances can mean that if it takes 10 to 15 years to field a system, the equipment may be outdated even before it is installed. Inevitably we are going to need to put in place a more streamlined evolutionary C^2 acquisition process that can keep pace with the major technological breakthroughs.

There is an associated problem, too: As we acquire new technology, we also must develop new ways of doing things—of *using* the technology. But, on the other hand, we cannot allow degradation of existing C^2 systems while we transition to new ones. There must be *evolutionary* introduction of new technology to upgrade existing command and control system capabilities rather than *revolutionary* replacement of entire systems. Our native enthusiasm for quick exploitation of technical possibilities must be reined in by leaders and staff people who can set aside fascinations and impulses, and evaluate potentials not in the abstract but in light of operational impact.

Future of Command and Control

The C^2 arena is dynamic, perhaps the most dynamic of all those in which leaders, planners and strategists must work. Systems requirements change continually as we evaluate and reevaluate the threat and our missions, which are now involved in outer space as well as on our globe. Technology continues to advance, creating opportunities, and we are faced, all the time, with the challenge of using our best technical capabilities to fashion better defenses. In an era of change, the future has to be shaped by leaders, and by deliberate, comprehensive plans that will allow a smooth transition from where we are in our C^2 systems today to where we want to be. The central principle is that command and control is a major link in the chain of deterrence.

It will take vision, discipline, and perseverance—and the unstinting efforts of commanders and operators who understand the capabilities and limitations of their systems, to envision what they need, and to engage intelligently with the processes that bring new systems on line.

My own view of history suggests that we will go forward with confidence in our ability to determine and build the systems we need, and that we will use them well whenever and wherever they are required. But as Mark Twain once said: "Even when you're on the right track, you'll get run over if you just sit there." The world of command and control may be the best illustration of the truth of that observation.

Strategic Command and Control and National Security

Bruce G. Blair

Since the beginning of the nuclear age, a very narrow view of nuclear capabilities has dominated the strategic debate in the United States. This view focuses on the number, technical characteristics and cost of nuclear weapons. It excludes the technologies and procedures created to manage those forces and barely acknowledges the seemingly obvious fact that if command and control (C^2) fail, nothing else matters. As a result of this exclusion, strategic C^2 has historically been neglected in defense budgets and arms control negotiations. The result is that deficiencies in the U.S. control system have been so severe for so long that developments in the areas of apparent importance—growing numbers and technical sophistication of weapons—have become practically irrelevant to the nuclear confrontation.

The Command and Control Problem

Overreliance on narrow measures of nuclear capability has both caused and obscured several fundamental problems for U.S. national security. First, it led to the comforting conclusion that the United States has always possessed the ability to destroy the Soviet Union in a second strike, even after absorbing a maximum Soviet attack. In fact, the U.S. retaliatory threat has been at serious risk since the 1960s; in all likelihood, Soviet nuclear strikes

**This article originally appeared in the March 1985 issue of SIGNAL.*

against U.S. command, control and communications (C^3) after 1966 would have severely degraded, and possibly blocked, U.S. retaliation.

That calculations based on weapons charateristics obscure this problem is illustrated by their underestimation of the vulnerability of the Minuteman force. According to standard calculations, a severe threat to the Minuteman force did not emerge until quite recently. An attack aimed at the Minuteman launch control centers instead of the missile silos, however, would have caused far greater damage. Using nuclear blast effects against a very restricted part of the U.S. nuclear control system, a Soviet ICBM attack circa 1966 offered, at least on paper, a good chance of disabling the bulk of the Minuteman force. The expected damage exceeded by a wide margin the damage that attacks on individual silos would have produced, at one-third the cost in terms of missiles expended. Since the launch control centers were also probably vulnerable to collateral nuclear weapon effects such as electromagnetic pulse (EMP), a Soviet attack might have neutralized the Minuteman force.

During the same period, the risk was also high that a Soviet attack aimed at a relatively small number of key control nodes would have effectively destroyed the U.S. control system and severed links to Strategic Air Command (SAC) and Navy forces. For instance, Soviet nuclear strikes against a few Navy command posts and shore based radio transmitters would have isolated the U.S. ballistic missile submarine (SSBN) force from higher authority. In the mid-1960s, the chain of command and parallel communications for execution of the Navy's portion of the strategic war plan ran from the national command authorities to the Commanders-in-Chief Pacific (CINCPAC) and Atlantic (CINCLANT), and then to the executing SSBNs via radio transmissions by dedicated very low frequency (VLF) shore stations controlled by CINCPAC and CINCLANT. It is likely that the destruction of the two soft command headquarters and the handful of dedicated radio transmitters would have produced so severe a concussion that positive control over the SSBN force could not have been effectively exercised.

Due to these threats to the ground portion of the nuclear control system, the United States expanded the back-up airborne system by deploying a diverse array of aircraft whose missions ranged from relay of Single Integrated Operational Plan (SIOP) messages to SSBNs, to airborne launch of the Minuteman force. The full system was in place at the end of 1969. But by then a unique threat to the airborne control network had emerged with the Soviet deployment of *Yankee* class submarines in patrol areas within striking range of the aircraft, all but two of which were normally maintained on 15 minute ground alert. Aircraft at normal alert readiness could not have coped with surprise attacks at the time because early warning of Soviet submarine launched ballistic missile (SLBM) attack was deficient and because the flight time of attacking missiles was shorter than the reaction time of most aircraft.

Among the aircraft that became vulnerable to sudden destruction by Soviet SLBMs was the National Emergency Airborne Command Post (NEACP)

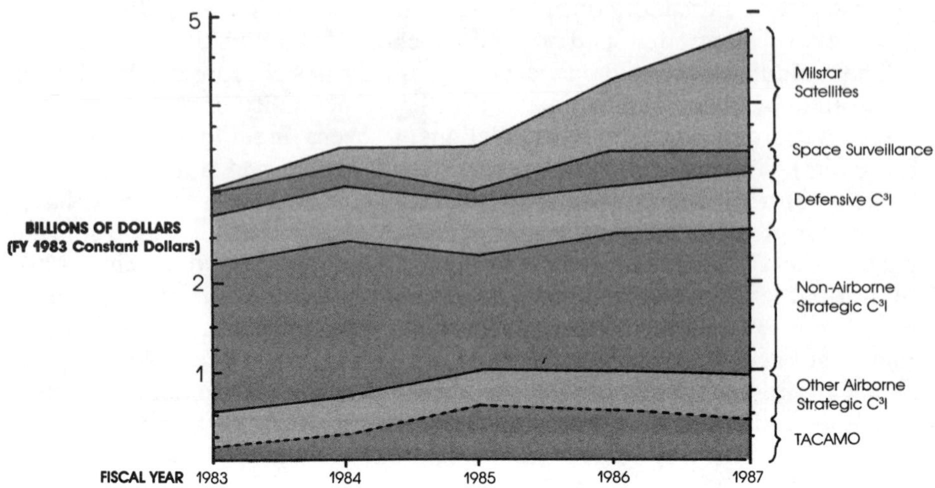

Figure 1. Strategic C³I trends.

aircraft stationed at Andrews AFB, Maryland, on ground alert. Depending on the exact location of *Yankee* subs at the time of attack, the flight time might have ranged from 13 minutes (900 nautical miles from the coast) down to as little as seven minutes (300 nautical miles from the coast). In either case, the prelaunch vulnerability of NEACP was acute, raising the specter of an attack that would destroy the most survivable component of the national military command system. Decapitation became the most significant threat to U.S. nuclear capability, yet standard assessment of the strategic balance continued to treat force vulnerability as the prime concern.

In recent years, efforts to strengthen C² have concentrated on deploying early warning and communications satellites and on installing stringent physical safeguards on long-range bomber and ICBM weapons to prevent their unauthorized use. Nonetheless, the U.S. nuclear control system remains vulnerable even to collateral nuclear weapons effects. One of these effects is scintillation. A nuclear burst at high altitude produces plumes that extend hundreds of miles horizontally and vertically. This plasma divides into filaments or striations that align themselves with the Earth's magnetic field. Radio waves that propagate through the disturbed region undergo random fluctuations in signal amplitude and phase as the striations change position and shape. These random fluctuations can degrade communications severely for hours, over very large areas, and the effect is most pronounced for the ultra high frequency (UHF) band used by the current strategic communications satellites.

Another collateral effect is magnetohydrodynamics (MHD). High altitude explosions create MHD effects that generate anomalies in the Earth's magnetic field, which in turn can "short out" long terrestrial cables such as those that the U.S. control system relies on to carry sensor data, to conduct emergency conferences between senior decision makers and to deliver emer-

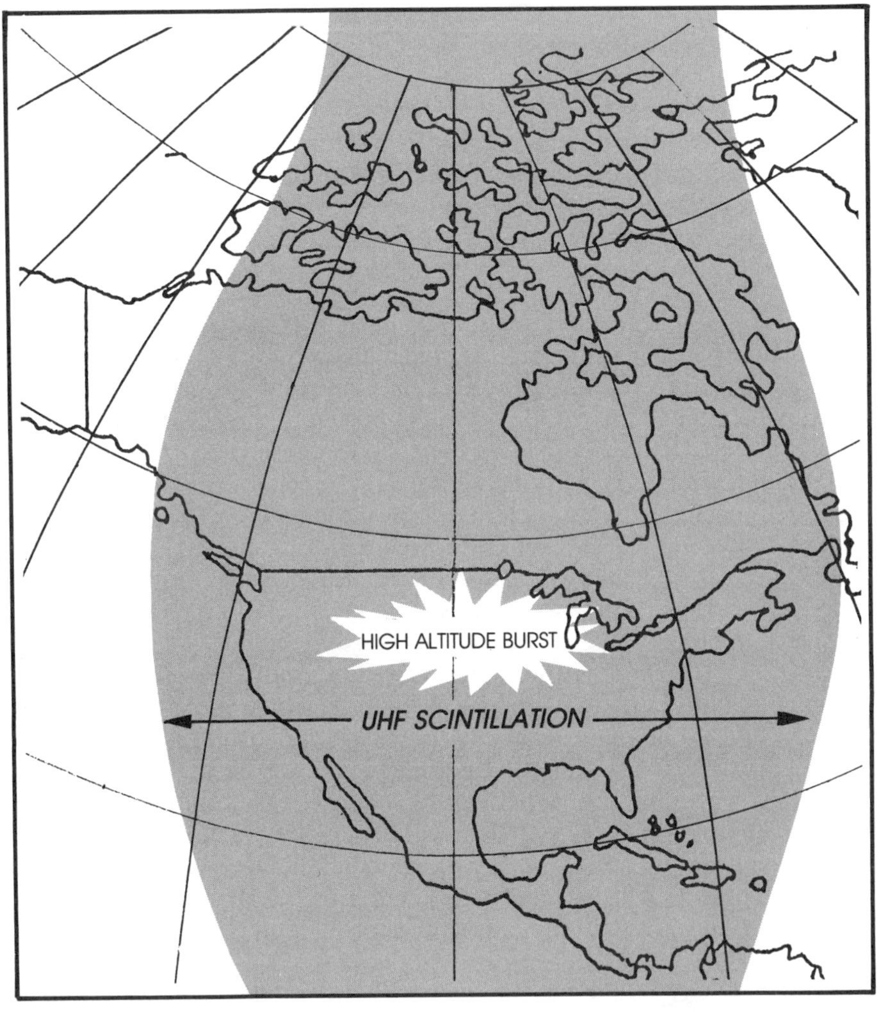

Figure 2. Area of scintillation effects.

gency messages to nuclear units during the critical pre-impact phase of a Soviet attack. This effect has stimulated concern that high altitude explosions, which could occur five or 10 minutes before other incoming weapons arrive, could effectively paralyze the control system and delay the time-critical launch of ground alert command aircraft and bombers.

Of even greater concern is the continuing gross disparity between the number of significant ground based control nodes and the number of Soviet nuclear weapons available for use against them. Since the mid-1960s, the Soviets have possessed more weapons than were needed to attack every important U.S. node and effectively destroy the ground portion of the U.S. control system. Today, there are still only a few dozen such targets. Primary, secondary and tertiary control targets comprise a list of no more than 450.

Moreover, questions remain about the performance of the airborne network despite the recent transfer of NEACP to an inland base (Grissom AFB, Indiana) in an effort to alleviate the threat of sudden decapitation. Prelaunch vulnerability, airborne communications deficiencies and limited endurance have combined to raise fundamental doubt about the airborne network's ability to direct the forces to coherent national purposes. A longstanding risk of decapitating attacks on the National Command Authorities, the National Military Command System, or both, heightens this concern over airborne performance as it adds uncertainty about the source of its instructions.

The second problem caused by the narrow view of nuclear capabilities is the underestimation of crisis instability. The vulnerability of the U.S. control systems long ago created a bias in the national command system toward rapid and early execution of the SIOP, while the system was still intact. This is manifest in the operational posture of the U.S. forces and control system. It is expected that retaliatory authorization would be released within very few minutes after the initial detection of Soviet missile launches. That expectation is so deeply ingrained that the nuclear decision process has been reduced to a drill of sorts—a brief, emergency telecommunications conference between senior military and political authorities whose purpose is to extract presidential authorization to retaliate in the shortest possible time.

This time pressure on senior military and political authorities undermines rational assessment and promotes miscalculations that could produce inadvertent war. In the short time available for decision, national policy officials are also bound to focus on a single, one-time plan of retaliation. This restriction encourages comprehensive retaliation. Whatever the SIOP may be and however invulnerable the forces themselves may be, the U.S. control system thus has been geared historically to a spasmodic, indiscriminate response. This sharply conflicts with national policy requiring the ability to ride out an attack and respond in a manner that is proportionate to the scale of provocation.

The third problem fostered by the narrow view of nuclear capability has been a dangerous interpretation of, and response to, the Soviet strategy. The overemphasis on the numbers and technical characteristics of weapons leads to the expectation that Soviet nuclear attack would be concentrated on, if not restricted to, U.S. strategic forces. Soviet strategy is thus seen as designed for limited attacks to gain exploitable bargaining leverage during a protracted conflict. This alleged incipient threat of Soviet nuclear blackmail has provided the main rationale for the President's strategic budget, the lion's share of which is earmarked for modernization of the forces, not the command system. In reality, Soviet planners would not spare the U.S. control system if damage limitation is the predominant objective of their strategy, and that means that nuclear war would become uncontrollable at the outset. I would go even further and argue that regardless of Soviet targeting doctrine, a nuclear war could not be controlled. The rudimentary design and short endurance of the U.S. control system nullifies the entire conception of multiple, time-phased strategic campaigns undertaken in pursuit of bargaining advantage. Such campaigns are purely intellectual constructions, metaphors really, that have no place in serious threat assessment and planning.

Lastly, the narrow view of nuclear capabilities places the entire burden of war avoidance on the threat of retaliation. But the prevention of nuclear war requires negative as well as positive control. The ability to prevent unintentional attacks, including unauthorized launches, is no less important than the ability to implement authorized strike plans.

Solving the Problem: Modernization

In 1982, President Reagan unveiled a plan for strategic modernization which seemed to depart from tradition by assigning the highest priority to C^3. So deeply ingrained is the narrow view, however, that not even the spirited promotion of C^2 by the President could dislodge it. Strategic C^2 remains underfunded. Planned expenditures for command modernization are following a modest growth curve in real terms. While the out-year accounts show sharp upward slopes, uniformed services tend to flatten them when the future arrives. During President Reagan's first term, the military departments have not failed in their efforts to postpone, delay, or eliminate important programs.

A prime example is the Navy's decision in FY 85 to postpone the initial acquisition of new SSBN communications relay aircraft (while vigorously pursuing a 600 ship fleet) that are key to the modernization of communications with the sea based nuclear forces. The jumbo jet E-4 aircraft once slated to replace the outmoded EC-135 aircraft used in two key roles—NEACP and SAC Airborne Command Post—met a similar fate. Shortly after the President assigned top priority to C^2, the Air Force trimmed the E-4 fleet down to four and cancelled plans to acquire these modern aircraft to support SAC. (In the early 1970s, then Deputy Secretary of Defense David Packard, undoubtedly the strongest senior level advocate of improved nuclear C^2 in Pentagon history, considered the Department's highest priority to be the deployment of a large fleet of E-4 aircraft.)

Major improvements in overall C^2 capabilities are still needed. The immediate concern of planners should be to create a system that provides for assured SIOP retaliation in the wake of a Soviet first strike against the U.S. nuclear control system. To accomplish this aim, budgetary support for programs in the five year defense plan must be sustained, and new programs must be initiated. For approximately \$30 to \$50 billion, over and above planned expenditures, the United States could deploy a high confidence command, control, communications and intelligence (C^3I) network capable of supporting traditional SIOP missions. Besides the SSBN communications relay aircraft, key C^3I programs that ought to be better protected during annual budget review include Low Frequency (LF) radio mini-receive terminals for the bomber force, which are essential to delivery of the "go-code" to outbound aircraft; Ground Mobile Command Centers for senior political

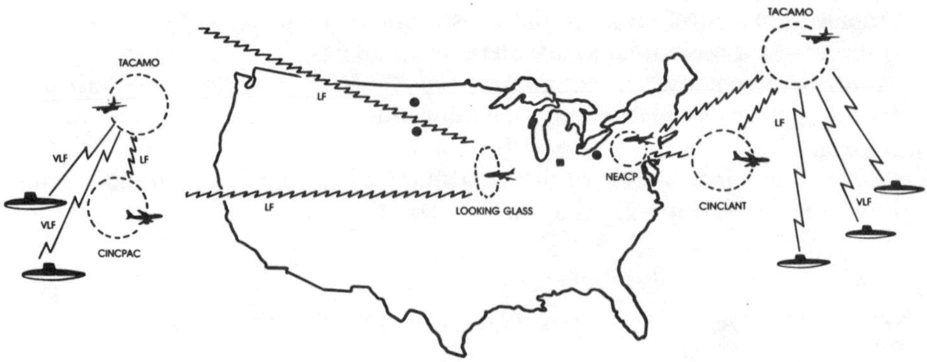

Looking Glass—SAC Airborne Command Post
NEACP—National Emergency Airborne Command Post
TACAMO—SSBN Communication Aircraft
CINCPAC—Commander-in-Chief/Pacific Airborne Command Post
CINCLANT—Commander-in-Chief/Atlantic Airborne Command Post
LF—Low Frequency Radio Communication
VLF—Very Low Frequency Radio Communications

Figure 3. Airborne submarine communications.

and military authorities, which would augment command posts; and Mobile Ground Terminals for survivable readout of tactical warning data from early warning satellites. These programs along with the SSBN communications relay aircraft replacement program should be expanded and accelerated.

A substantial increase in the C^3I budget is needed to modernize the airborne command network on which the United States relies for SIOP retaliation. The current outdated inventory of EC-135 aircraft that form the backbone of this network should be replaced by modern aircraft designed to resist collateral nuclear weapon effects—particularly EMP effects—and operate from austere airfields with short runways. Additional funding also is needed to install modern communications suites on all C^3I aircraft. There is almost no money in the budget to replace or upgrade the old, unreliable but critical LF/VLF radio systems carried by C^3I aircraft. New equipment is needed for effective communications in a jamming and nuclear weapons environment. Other needed but unfunded improvements to the network include modern automatic data processing; onboard sensors for detecting airborne nuclear radiation and turbulence; sensors to identify safe landing strips; and two-way radio links between airborne launch control centers and Minuteman missile silos.

Modernization of the ground segment of the nuclear control system is also underfunded. The prelaunch survival of C^3I aircraft and long-range bombers on ground alert depends upon timely detection of an in-progress attack, rapid decision making and speedy transmission of take-off instructions. These vital functions are performed by ground command posts and terrestrial communications whose operations could be severely disrupted by high altitude nuclear detonations. During the critical pre-impact period of missile attack, the United States can ill afford disruption of the means by which it alerts nuclear units to impending attack. Such degradation compromises both negative control—prevention of inadvertent war—and positive control—authorization and implementation of effective counterstrikes. To bolster both, additional hardening of ground facilities and the commercial landline network, emphasizing fiber optics, at a cost of several billion dollars, should be undertaken.

Two emergent threats to future C^3I systems are not being adequately addressed. These threats are modern Soviet cruise missiles and advanced Soviet antisatellite (ASAT) weapons. Cruise missiles are much slower than ballistic missiles, but they are much harder to detect. They could be used effectively in a decapitation attack, especially if launched from delivery systems located close to U.S. borders. For these reasons, development and deployment of Over-the-Horizon radar and other sensors capable of detecting a cruise missile attack should have higher priority and funding support.

Of equal or greater concern are Soviet ASAT weapons that could pose a major threat to high altitude satellites during the 1990s. Although deployment of such weapons as space based lasers or neutral particle beam weapons is not expected in the near future, they are plausible threats in the medium term (five to 10 years), and yet satellites on which the United States will heavily rely during this period and beyond are not being designed to cope with them. Since space is becoming an arena of intensifying military competition, it seems only prudent to anticipate the emergence of new threats to satellite constellations such as MILSTAR, which is the centerpiece of the current program of command system modernization. There is no doubt whatsoever that with concentrated effort the Soviet Union could deploy ASAT weapons that would render MILSTAR acutely vulnerable to attack.

The United States eventually must confront the issue of command endurance. While the immediate concern is to modernize C^2 so that it can support traditional SIOP missions, American planners cannot evade very much longer the responsibility to provide the President with the option to withhold all or part of the SIOP forces in the wake of a nuclear attack. If deterrence fails, there would be a continuing necessity to deter an aggressor's residual forces and to maintain strict national control over U.S. forces. In such grievous circumstances, national policy officials surely will need more than a few minutes, or even a few hours, to determine the national security interests of the nation and the role of nuclear weapons in pursuing those interests. If providing for assured SIOP retaliation is difficult and expensive, how much more so is it to provide adequate endurance? The U.S. supposed commitment to effective nuclear C^2 will truly be tested, as it would force very hard choices between investment for C^2 modernization and weapon system modernization.

Solving the Problem: Arms Control

A broader view of nuclear capability and national security lends added importance to arms control measures that serve to protect control systems and to reduce risks of misperception in a time of crisis. A number of measures could alleviate current problems:

A ban on antisatellite (ASAT) testing and deployment. The ASAT negotiations with the Soviet Union suspended by the United States after the invasion of Afghanistan in 1979 should be resumed. A Soviet offer to impose a mutual moratorium on ASAT launches provides a basis for initial discussion that could lead to a total ban on ASAT testing and deployment.

Although the United States is conducting initial tests of an F-15-based ASAT weapon, which is superior to ASAT weapons in the Soviet inventory, the long-term costs of unrestrained ASAT competition could vastly outweigh any near-term advantages the U.S. may hope to gain. Space based assets play important, in some instances vital, roles in national defense. Both sides have a strong interest in preserving the accepted legitimacy of these activities—notably, worldwide communications and intelligence collection—but the stakes are especially high for the United States. Its strategic forces are more far-flung and hence more dependent upon satellite communications. Furthermore, the United States is becoming increasingly dependent upon satellites for strategic communications and tactical early warning. Under the provisions of the current C^3I plan, MILSTAR communication satellites and early warning satellites will form the backbone of the future U.S. C^3I network. The emergence of an ASAT threat to these systems, a development that should be anticipated in an environment of unrestrained ASAT competition, could derail current plans for command modernization.

A prohibition on stationing missiles close to enemy borders. Both sides currently operate nuclear ballistic missile forces at forward locations; these missiles can strike enemy command facilities within a few minutes after launch. U.S. and Soviet submarines are now positioned for such short warning attacks. Pershing II deployments in western Europe will, from a Soviet standpoint, aggravate the problem because they could strike key Soviet command facilities to the west of Moscow within 10 minutes after launch.

An arms agreement that restricts nuclear deployments to zones that preclude short-range attack would mitigate the threat of sudden command decapitation. Stipulations of this sort would promote crisis stability and ought to be a subject of negotiation. Since the pertinent deployments to be traded off include weapons that cut across both INF and START—Pershing II and SSBNs, to cite the prominent examples—the two sets of negotiations would have to be bridged for this purpose.

A prohibition on tests of SLBM "depressed trajectories." The flight time of SLBMs can be substantially reduced by lowering the missiles' trajectory. To date, neither side has conducted tests of depressed trajectories. Were they to do so, the threat of command decapitation would become more pronounced, thereby undermining crisis stability. Therefore an agreement that prohibits tests of depressed trajectories should be negotiated.

A ban on forward deployment of cruise missiles. To mitigate the threat of precursor cruise missile attacks on command facilities, the parties could agree to prohibit close-in deployment of cruise missile carriers—ships, sub-

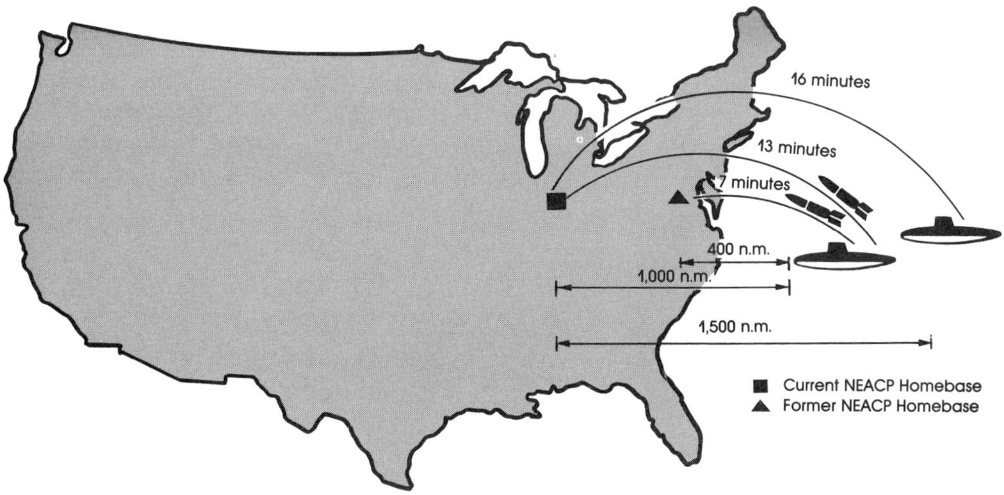

Figure 4. Flight times of Soviet SS-N-6 SLBMs.

marines, aircraft and ground bases. Due to the extreme difficulty of verification, however, a far superior solution, albeit a politically infeasible one, is to ban cruise missile deployments altogether.

In making command improvement the pivotal component of its plan for strategic modernization, the Reagan Administration formally acknowledged the importance of command performance to national security. More than that, for the first time in 25 years the U.S. government defined command modernization as the overriding aim of U.S. nuclear policy, putting it at the top of the strategic agenda. Underneath the encouraging rhetoric, however, lies the sobering reality that the United States is not much closer to answers to the central command questions than it was a decade ago.

The crucial first step, which has yet to be taken, involves major legal and institutional change. Implementation of any plan that elevates the priority of command development to the same level as weapons systems requires that a powerful C^3I advocate be created within the defense bureaucracy. Advocates have historically lacked the clout necessary to alter substantially the logic and pattern of strategic investment. Unless a countervailing power base is created, weapon programs will continue to take precedence, regardless of the prominence of command modernization themes in presidential oratory and/or defense guidance.

Nevertheless, the climate for new conceptualization and new initiative in this area has never been more favorable. Defense professionals must reemphasize that command performance is a central question of national security and must actively support the search for answers. This search needs to be conducted without the emotional extremes of despair or wildly inflated expectations. With concentrated and sustained effort, the United States could at least begin to come to grips with one of the most important security dilemmas of our time. Given current technological constraints, all promising avenues must be explored.

This article is an adaptation of an article which appeared by the author in Arms Control Today, *January 1985, a publication of the Arms Control Association. The artwork is reprinted by permission from* Arms Control Today, *January 1985.*

Reflections by a Soldier on Command, Control and Leadership

H. F. Zeiner-Gundersen

Command and control (C^2) in relation to NATO and European operations is, in essence, not very different from that in any one nation, but it can require more understanding and tact. C^2 is a means of exercising leadership, but it is not the only way.

Keeping these points in mind, I would like to discuss four major C^2 topics: NATO and national considerations; command; control; and leadership.

NATO and National Considerations

When working in NATO it is important to remember and always to be aware that NATO consists of 16 independent countries joined in an alliance of defense—it is not just the United States and Western Europe. As it happens, two of the member countries (with about 250 million inhabitants) are situated in North America. Fourteen of the countries (with about 350 million inhabitants) are situated in Europe and, for reasons such as history, are even more different from each other than they are from the two North American countries. Whereas citizens of the United States and those from most of Canada speak the same language, the 14 European countries speak 12 different languages. All 16 countries are independent and are proud of their nation and their language. This pride and the fact that all of the different

This article was written especially for **Principles of Command and Control.*

languages convey thought equally well are important considerations, and national aspects and characteristics must be remembered when exercising C^2.

National characteristics usually are difficult to change in the short run, but they may change over time. These characteristics develop at an early age and usually have roots in history and tradition. They also can be influenced by the homogeneity of the population.

As already mentioned, each one of the 16 nations has tremendous pride in itself, although, admittedly, in different ways and on different subjects. The general and deep-set feelings of any nation are that its own educational and governmental systems are the best, and its own ways of doing things are the most effective. This applies to all 16 NATO countries, rightly or wrongly, irrespective of population size.

In my opinion, three points are important when exercising leadership within NATO: first, be a willing and, preferably, a polite listener, and establish confidence through listening; allow those from other nations to choose their own ways and methods as long as the aim is attained; and finally, try to persuade and influence, but do not force your way. People will resent being forced.

Realize that some members of the armed forces want direct orders, which take responsibility away from their soldiers. Others must be convinced or persuaded, which acknowledges their intellectual capabilities or their sense of freedom and independence. Some forces like to decide for themselves and they despise orders but tolerate broad directives.

In working with NATO nations, recognize the differences in native languages and in their traditional use. Sentences may not mean what is intended when translated literally. Another more important difficulty lies in the knowledge of languages and the perceptions of the person who receives the order. Personnel from the larger nations should not consider those from other nations as lacking in knowledge and intelligence. Such prejudices will be felt by "recipients" and are obviously harmful. Intelligence and knowledge do not necessarily follow size or wealth. Smaller language groups and/or very different language groups are particularly vulnerable in this connection.

Whereas setting the aim is always the privilege and the right of the group exercising leadership, the local people always have the privilege of the perception of leadership and are correct when it comes to judgment of the execution of such leadership. Traditions are important to adhere to, more so in some countries (or regions) than in others. Such traditions will vary as do the social structures among countries. In all countries some harmony must exist between the execution of leadership and the social order of the day.

Lastly, and as a general consideration, remember that when you deal with your allies and they believe their national pride has been hurt, they will ride to the end of the course.

Exercising Command

To me, the concept of command has a number of faces:

- To be in command is a responsibility given to one or more persons.

- To command is to exercise leadership, be it through instructions, orders, etc., or in verbal, written or electronically transmitted form.
- A command may be a particular directive or an order.
- Command must provide inspiration because it requires understanding from the recipients.
- The lines (channels) of command always must be clear.
- Any command given must be clear without the possibility of it being interpreted in any other way than intended.

Because of these rather different faces, it is not easy to formulate a brief definition, but a general definition might be as follows: Command is one way of exercising leadership. A command transfers thoughts from the person in command. It could be given in the form of an order, an instruction, etc. It would furthermore be verbal, be in writing or be electronically transmitted.

If one chooses to exercise command in the traditional way, then orders usually are issued. Orders also may be given in industry and business, although their format will differ and possibly be less direct than in the military services. The reasons behind such orders are, however, the same as in the services, with some minor adjustments. I therefore believe deliberations on command will be broadly the same in the military services and in the civilian sector, during peace and wartime. This also would be the same within the NATO organization.

While terminology may vary, as a retired soldier, I will use the word "orders" for conveying command decisions. Orders will be given when minds are made up and decisions taken. They can be broad or limited in scope, but they always must inspire and direct toward attainment of the aim.

The Aim

Generally speaking, an aim may be given to people or it may be decided by the people themselves. In the latter case, it is really a sub-aim, that is, an undertaking within the limits of some wider, more far-reaching aim. Aims may be realistic and easy or difficult to attain. Whatever they are, aims always are determined in consideration of each person's strengths, weaknesses and differences. Such, also, is the case with those who are given an aim or who set a sub-aim. While only with difficulty can you influence an aim you receive, it is always possible to influence your own choice of sub-aim. Human factors, such as sense of realism, creativity, willingness to accept risks, etc., will color the choices.

Most aims given or personally set are realistic and may be attained through hard work and some luck. They may need some extra tools, in which case such tools should be requested. After that, the job must be accomplished: It does not matter if the aim is to save a failing company or to take the next hill.

Occasionally, someone is given or sets an unrealistic aim. This can happen at any level. When receiving the order, you may be tempted to pass it on, only issuing the logical and necessary instructions. While you then could blame the subordinate for the failure, this is obviously a dishonest and unethical behavior. Furthermore, in the end, such behavior will result in loss of trust and credibility. It is the loser's behavior.

A much better choice is to use your judgment, to speak your mind and to take the consequences. In most cases, doing this will change things; however, if instructions and orders are not changed, they must be followed. In both cases the next day you are able to look yourself in the eyes in the mirror.

General Comments

Orders will be passed to many types of subordinates: some are independent minded, some are not; some are rash, others, thoughtful; some are strong, others, weak. In NATO some may have national characteristics that vary from your own. At times, fast action is required; other times, speed is unimportant. Orders will be given in different ways. The choices are through personal appearance or by portraying the decisions reached through written orders, messages, etc. The choice is conditioned by whether or not a personal appearance is possible (time, space, etc.) A decision on this will require considerable reflection on factors such as:

- The importance of each case.
- A realistic evaluation of what influence, particularly in cases of criticality, your own person could have.
- Evaluation of the morale effect of not appearing personally.
- Need of your presence elsewhere and/or of availability.
- The need to stress the importance but also of not causing panic.
- The need for speedy reaction.
- The need for closer explanation of the aim.
- The level of command worked at.

Quite obviously, there is not one solution. It varies with the situation but one or two factors always are decisive, and they vary. By and large I favor personal appearance and explanation wherever it is physically possible. This is particularly necessary if you deal with multinational commands, which is often the case within the NATO organization. Otherwise, the complaint that "somebody works or sits in ivory towers" is heard.

However, be cautious of personal preferences, wishes, etc., that can blur your judgment with regard to making an appearance particularly in awkward cases. Similar factors apply when considering whether or not to use your own voice when telecommunicating.

Having decided on whether to make a personal appearance and/or to use your own voice when communicating, you must decide among the different technologies. Today's technologies (telexes, telefax, television, etc.) are faster than in the days of the trumpets. Experts should be used to choose the media, and their choices should be used to the fullest extent.

All transmitted messages must inspire, be concrete, be complete, allow room for initiative and ample time for execution. But, technical means can make a situation volatile. This is dangerous, particularly in crises, as telecommunications systems may be overloaded. Besides, the more words transmitted the greater the risk of contradicting yourself. Remember, it is the interpretation that matters, not what you think you said. This is particularly important when considering (and drafting) orders to someone from another nation. Remember that English is not always a precise tool. It is so easily misinterpreted.

Whatever is transmitted should show the influence of the decision maker, so as to inspire, but not to dramatize. As electronic means work fast, there is always the tendency toward back seat driving. This happens in industry, business and the Service, during peacetime and wartime. This tempting tendency must be resisted, particularly when dealing with personnel from other nations. When put to improper use, this tendency will undermine your allies' sense of responsibility, initiative and self-confidence. The original order should give enough room for initiative and should be changed only when major new information is available.

The Control Aspect

I think the control aspect easily stands alone. Furthermore, it really only has one face—namely that of controlling the actions undertaken to reach the aim. In today's practice, control often is used as a substitute for the word "command." Many ways of controlling easily develop into command methods. I define control as follows: Control is keeping an eye on the actions executed to attain the aim. This is in order to assist executors and to change your own commands to fit realities.

Whoever is in charge—the corporate chairman of the board, the executive president, the military commanding officer—wants to be certain his or her aim is attained through orders and instructions being followed. This, in principle, is done through control. The definition of control may seem to be limited in scope, but it must be viewed in light of current circumstances and means.

Generally speaking, development in most spheres goes much faster today than in earlier times, allowing rapid change in circumstances. This can be an advantage, and it can necessitate the higher echelons being "in the detailed picture" all of the time. Such involvement will make it easier to adapt orders and instructions to suit circumstances.

The development of telecommunications has made a more continuous form of control possible and has enabled those in charge to have real time information all of the time.

In the military services, it is necessary to have continuous and detailed control. The increase in speed also makes it necessary to pass relevant information rapidly to subordinates both civilian and military. This allows them to adapt to whatever situation develops.

Having established the need for control, the major questions are: how detailed should it be? and how should it be undertaken? I believe it is important that all concerned realize that good control always allows valuable information to travel both ways. Control should therefore be considered an asset, not a nuisance.

New information obtained through control may lead to a change of instructions and orders, but then command and not control is exercised. It is easily forgotten that there are two elements of control: the controlling agency and the controlled agency. When different nations and different natural languages are involved, other problems can develop.

The group in control will continuously want to be up to date; the group being controlled will prefer not to be bothered with what it considers unnecessary work. Irrespective of sphere and level, such a situation must end with a compromise.

The people being controlled should understand the necessity of control. They must be aware that it gives higher authorities valuable information and must recognize and accept control as a benefit, because information obtained through control will enable higher authorities to adapt their instructions and orders if the information warrants it. On the other hand, the controlling agency must remember to consider feelings and sensitivities. With multinational forces, this is particularly important. Pride can be damaged easily.

Occasionally, there may be a tendency at some levels to ask for reports that are not necessary and are additional to the routine reports. Prior to asking, remember that such work often will distract minds and will be undertaken at the cost of other, possibly more important, work.

At other times, you may find that staffs ask for reports in order to be able to give clearer upward briefings. These are borderline cases where necessity has to be carefully weighed to avoid misuse. Additional valuable information might be available from people who are so busy they forget to pass the information on. Very rarely do you find someone who deliberately holds information that might influence a wider picture. Such behavior is unforgiveable. All valuable information must be passed to those it might influence. This could be accomplished through a routine process or, when the information is time critical, through special means. In military life, minute-to-minute control exists and is often a necessity.

The best way of avoiding wrongs and ill feelings is when controller and executor know each other personally. This is particularly important when they do not share the same native language. Being personally acquainted with those you control will prevent control from being misused. Trust is necessary, and it develops easier between those who know each other and between those people who share the same native language.

Delegation of control authority exists within all staffs and headquarters. Today, when speed and urgency dominate, no single person can sit at the controls throughout. Although obviously one person is in charge at the controlling agency, when it comes to the continuous exercising of control, that person will have to delegate to persons who then act on his or her behalf. The choice of the right people to execute control is therefore a major issue. The following are the major controller factors to consider:

- Does he or she have the necessary knowledge?
- Do you trust him or her?
- Is he or she able to see the consequences of his or her actions?
- Do subordinates and parallel authorities trust him or her?
- Does he or she use simple, direct and clear language?

I cannot provide a more detailed list. However, about 10 or 15 years ago, I witnessed an example that illustrates the control situation. The scene was the Norwegian Sea, where an air control exercise was underway. The land based Norwegian authority and the sea based allied units had agreed on certain boundaries between their air control areas. On two occasions, the sea based authority trespassed and exercised control of aircraft within the land based authority area. The land based controller immediately flashed the following signal to the sea based authority: "Stay out of my area; I do not like your interference." Then, of course, he signed the signal. The reply came back at once: "Thank you, acknowledged." That signal was also signed. After this, there were no more interferences.

To all who know these matters, this is a perfectly normal procedure. I believe it shows that both parties had the necessary knowledge; both parties trusted each other; both used simple and direct language; and the land based controller knew he had the confidence of his superiors. Both signals were framed in glass and pinned on the wall at the headquarters. They still may be there.

Leadership

For some time, I have tried to follow the current management style debate in military publications as well as in industrial and business journals. It appears that the debate has tended to develop catchwords: in industry and business the word is leadership; in the military services the words are command, control and information.

In some of the writings, "leadership" tends to develop into "rules for stage performance," or, even more bluntly, into how to develop the appearance of persons and how to master the "tricks of the trade." These often are of little substance but of much image. Less attention seems to be given to developing the sincerity of the same persons or to teaching the practical ways in which to lead. This rather critical, and hopefully unfair, comment is based mainly on reading, as I regrettably have not attended any leadership seminars.

In other writings, the words command and control (with information as an addendum) seem to focus on the tools (to command and control) and preferably the electronic tools.

I believe these catchwords are not entities in themselves, as they do not present the whole picture, nor do they exclude each other. But stressing any of them exclusively creates the danger of developing superficiality. I also think a close relation exists between each catchword and what it really stands for. I also believe the real content must not be treated superficially.

Taking a basic approach, I would like to put the relations between the catchwords in one rather simplified phrase: Aims are attained through leadership, which again is practiced mainly through C^2, be it in the military services or in the civilian sector.

The words command and control often mean different things to different people. However differently they are practiced—for instance in the Services as opposed to industry—their content is the same. The danger with a brief definition of leadership is that, in simplifying, you gloss over some facts. However, leadership can be defined as inspiring and directing actions to attain an aim. This may be agreed upon by several or decided upon by oneself. Leadership may be practiced in different ways.

Leadership, therefore, serves the aim. It may be executed in many different ways and will inspire actions rather than force them. When exercising leadership, you should look at it as providing a service to both the aim and the organization. It must also be viewed as a service to those being led in that it must provide them the inspiration needed as well as the clearcut directives and/or orders. Such leadership can be exercised by one person or by more than one when they work together in harmony. The object is always to attain the aim.

General Reflections

Leadership, which is exercised in many spheres, in differing circumstances, at a multitude of levels and by many different people on differing audiences, must always inspire. The basis for exercising leadership is knowledge, and the internal ingredient of leadership is C^2. Modern times have highlighted the need for providing information on what leadership really does. Such information is partly needed internally and partly needed externally.

Modern telecommunication with its ever-increasing speed of transmission and its efficiency will help in the execution of leadership. But it also will put additional burdens, particularly in the information field, on those who exercise the leadership.

It is natural to question if a general recipe for leadership exists, and if it does, will such a recipe ensure success? These obviously are important questions to anyone who aspires to lead. They even can be important for current leaders.

Prior to answering such questions, and the answer may seem to be "no" to both questions, let us discuss some of the different conditions a leader might have to address. And this applies to all leaders at any level, in the Services or in the civilian sector, whether introverted or extroverted.

Peace and War

In industry and business not much difference exists between exercising leadership during peacetime or during wartime. There naturally are sizeable differences in the military Services, which varies with the levels.

Overall, leadership in war:

- Requires personal courage (moral and physical).
- Must seriously consider matters of losses and casualties and their consequences. These are factors that hardly exist in peace, but that will influence methods chosen in war.
- Must be more absolute. There must never be any doubt as to who is in charge.
- Must act faster as situations creating opportunities will occur suddenly.

The latter of these, of course, varies and depends upon the level concerned.

To some extent, these factors, particularly when viewed in relation to other leadership considerations, work at cross purposes. This obviously does not make it easier. On higher levels, the political considerations influencing leadership in peace partly disappear. But other considerations must be faced. The military traditionally plans and prepares for the worst case, thus potential leaders are trained for handling war. This is an advantage when related to wartime conditions, but it might not be an equally good basis for exercising peacetime leadership. And as not all human beings master both conditions, this could and sometimes will create problems both in peacetime and in wartime.

Those who exercise any level of leadership must be aware of this fact, consider it and remember it when they issue their orders or select their subordinates.

Other Differences

At least one major difference exists when considering industry and business versus the military Services. In industry and business, both in peacetime and wartime, corporate leadership is necessary only during an eight hour day. Society has the duty of leadership during the other 16 hours. In the Services, leaders are responsible for providing leadership 24 hours a day, including leadership on all ordinary matters of society, such as health and welfare. Therefore military leadership appears to be more demanding.

In some countries, all members of the Services are career employees. Their solid military knowledge can result in fewer and probably briefer orders and fewer explanations of orders. Soldiering to them is a job for which they are paid. In other countries, although officers and non-commissioned officers (NCOs) might be career employees, the other members of the Services are not. In such cases, young men (and sometimes women) are conscripted for service. They have less military knowledge than career military personnel, requiring a leadership that explains and tries to convince. This again means it will be more volatile. And, in a third group of countries, a mixture of career personnel and those fulfilling their national duty exists.

Other factors include the amount of education, and, on a more sensitive level, intelligence. In countries of national service (or conscription) leaders know they will deal with the average youth in terms of their education and intelligence. With career personnel this theoretically might not be so. An indication of this is the extensiveness and types of non-military instruction in the Services where you find career only. Such instruction might be based on the desire to overcome weaknesses, particularly in the educational field. Of course, it also might be caused by the desire to prepare for civilian life.

My intention here is to show that these differences influence the ways of executing leadership. My intention is not to argue the advantages/disadvantages of career personnel versus national service personnel. Some military leaders prefer one way, and some prefer the other way.

Methods of Execution

While there are many different methods of execution, the methods chosen must recognize the attainment of the aim as the main point. Any method of leadership chosen must include both sides of the coin: inspiration and direction (orders, instructions, etc.). You may choose to lead through conferences, personal visits, broad instructions, exact orders, a stream of transmitted signals, etc., but you always must both inspire and direct.

Whereas directing is based on facts and needs, inspiration is mainly based on belief in you as a person and in the cause. And if you are not convinced, you will not convince others. In determining a method of execution, you must know how fast action is required. Sometimes this is not a factor; however it is necessary to act fast to catch the moment of opportunity. The development of tools, machines and production methods has made fast action, with little time to think, more important. This may be most noticeable within the military Services. I am particularly thinking of the development of the sharp-end tools and their weapon platforms.

The development in electronic transmission has had the same tendency and puts extra pressure on those who decide. This has to be watched carefully or commanders might find themselves being forced by their transmission agencies. I believe that knowing when speed is required and when it is not, as well as being able to convince others, may be one of the true criteria for exercising leadership.

Characteristics of Personality

Personal characteristics of leaders vary as do the characteristics of those they lead. Some leaders will readily be accepted, trusted and even loved by subordinates, while others may not be tolerated. It is difficult to explain this logically, but as we all have different limitations, there may be a connection with subordinates sensing these limitations. There also may be a connection with subordinates sensing dishonesty.

How then can a person be accepted as a good leader? Opinions on this vary, but I believe the first criterion would be knowledge, because it is appreciated automatically. I also believe that the spark of leadership and an ability to inspire are traits you are born with. You can be taught all of the "tricks of the trade," but if you do not have natural leadership qualities, using these tricks will not help you in the long run, although in the short run it might.

A leadership spark may be developed into a flame. As for personal behavior I believe the old virtues of knowledge, honesty, fairness, thoroughness and, not least, the personal example are really the right remedies. Always be yourself and do not attempt to portray yourself as somebody or something else, however tempting it may be. Therefore, setting a good personal example is the acid test of good leadership.

I believe the next important aspect is to appreciate the subordinates, the tools they have and their strengths and weaknesses. This is not always easy to do. Some subordinates are suited for lower levels of leadership, some for higher, some for none and some for all. You will probably have no idea how you are accepted. However, your subordinates and their subordinates will know. Your superiors may even know.

As advice on this is hard to come by, skepticism toward yourself is really the only tonic.

As a leader you should be acknowledged by others as being able to do (or having been able to do) what you ask others to do. You also should not ask anyone to undertake something you would not have taken on yourself had you been in their place.

Reaching Decisions

One important test of ability to lead is the ability to reach a decision. A decision has to be reached in order to continue with the job. In the end, it is usually one person who must take it and say the word.

In earlier days, the decision making definitely was a one-person undertaking and was built on that person's observations and evaluations. Not so anymore. Modern society coupled with modern technology requires other ways. In the first place, only a few people have the competence and knowledge needed to reach a reasonable decision by themselves. This mainly is because dimensions and sophistication of available means have grown. The modern way of leadership is to listen to advice and to have a broad spectrum of information available before making decisions.

Listening to advice is an art. You must not be swayed by those who talk well but who do not think too clearly. It is important to select your advisers, allowing for a fair distribution of qualities. It is so easy to choose the ones who say yes. But the ones who will say no are your real friends. They are willing to give an honest answer because they too want things to work correctly. Remember, you do not need to understand all things, but you must understand who does.

The next point, on allowing for a broad spectrum of information is important, yet difficult. Modern telecommunications has considerable capacity for information transmission—some of it important, some not; some correct, some doubtful; and some wrong. But you often do not know which information falls into which category. Still, decisions have to be reached.

In business, industry and the Services only a limited number of factors are decisive, although a much larger number may influence, but, in the end, have to be discarded. You and your advisors personally will have to do this sifting to reach a decision. And remember, it is on the ability to judge what is important and what is not that leaders are weighed, not on their ability to have considered absolutely everything.

It is really a paradox that improvements in signaling capabilities might overload commanders with information on every aspect of the decision-making process.

Does a Leadership Recipe Exist?

Earlier in this article, I indicated that the answer to this question might be no. Different circumstances under which leadership has to be executed and differences regarding those who are at the receiving end of the leadership have been discussed at some length. I believe this discussion has shown that my conditioned response of "no" is correct. I do, however, believe certain ground rules exist for whoever exercises leadership:

- Decide upon the aim and remember it.
- Appreciate and learn the characteristics, strengths and limitations of those you lead.
- Choose your method. It must inspire and direct. And do not let technical gadgets force you to choose wrong methods.
- Be cautious of experts; they often are wrong.
- Be yourself; be honest; and be fair. Portray your image, for good or bad, not that of somebody else.
- Do not let the temptation of popularity sway you.

This may not give you the success you are after, but I believe it will help you in obtaining it.

Summary

I can summarize the exercising of command, control and leadership as follows:

- Have knowledge

- Be honest
- Show fairness
- Have independence of mind
- Be courageous
- Bear in mind the aim
- Use simple and direct language
- Show understanding.

I cannot provide a solution to it all; however, the key words are mutual trust. That is something nobody can buy. It has to be earned through time and hard work. On the other hand, it can be destroyed in a second.

Human Information Processing Principles for Command and Control

Andrew P. Sage

Abstract

This chapter provides an overview of human information processing principles for command and control, and in particular for command and control systems design. Knowledge of the purpose and function of C^3I systems leads to the identification of three primary need areas for the development of efficient and effective support systems for C^3I functions: communication system hardware and software for command and control; human information processing concerns and how human information processing capabilities can be enhanced through use of effective knowledge support systems; and user requirements and leadership characteristics as they effect the need for and the use of support systems to aid in command and control functions. Satisfaction of the last need is necessary for satisfaction of the second need. These two needs, each of which relate to effective system design for human interaction, strongly influence the requirements for operational functionality of

This article was written especially for **Principles of Command and Control.*

appropriate hardware and software. In this chapter, we examine some of the human information processing principles necessary for successful command and control.

C^2 *Purpose and Function*

Command and control (C^2) is that integrated function which enables the establishment of objectives and associated selection of action alternatives for force deployment. The basic C^2 functions are information gathering, situation assessment, action selection, response planning and execution, and monitoring of the implementation of alternative courses of action. It must be possible to develop effectively and to convey efficiently commands from higher to lower levels within a management hierarchy, and to enable information feedback throughout the command and control system under extreme environmental conditions.

C^3 *Purpose and Function*

An effective communications system for command and control is necessary to support the information sensing, processing and transmission capability of military commanders in these tasks. This decision support system, often called a command control and communication (C^3) system, consists of: radars and other sensors and intelligence sources for gathering information; computers for information processing, storage and display; and communications equipment and networks to enhance information transfer among various command and staff levels.

C^3I *Purpose and Function*

Intelligence analysis is a major purpose of a C^3 system. The goal of intelligence analysis is to predict human intent and behavior on the basis of retrieved information. This includes: estimates of personnel and equipment locations; the size of these elements; and their capability, options available, and intent.

Human Information Processing in C^3I *Systems*

Human information processing tasks in C^3I environments are especially difficult due to the many factors which lead to information imperfection. The available information in a C^3 system is sometimes likely to be of very questionable validity and reliability. It also may be very incomplete and, especially since C^3I is inherently a group effort, often be inconsistent and possibly even be incoherent. There often are severe time constraints and personnel shortages. Rapid degradation of portions of a C^3I systems is a reality. Message summarization often is needed, as is the aggregation of summarized messages as they pass through the system. The individuals involved in the many C^3I activities necessarily have different experiential familiarity with various intelligence activities and situations. Although they may be most strongly influenced by doctrine, and the policy and alternatives suggested by doctrine, there still will be a multitude of perspectives through which intelligence information and decision situations are seen.

It is, therefore, easy to reach the conclusion that a C^3 system is a system of very large scale and scope [Andriole, 1986]. It is a distributed system of people and machines in which there exists a need to forecast the present and future intent of potential and real adversaries and to identify and communicate alternative courses of action that are in support of established objectives.

Decision and Problem Solving Activities in C^3I Systems

There are many ways to characterize human information processing and associated judgment and decision activities [Janis and Mann, 1977; Sage 1981a]. Regardless of the way in which an information processing, judgment, and decision process is characterized, all characterizations will necessarily involve an initial step in which some need to resolve an issue is identified. This, we will call stage 0:

0. *Recognition and assessment of a decision situation*—in which the need for a decision is recognized, information requirements for the decision situation are identified, and the meta-level decision of how to decide is identified.

After this is accomplished, there is an effort at problem resolution that involves three fundamental activities [Sage, 1977, 1981ab, 1982]:

1. *Formulation of the decision problem*—in which the needs and objectives associated with a perceived decision need are identified, and potentially acceptable alternatives, or options, are identified or generated;
2. *Analysis of the alternative decisions*—in which the impacts of the identified options are evaluated; and
3. *Interpretation and selection*—in which the options are compared by means of an evaluation of the impacts of the alternatives. The needs and objectives of the client group are used as a basis for evaluation. The most acceptable alternative is selected for implementation or further study in a subsequent phase of decision-making.

Without question, this is a formal rational model of the way in which judgment and decision is accomplished. Even within this formal framework, there is the need for much iteration from one step back to an earlier step when it is discovered that improvements in the results of an earlier step are needed in order to obtain a quality result at a later step of the effort.

Table 1. Three Fundamental Steps in Problem Solving and Nine Associated Activities

1. Issue Formulation
   1.1 Environmental Monitoring
   1.2 Environmental Understanding
   1.3 Identification of Information Needs
   1.4 Identification of Alternative Courses of Action
2. Issue Analysis
   2.1 Identification of the Impacts of Alternatives
   2.2 Fine Tuning the Alternatives for Effectiveness
3. Issue Interpretation
   3.1 Evaluation of Each Alternative
   3.2 Selection of a "Best" Alternative
   3.3 Implementing the Selected Alternative

The aforementioned four steps are pre-option selection activities. There are three primary post option selection activities that are involved in the process of decision making:

4. *Specification of Decisions, Plans, and Architectural Structures*—in which the selected alternative is further developed or expanded conceptually such as to yield specifications for activities to be undertaken;
5. *Identification of Operational Priorities*—in which the conceptual plans and their specifications are decomposed to yield specific operational activities, such as route specifications; and
6. *Forecasting, Explaining, and Describing to Others*—such that those who need to know information about a selected course of action are provided this information.

This framework does not emphasize the different types of information and different types of support that are needed within each of the identified activities. The iterative nature of the steps also is not emphasized. A number of constituent activities may be identified for each of the identified activities. For activities 1 through 3, which are the primary problem solving activities, we might identify the 9 activities shown in Table 1:

Military commanders in a C^2 environment perform each of these activities. Some of them may be performed at an intuitive level which is made possible because of much experiential familiarity with particular task requirements. Some should be performed in a formal analytical manner because they are initially unstructured and unfamiliar. We could now proceed to describe each of these nine activities in considerable detail. Each of the descriptions would provide ample testimony to the critical importance of information in all of them. Information is needed about the external environment such as to enable understanding of the environment. This leads to identification of a set of information needs relative to already identified command and control objectives. Parallel with much of this, some preliminary identification of potential alternative courses of action are identified, and these further act to frame the information needs for proper analysis and evaluation of these alternatives. All of these are accomplished as part of issue formulation. The issue analysis and issue interpretation steps are equally rich in terms of their need for information.

Decision-making is an activity involving iterative *hypothesis, generation* and *test* of alternative courses of action. The hypothesis generation activity, which is a part of issue formulation, involves primarily inductive skills which are generally based upon experiences that enable generation of alternative courses of action. These are evaluated through the primarily deductive activity of evaluation or testing. An initial hypothesis is often not acceptable. When it is rejected, generally through evaluation, iteration back through the hypothesis generation effort enables modification of the identified alternative, and reevaluation so as to, hopefully, lead to a successful alternative course of action that will resolve the command and control issue under consideration.

This C^2 problem solving methodology is somewhat incomplete in that it does not formally recognize the stimulus that leads to hypothesis generation. There are at least two types of hypotheses generated in a typical command and control effort. Hypotheses concerning situation assessment, which includes the intent of a potential adversary, represent the first type. Other types of hypotheses relate to the alternatives or options that are available for choice. Uncertainties and information imprecision abound relative to each of these hypotheses. The first is information input imprecision. This creates the need for hypothesis generation regarding situation needs and/or intent of a potential adversary. The second is consequence of option implementation imprecision. This creates the need for generation of hypotheses relative to alternative courses of action. In a similar way, we might talk about stimuli that lead to hypotheses about situation needs, and to hypotheses about available options and their impacts. The first stimulus is generally the result of requirements identification or specifications for a specific C^2 effort. The second is the result of imperfect knowledge of the environment, and the effect of a given option on ameliorating the effects of a real, or impending, problem solving need.

Thus, we see that typical command and control information processing efforts will involve iteration and feedback, and hopefully learning as well through a systems and information technology process that leads to implementation of a set of action alternatives. Such a process may be depicted as shown in Figure 1.

While models such as this illustrate the iteration and feedback involved in the crisis management process, they fail to show the activities of the many people involved in these activities, or interactions with the external environment. Nor do they indicate the activities that precede problem solving, stage 0 efforts, or post problem solving, stage 4, 5, 6 efforts. Often the stage 0 efforts occur as the result of a previous problem solving effort at a higher level, and the stage 4, 5, 6 efforts are part of succeeding efforts in a different phase of problem solving. For technological processes, these phases are often given such names as planning, development, design, . . . and other activities in a systems life cycle that culminate in system modification, retrofit,

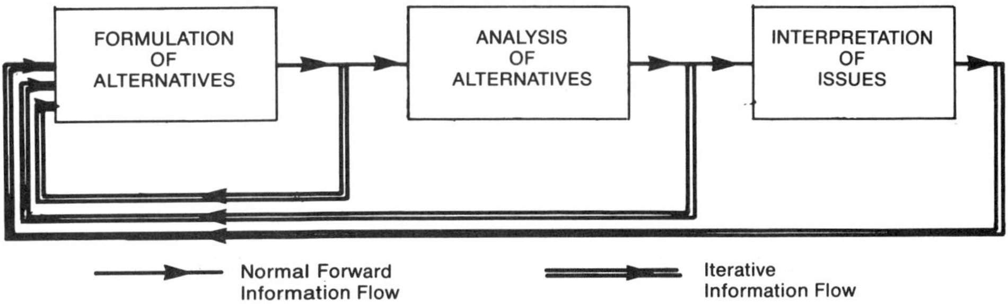

Figure 1. Iteration and Feedback in Command and Control Problem Solving

or phase out [Sage, 1977]. In a management situation, names typically given to the phases are [Anthony, 1965]:

- *Strategic Planning Decisions*—which are decisions related to choosing highest level policies and objectives, and associated resource allocations.
- *Management Control Decisions*—which are decisions made for the purpose of assuring effectiveness in the acquisition and use of resources.
- *Operational Control Decisions*—which are the decisions made for the purpose of assuring effectiveness in the performance of operations.
- *Operational Performance Decisions*—which are the day to day decisions made while performing operations.

Within each of these types of management decisions, or within each of the phases of a technological system life cycle, there would be problem solving activities, such as in Figure 1. For our purposes, we simply will augment Figure 1 with the four pre- and post-problem solving efforts to obtain the command and control problem solving efforts shown in Figure 2.

Different types of information and different types of support are needed within each step of a C^2 process. During the issue formulation step, the support that is needed tends to be of an affective, perceptive, or gestalt nature. Intuition based experiential wisdom will play a most important role in this. During the analysis step, the need is typically for quantitative and algorithmic support, perhaps through the use of the formal models of oper-

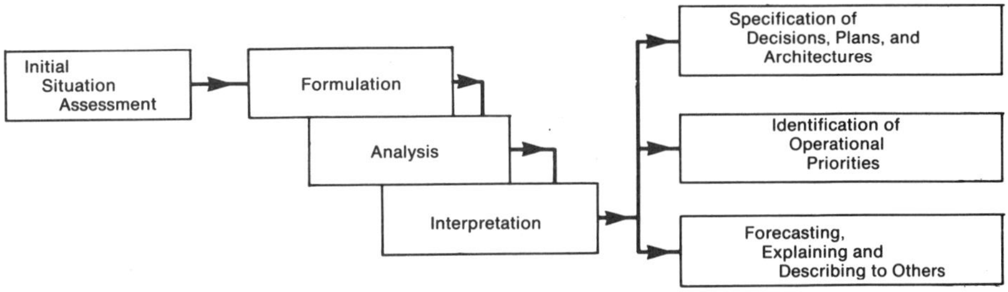

Figure 2. Role of Pre- and Post-Problem Solving Activities

ations research and management science. In the interpretation step, the needed effort shifts to a blend of the perceptive and the analytical. It does not emphasize or take explicit account of the various perspectives on judgment and decision making that will follow as a result of implicit identification of the contingency task structure on the part of the decision-maker. For it is this contingency task structure identification that influences both the determination of information requirements as well as identification of the meta-level decision of "how to decide how to decide." There are many attributes that affect information choice factors. We now turn our attention to a discussion of some of these.

Information and Decision Attributes

There are many attributes of information and associated decisions. Simon [1960] has described decisions as structured or unstructured depending upon whether or not the decision making process can be explicitly described prior to the time when it is necessary to make a decision. This taxonomy would seem to lead directly to that in which expert skills [holistic reasoning], rules [heuristics], or formal reasoning [holistic evaluation] are normatively used for judgment. Generally, operational performance decisions are more likely than strategic planning decisions to be pre-structured. It is important to note that expertise is a relative term which depends upon familiarity with task and the operational environment into which it is imbedded. Since decision environments do change, and since novices become experts through learning and feedback, it is clear that there should exist many areas in which the proper form of knowledge base support for command and control decisions is a hybrid of an "expert system" and a "decision support system." This suggests that there will be a variety of decision-making processes in practice and that an effective support system should support multiple decision processes. In a similar way, the information requirements for decision-making can be expected to be highly varied, and an effective support system should support a variety of knowledge base management needs.

There are many variables that will affect the information that is, or should be, obtained relative to any given decision situation. These variables are very clearly task dependent. Keen and Scott Morton [1978] identify eight variables:

- *inherent accuracy of available information*—Operational control situations often will deal with information that is relatively certain and precise. The information in strategic and tactical planning situations often is uncertain, imprecise and incomplete.
- *needed level of detail*—Often very detailed information is needed for operational type decisions. Highly aggregated information often is desired for strategic decisions. There are many difficulties associated with information summarization that need attention.

- *time horizon for information needed*—Operational decisions typically are based on information over a short time horizon, and the nature of the control may be changed very frequently. Strategic decisions are based on information and predictions based on a long time horizon.
- *frequency of use*—Strategic decisions are made infrequently, although perhaps they are refined fairly often. Operational decisions are made quite frequently, and are relatively easily changed.
- *internal or external information source*—Operational decisions often are based upon information that is available internal to the organization, whereas strategic decisions are much more likely to be dependent upon information content that can only be obtained external to the organization.
- *information scope*—Generally, operational decisions are made on the basis of narrow scope information related to well defined events internal to the organization. Strategic decisions are based upon broad scope information and a wide range of factors that often cannot fully be anticipated prior to the need for the decision.
- *information quantifiability*—In strategic planning, information is very likely to be highly qualitative, at least initially. For operational decisions, the available information is often highly quantified.
- *information currency*—In strategic planning, information often is rather old, and it often is difficult to obtain current information. For operational control decisions, very current information often is needed.

Critical to success in the design of support systems to enhance the efficiency and especially the effectiveness of command and control decisions is an understanding of the way in which humans formulate issues, identify possible alternative courses of action, analyze the impacts of these alternatives and integrate these impacts in accordance with a value system. It now is generally recognized that experiential familiarity with the perceived task at hand is the dominant influence upon the way in which the task requirements are cognized, and the way in which human information processing and associated problem solving activities proceed.

Conceptual Frameworks for Information Processing and Problem Solving

An appropriate conceptual framework for human information processing and judgment must be set within the context of real command and control decisions made by real people in real decision situations. The terms judgment and choice refer to human cognition over a continuum of perspectives that range from formal analytical thought at one end of the spectrum to holistic intuitive thought at the other.

The difference between analytical and perceptual thought processes is rooted in the writings of many authors. Piaget's theory of intellectual development [Ginsburg and Opper, 1979; Sage, 1981] was among the first to identify relevant variables that influence selection of judgment and choice style: maturation; experience; education; and "self-regulation." Brunswick [1952, 1956] describes perception as an intuitive, continuous and rapid process; as contrasted with analytical reasoning, which is typically deterministic, discontinuous and slow. Cyert and Simon [1983] describe "behavioral rules" as modes of behavior that an individual or an organization develops as guidelines for decision-making in complex environments characterized by uncertainty and incomplete information. Cyert and March further posit that these rules incorporate the decision-maker's assumptions about both the nature of the internal environment of the organization and the external environment of the world surrounding the environment. In this model, judgment is an entity that is capable of being decomposed into a set of behavioral rules that change and move closer to those described by simple known situations as uncertainties give way to certainty and as knowledge increases.

One particularly useful taxonomy due to Rasmussen [1983] conceptualizes three distinct types of problem solving behavior, or reasoning:

FORMAL KNOWLEDGE BASED BEHAVIOR
RULE BASED BEHAVIOR
SKILL BASED BEHAVIOR

The choice of which type of reasoning to employ is made by the problem solver on the basis of experiential familiarity with the task at hand, and the environment into which this task is imbedded.

Often, there is a mismatch between the problem solving behavior a particular human might wish to use in a given situation, and the behavior a machine designed to support a human in judgment and choice activities might be programmed to emulate. Expert systems often attempt, for example, to use rule based representations of skill based knowledge, or alternately use rules when an experienced expert would use a more skill based representation of knowledge. This knowledge initially may be expressed in any of several affective or intuitive forms, that have been learned by the expert on the basis of much relevant experience. Although the expert may well have learned how to be an expert through formal means that require an absorbed and alert monitoring of the task components, experts no longer may be formally aware of all of their monitoring actions, and therefore may give incomplete information concerning their activities. This simply reflects the view that expertise develops through perceptual learning *and* through the acquisition of a larger number of rules. Experts are not just faster and more accurate with their use of a larger number of rules; but also can perceive situations, similarities, and task objectives with considerable clarity. Dreyfus [1982] and Klein [1980] have stated very useful models of human problem solving activity based on these observations.

It is especially necessary that a knowledge representation scheme be capable of coping with the types of expertise, and the reasoning perspectives that reasonably can be expected to exist among the users of the knowledge based system. In this way, we will be much better able to accomplish needed activities that involve learning and discovery such as the diagnosis of faulty theories, the proper assimilation of new knowledge, and useful frameworks in which to pose questions so that they are understandable and interpretable in the way in which the questioner (should have) intended. Clearly, all of this has major implications for subjects such as decision-making in general and such important subareas as command and control decision aiding.

A major conclusion from this is that it would be very desirable to develop predictable interrelations among human cognitive processes for human information processing, associated judgment and choice and environmental characteristics. While these will depend upon the contingency task structure of decision task, surrounding environment and decision maker experiential familiarity with these, it should be expected that *most* knowledge support processes will tend to encourage use of formal analytical thought processes by both novices and experts. Several observations support this:

- A support system encourages people to justify their judgments, and this has been shown to encourage formal analytical reasoning [Hagafors and Brehmer, 1983]; and
- People tend to test and potentially change their initially favored hypotheses when provided with relevant information feedback [Batra and Ray, 1983; Fiske and Kinder, 1981; Fiske, Kinder and Larter, 1983; Chaiken 1980].

It will be very important to determine the extent to which judgment and choice is improved through the use of support systems brought about by the likely encouragement to use formal operational thought processes. One role for a knowledge support system is to assist the user in minimizing errors between his perceived knowledge level relative to a particular task and his actual knowledge level. When both perceived and actual knowledge levels are at the level of "master" for example, then skill based knowledge generally is appropriate for judgment and choice tasks. When the perceived and actual knowledge levels are that of "novice," then the knowledge support system user is generally aware of the need for support. When the perceived knowledge level is that of "master" and the actual knowledge is that of "novice," perhaps due do an unrecognized change in environment, then it is very likely that acts of judgment and choice will be associated with self deception. The task for a knowledge support system, in this regard, would be to alert the decision-maker to the potential difficulties of skill based behavior in an appropriate manner. A very important role for a knowledge support system, therefore, is that of alerting the decision-maker to the information requirements appropriate for the decision-maker, the task at hand, and the environment surrounding task and task requirements. The situation is much like that shown in Figure 3. If it is correct that use of a support system will encourage people in quadrant 1 and 2 to use formal reasoning based knowledge, then the change from quadrant 2 is appropriate, whereas that from quadrant 1 is not necessarily appropriate. This suggests the addition of a third dimension to Figure 3, necessary to reflect the influence of a support system, and

	Actual Expert	Actual Novice
Perceived Expert	Uses Skill Based Knowledge as Appropriate (1)	Uses Skill Based Knowledge Should Use Formal Analytical Reasoning (2)
Perceived Novice	Uses Formal Analytical Reasoning Should Use Skill Based Knowledge (3)	Uses Formal Analytical Reasoning as Appropriate (4)

Figure 3. Actual knowledge perspective used versus appropriate knowledge that should be used.

extension of experiments reported by Lehner and Zirk [1987] concerning a consistency of human and support systems models.

Knowledge Representation in Humans and Machines

Among the characterizations that have been proposed of the ways in which humans represent knowledge are the inquiry system models of Churchman [1971], the multiple perspectives models of Linstone [1984], the decision framework and organizational choice models of Allison [1971], Steinbruner [1974] and others. Summary discussions of many of these models, which have particular relevance to purposeful design activities, are contained in Sage [1981] and in several chapters of Nystrom and Starbuck [1981]. The central message of essentially all of these studies is that the inquiry system that one adopts and the type of rationality employed is very much a function of the background and experiential familiarity that is brought by the problem solver, or the organization involved in problem solving, to the issue under consideration. Included also within the elements affecting decision or problem solving behavior are such "stress" related elements as time and resource availability. The stress based model of choice due to Janis and Mann [1977] is one particularly useful model that explicitly incorporates these considerations into a descriptive model of judgment and choice. This model suggests that judgment "style" will range from unconflicted adherence to a present pattern or change to a new but familiar option, to hypervigilance or panic, to decidophobia, to the vigilant information processing of the economically rational actor. The dominant variables influencing this choice of style are experiential familiarity with the task, the significance of the decision, and the time allowed to exercise judgment and choice.

There are several different knowledge representation schemes that have been proposed for use in computers. These include: logic; procedural representations; semantic networks; production systems; direct (analogical) representations; semantic primitives; frames and scripts; and schemas. Each of these may be used to describe the four different types of factual knowledge elements that may be captured in a knowledge base that is used for command and control purposes: objects; events; performance; and meta-knowledge. These representations may be used to describe knowledge retrieval, reasoning, acquisition of new knowledge and relating this new knowledge to already known knowledge. New knowledge can cause any of several changes in a knowledge base. It can cause expansion of the knowledge base to encompass new knowledge. It can cause contraction of the knowledge base because of rejection of old knowledge as currently inappropriate.

Our interest here is primarily with respect to the information processing perspective concerning human representation of knowledge. The purpose of a particular knowledge representation is to enable ultimate human use of the knowledge for purposes of problem solving. Typical tasks include: retrieving factual information from the knowledge base that is judged relevant to the task at hand; reasoning about these facts in the search for a resolution of the task requirements; and acquiring more knowledge.

There are various mechanisms that can be used to represent the organization of declarative and procedural knowledge. The most common, and simplest, representation is that in which knowledge is structured as a set of facts, with each fact related to one or more facts in a causal type of inferential relationship. The basic idea behind a production system is that there exists a set of productions, or rules, in the form of various condition-action pairs, generally in the form of "If-Then" combinations. Initially these were exclusively explicit rules, although there is much current interest in incorporating fuzziness and imprecision into production rule concepts. Generally production rules are heuristic in nature, and may be appropriate or inappropriate to the task at hand.

The basic concept of having nodes that represent elements in the universe and links that represent the contextual relations between these elements is a very appealing form of knowledge representation, known as a semantic network. A semantic network is intended to represent concepts expressed by natural-language words and phrases as interconnected nodes, connected by a particular set of arcs called semantic relations. Concepts in this system of semantic networks are word-sense meanings. Semantic relations are those which the verb of a sentence has with its subject, object, and prepositional phrase; in addition to those that underlie common lexical, classificational, and modificational relations.

A semantic network is a convenient computational representation, that is readily implementable on computers, knowledge which is expressed in natural language. The extent to which this type of knowledge representation can be effectively utilized for inference processes depends on the existence, or lack thereof, of a well structured and sophisticated set of rules. The antecedent—consequent type of rule is commonly used. Structurally the left-hand side of the rule, or antecedent, represents the set of assumptions or conditions appropriate for use of the rule. The right-hand or consequent side of the rule represents the set of end results. Conjunctions, disjunctions, and negations can occur on either side of the typical relational rule. The contemporary frame and script concept, in which the structure and framework within which new information is interpreted in terms of the concepts that have been acquired through previous experience, has evolved from semantic network concepts. Again we see the strong contextual dependency, or expectation driven processing, of scripts and frames such that one looks for things based on the context that one believes exist.

A well designed semantic network is able to represent physical characteristics of actions and the purposeful aspects of these actions. A semantic network representational scheme must, as a consequence of this requirement, include the *goals* or objectives that are obtained as a result of actions, those *scripts* that describe scenarios in simple stereotyped situations, the *plans* that allow for flexible description of potentially adoptable action-impact pairs, and the *themes* which allow for such environmental descriptions as the occupations and aspirations of actors involved in the issue under consideration. In this way, a semantic network may be a descriptive as well as a prescriptive mechanism for aiding the designer through knowledge support.

There are bound to be difficulties when the size of the semantic network, in terms of the associated data-base, becomes sufficiently large such that it can represent a nontrivial amount of knowledge. The computational difficulties in processing the network and the cognitive difficulties in coping with the associated complexity may be overwhelming. This leads to the need for network aggregation in order to obtain "summary" representations that are efficient and effective. These aggregate networks are called *frames*. Thus the concept of a frame eliminates many of the defects of semantic networks. In human information processing, experimental familiarity with a specific domain often will allow this "chunking" of knowledge to be accomplished satisfactorily.

There are difficulties associated with the semantic network concept. There are, for example, difficulties associated with representing time, as structural models are basically static devices. There are difficulties associated with maintaining a distinction between "facts" and "values," and in incorporating concepts of uncertainty, fuzziness, and imprecision. Situations often arise, for example, in which reasoning must be accomplished using incomplete, inconsistent, and perhaps even contradictory data. These may arise when collecting data from information summaries, or from imperfect and/or distributed sources. Much research is being done in semantic networks to improve their usefulness as representational schemes for general knowledge [Findler, 1979]. Some contemporary applications illustrate the use of semantic networks in constructing the knowledge base of programs which exhibit some aspects of understanding—paraphrasing, abstracting, and classifying a corpus of text, answering questions on the basis of common sense reasoning, drawing deductive and inductive inferences and obeying commands.

A much simpler network based representation is one in which knowledge is structured as a set of concepts; with each concept related to one or more facts by a single causal type of relationship. Needed to accomplish a structural model of this sort is a theory of psycho-logic [Axelrod, 1972, 1976] or structural model in which relations may have enhancing (+1), inhibiting (−1), or neutral (0) causal influences on other relations. A number of applications of the resulting cognitive maps, as the resulting structural models are often called, have been reported. In its more general form, a cognitive map is the result of an attempt to capture an individual's view of the world with respect to a particular issue.

Unlike semantic networks that are typically multi-relational structures and as such require a sophisticated set of production rules or grammars for representation and interpretation, a cognitive map is generally based on a single specific contextual relation, such that any given element will have enhancing, inhibiting or neutral causal impacts on each other element in the cognitive map. Thus the representation and analysis of a cognitive map usually will be simpler than is the case for semantic networks. The simple contextuality of the cognitive map may make it difficult for such a map to replicate a complex belief structure. However, the elements which represent concepts in a cognitive map are variables that can take on different values. The linkage, or contextual relation, among concepts may represent causal assertions and perceptions concerning how one concept variable affects another concept variable. Since the dynamics of the reasoning mechanism in a cognitive map is imbedded in structural considerations, it is a simple matter to simulate the reasoning process of a person if we can assume that the cognitive map has been faithfully constructed.

Minsky [1975] presents a rather different approach to semantic networks that eliminates some of their defects. He advocates the use of *local procedures* within a "frame" in order to represent structured knowledge. A frame is, as we have noted, a chunk of knowledge for representing a stereotyped situation. Attached to each frame are several kinds of information. Some of this information concerns how to use the frame and some of the information may concern what can happen next. A frame then can be represented as a hierarchical network of nodes and relations. The top level element in the frame will represent facts that always are assumed to be true about the generic situation at hand. The lower levels of the frame will have many empty terminal slots that must be filled by the specific context-dependent information about the frame and the situation at hand.

The concept of scripts and frames is related to that of schema representations of knowledge. Fundamentally, schema are conceptual structures that explain the relationship between information acquired, task variables, and decision making performance. Their use is suggested whenever decisions are made based on what has worked before in previously encountered situations believed to be similar. There are three layers to a schema: a feature slot layer; a constraint layer; and an inference and action layer. Slots contain features of situations that are believed relevant to a given issue. Features may be spatial relationships or other characteristics of a situation. When enough slots are filled, then a schema becomes activated and may be used to understand a decision situation. The constraint layer defines the degree to which features with different characteristics fill a slot. The inference and action layer contains inferences about decision situations that are associated with schema, and alternative courses of action or design options that are believed appropriate to the particular decision situation that is associated with a schema.

Schema have two essential functions: They enable situations, objects and concepts to be recognized and classified; and they specify actions and inferences that are believed appropriate to entities that are associated with a schema. Schema theory argues the prevalence of decisions based on recognizing that a situation corresponds to a particular schema. It does not assert, however, that all decisions are of this type. When situations are ambiguous and correspond to several schema, or when the lack of experiential familiarity with a situation does not suggest any schema, then alternative methods to make a decision are used.

Fundamentally, a schema is an abstract cognitive structure that specifies both the defining features and the relevant attributes of some stimulus domain, as well as the interrelation among the features and attributes [Crocker et. al. 1984]. It is through schema that humans accomplish the encoding and representation of functions, as well as various inferential and interpretive functions. It is through schema that humans are able to structure experiences. A persons schema relative to some purpose determines the information that will be encoded or retrieved from memory. Further, it is a persons schema that determines the speed of information processing and associated problem solving. It is through schema that humans are able to fill in data that are missing from an input stimulus condition. Schema provide a basis for evaluating the relevancy of past experience to a present problem, as well as for goal setting and anticipating the future, and other efforts associated with making plans and solving problems.

Analogies and analogous inference play a very important role in human judgment. Philosophers of science often claim that reasoning by analogy is the basis for hypothesis formation and identification in science. Often analogous reasoning is used when there exists uncertainty and imprecision associated with the judgment task at hand. Learning is a process whereby we are able to do things more efficiently, or effectively, or more efficiently *and* effectively the next time we do them. Thus learning can involve rote memory and direct implementation of instruction, reasoning by analogy and also learning through discovery and observation, of a concrete or formal nature.

When there is a lack of explicit, certain and concrete knowledge about a specific issue, reasoning by analogy often will be used. In such cases, one searches for similarities between the task situation extant and a previously experienced and familiar situation. When the situations are sufficiently analogous that one can see "parallels" between elements in one situation and elements in the other, then reasoning by analogy becomes possible.

Silverman [1983], in an extensive review of research concerning analogous inference in systems management, identifies a taxonomy that facilitates the development of procedures and protocols for the identification and correction of pitfalls in this form of reasoning. Sternberg [1977] also presents a descriptive theory of analogous reasoning and experimentally identifies sources of errors that may occur during each of the operations that constitutes his paradigm for analogous reasoning. The basic operations in Sternberg's model are encoding, inference, mapping, application and response. This research provides a great deal of descriptive information that has motivated the development of normative theories of concept formation and issue formulation.

There have been many studies of the decision styles associated with judgment and choice activities. These studies and related research and associated findings concerning the major role of leadership attributes [Bass, 1981; Fleishman and Quaintance, 1984] in influencing support needs have the strongest of descriptive, normative and prescriptive implications for the design of computerized support systems that aid in various command and control functions. Thus, fundamental studies concerning information processing and decision behavior as a function of leadership level and the contingency task structural perspective that leaders bring to specific decision situations is a major need. This will enable both the evaluation of existing models of information processing and decision behavior, and the formulation of improved models of these important functions.

Two distinct phases can be identified in the process of knowledge acquisition. The first is a generative phase in which the need for knowledge acquisition is recognized and some requirements concerning the scope of this knowledge are identified. Next, there is a cognitive validation phase which is concerned with degrees of validity and confidence that should be attributed to the present knowledge and level of expertise of the decision-maker(s) relative to tasks at hand. During the various stages of a decision-making process, each of these phases may be repeated several times. Initially, a preliminary set of requirements concerning the scope of available information requirements is identified. This set of requirements leads to an estimate of validity and confidence relative to the expertise associated with the task at hand. This, in turn, leads to the identification (implicitly of course) of the decision perspective that will be used for selection of an alternative course of action. In some cases, the associated information processing behavior is "open" in the sense that the decision-maker actively searches for potentially disconfirming information. At times it will be "closed" in the sense that the decision-maker will ignore information that is potentially disconfirming to already held beliefs. This open and closed information processing behavior has been observed by several researchers. Kruglanski and his colleagues [1983ab] conceptualize these tendencies as consequences of epistemic motivations and uses the terms "freezing" and "unfreezing."

To examine conditions under which one of these information processing behavior patterns is evoked should yield much insight into the variables that influence meta-level decisions concerning how to decide how to decide. In part, this might be accomplished by identifying classes of internal and external factors that influence various "freezing—unfreezing" tendencies. Three types of epistemic motivations which influence "freezing–unfreezing" tendencies have been suggested: fear of invalidity; sensed need for structure; and need for a desired conclusion. Several external or environmental factors that influence these may be identified: availability and saliency of information; time pressure; and group and organizational factors.

Information Needs in C³I Systems

Judgments, at least prudent judgments, are seldom made without information. Information is often defined as data of value for decision-making. Activities associated with acquisition, representation, storage, transmission, and use of pertinent data are generally associated with information processing. The task of information requirements determination is associated with each of these. Initially, it might appear that this would be associated only with an information acquisition effort. But since information acquisition is necessarily related to other activities associated with information processing, so also is information requirements determination.

There are many ways in which information may be characterized. Among attributes that might be used are: accuracy; precision; completeness; sufficiency; understandability; relevancy; reliability; redundancy; verifiability; consistency; freedom from bias; frequency of use; age; timeliness; and uncertainty. It is also possible to define information at several levels. At the *technical level,* information and associated measures are concerned with transmission quality over a channel. At the *semantic level,* concern is with the meaning and efficiency of messages. At the *pragmatic level,* information is valued in terms of effectiveness in accomplishing an intended purpose. Clearly, this relates directly to the concept of *value of information* and indicates that this concept is very contingency task structure dependent. The mix of task at hand, the environment into which the task is imbedded, and the problem solver's familiarity with these interact to determine both the perceptions and the intentions of the problem solver. There are at least three human limitations associated with information processing that may significantly affect the requirements for information that are likely to be identified in operational C³I situations: limited information processing capabilities; potential bias effects in the selection and use of information; and limited knowledge of appropriate problem solving strategy which, in part, may result from incorrect assessment of the contingency task structure associated with a given issue.

Information obtained from any of many approaches must be capable of representation in a knowledge support system. A purpose of a support system should be to determine possibilities of insufficient and/or inappropriate information; that is information that is sufficiently imperfect such as to make the likelihood of choosing an effective alternative course of action low. The support system should then be able to determine the nature of the missing or otherwise imperfect information, and suggest steps to the user to remedy this deficiency. This suggests that a support system should be capable of detection, diagnosis and fault correction in a set of information obtained for an issue.

C^3I *Knowledge Support Needs*

There are three primary need areas for the development of efficient and effective support systems for C^3I functions, including training of potential users of these systems: system hardware and software; human information processing concerns and how human information processing capabilities can be enhanced through use of effective knowledge support systems; and user requirements and leadership characteristics as they effect the need for and the use of support systems to aid in command and control functions, and in training for command and control. The last of these provides major implications to the second need. These two needs, each of which relate to effective system design for human interaction, strongly influence the requirements for operational functionality of *appropriate* hardware and software.

Human information processing tasks in C^2 environments are especially difficult due to the many factors which lead to information imperfection. The available information in a C^2 system is sometimes likely to be of very questionable validity and reliability. It also may be very incomplete and, especially since C^2I is inherently a group effort, often inconsistent and possibly even incoherent. There often are severe time constraints and personnel shortages. Message summarization often is needed, as is the aggregation of summarized messages as they pass through the system. The individuals involved in C^2 activities necessarily have different experiential familiarity with various intelligence activities and situations. Although they may be most strongly influenced by doctrine and the policy and alternatives suggested by doctrine, there still will be a multitude of perspectives through which intelligence information and decision situations are seen.

A major inherent characteristic in command and control decisions is the need to deal with information imperfection [Sage, 1987ab]. This can represent information input equivocality which creates a need for hypothesis generation and evaluation; and it can also represent imprecision associated with consequence of action implementation. This creates the need for option generation, analysis, and evaluation. To aid humans in these endeavors is a primary goal of a C^3I system.

This discussion suggests three related areas of primary importance for broad scope understanding, and effective design and use, of C^3I systems:

- The technological design of communication systems hardware and software for command and control;
- Human information processing concerns and how human information processing capabilities can be enhanced through use of effective knowledge support systems; and

- User requirements and leadership characteristics as they effect the need for and the use of support systems to aid in command and control functions.

Each of these is important and to neglect any is to invite failure. Of particular importance is the role and effect of leadership characteristics and experience in command and control decisions.

Clearly these areas are related. It often will be the case that C^3I system users will be unable to express explicit requirements for an appropriate system architecture without actual experimentation with an evolving system. This suggests the need for rapid prototyping capabilities such that it becomes possible to quickly evolve tentative software for these systems such that evaluation of prototype C^3I system qualities can be efficiently and effectively accomplished.

Bibliography

1. R. Abelson, "The Structure of Belief Systems," in Schank, R. C., and Colby, K. M. (Eds.), *Computer Models of Thought and Language,* Freeman, 1973.
2. S. J. Andriole, *Handbook for the Design, Development, Application, and Evaluation of Military Decision Support Systems,* (in Press 1986).
3. R. N. Anthony, *Planning and Control Systems: A Framework for Analysis,* Harvard University Press, 1965.
4. R. M. Axelrod, (Ed.), *Structure of Decision: The Cognitive Maps of Political Elites,* Princeton University Press, 1976.
5. R. M. Axelrod, *Framework for a General Theory of Cognition and Choice,* Institute of International Studies, University of California at Berkeley, 1972.
6. B. M. Bass, *Stogdill's Handbook of Leadership: A Survey of Theory and Research,* Free Press, 1981.
7. R. Batra, and M. L. Ray, "Advertising Situations: The Implications of Differential Involvement and Accompanying Affect Responses," in Harris, R. J. (Ed.), *Information Processing Research in Advertising,* Erlbaum, 1983.
8. E. Brunswick, "The Conceptual Framework of Psychology," in *International Encyclopedia of Unified Science,* Vol. 1, No. 10, University of Chicago Press, 1952.
9. E. Brunswick, *Perception and the Representation Design of Psychological Experiments,* University of California Press, 1956.
10. S. Chaiken, "Heuristic Versus Systematic Information Processing and the Use of Source Versus Message Cues in Persuasion," *Journal of Personality and Social Psychology,* Vol. 39, 1980, pp. 752–766.
11. J. Crocker, S. T. Fiske, and S. E. Taylor, "Schematic Basis for Belief Change," In Eiser, J. R. (Ed.), *Additudinal Judgment,* Springer Verlag, 1984.
12. R. M. Cyert, and H. A. Simon, "The Behavioral Approach: With Emphasis on Economics," *Behavioral Science,* Vol. 28, 1983, pp. 95–108.
13. H. L. Dreyfus, and S. E. Dreyfus, *Mind Over Machine: The Power of Human Intuition and Expertise in the Era of the Computer,* Free Press, 1986.
14. S. E. Dreyfus, "Formal Models Versus Human Situational Understanding: Inherent Limitations in the Modeling of Business Expertise," *Office, Technology, and People,* Vol. 1, 1982, pp. 133–165.
15. C. Eden, T. Smithin, and J. Wiltshire, "Cognition, Simulation and Learning," Journal of Experiential Learning and Simulation, Vol. 2, 1980, pp. 131–143.
16. C. Eden, S. Jones, and D. Sims, *Thinking in Organizations,* The McMillan Press Ltd., London, 1979.
17. N. V. Findler, (Ed.), *Associative Networks: The Representation and Use of Knowledge by Computers,* Academic Press, 1979.

18. S. T. Fiske, and D. R. Kinder, "Involvement, Expertise, and Schema Use: Evidence from Political Cognition," In N. Cantor and J. Kihlstrom (Eds.), *Personality, Cognition and Social Interaction,* Erlbaum, 1981.
19. S. T. Fiske, D. R. Kinder, and W. M. Larter, "The Novice and the Expert: Knowledge Based Strategies in Political Cognition," *Journal of Experimental Social Psychology,* Vol. 19, 1983, pp. 381–400.
20. E. A. Fleishman, and M. K. Quaintance, *Taxonomies of Human Performance,* Academic Press, 1984.
21. H. Ginsburg, and S. Opper, *Piaget's Theory of Intellectual Development,* Prentice Hall, 1979.
22. R. Hagafors, and B. Brehmer, "Does Having to Justify One's Judgments Change the Nature of the Judgment Process," *Organizational Behavior and Human Performance,* Vol. 31, 1983, pp. 223–232.
23. R. Hakathorn, and P. Keen, "Organizational Strategies for Personal Computing in Decision Support Systems," *MIS Quarterly,* Vol. 5, No. 3, 1981.
24. G. P. Huber, "Cognitive Style as a Basis for MIS and DSS Designs: Much Ado About Nothing," *Management Science,* Vol. 29, No. 5, May 1983, pp. 567–579.
25. I. L. Janis, and L. Mann, *Decision Making: A Psychological Analysis of Conflict, Choice, and Commitment,* Free Press, New York, 1977.
26. P. G. W. Keen, and M. S. Scott Morton, *Decision Support Systems: an Organizational Perspective,* Addison Wesley, Reading, Massachusetts, 1978.
27. G. A. Klein, "Automated Aids for the Proficient Decision Maker," *Proc. 1980 Conf. on Systems, Man, and Cybernetics,* Boston, Massachusetts, October 1980, pp. 301–304.
28. A. W. Kruglanski, and I. Ajzen, "Bias and Error in Human Judgment," *European Journal of Social Psychology,* Vol. 13, 1983a, pp. 1–44.
29. A. W. Kruglanski, and T. Freund, "The Freezing and Unfreezing of Lay-Inferences: Effects of Impressionable Primary, Ethnic Stereotyping, and Numerical Anchoring," *Journal of Experimental Social Psychology,* Vol. 19, 1983b, pp. 448–468.
30. P. E. Lehner, and D. Zirk, "Cognitive Factors in User/Expert System Interactions," *Human Factors,* 1987, in press.
31. H. A. Linstone, *Multiple Perspectives for Decision Making: Bridging the Gap Between Analysis and Action,* Elsevier North-Holland, New York, 1984.
32. M. Minsky, "A Framework for Representing Knowledge," *The Psychology of Computer Vision,* P. A. Winston (Ed.), McGraw Hill, New York, 1975.
33. J. Rasmussen, "Skills, Rules, Knowledge; Signals, Signs, and Symbols; and Other Distinctions in Human Performance Models," *IEEE Transactions on Systems, Man and Cybernetics,* Vol. SMC 13, No. 3, May 1983.
34. J. Rasmussen, *On Information Processing and Human-Machine Interaction: An Approach to Cognitive Engineering,* North Holland, 1986.
35. A. P. Sage, *Methodology for Large Scale Systems,* McGraw Hill Book Company, 1977.
36. A. P. Sage, "Behavioral and Organizational Considerations in the Design of Information Systems and Processes for Planning and Decision Support," *IEEE Transactions on Systems, Man and Cybernetics,* Vol. 11, No. 9, September 1981a, pp. 640–678.
37. A. P. Sage, "A Methodological Framework for Systemic Design and Evaluation of Computer Aids for Planning and Decision Support," *Computers and Electrical Engineering,* Vol. 8, No. 2, 1981b, pp. 87–102.
38. A. P. Sage, "Methodological Considerations in the Design of Large Scale Systems Engineering Processes," in Y. Y. Haimes, (ed.), "Large Scale Systems," North Holland, 1982, pp. 99–141.
39. A. P. Sage, B. Galing, and A. Lagomasino, "Methodologies for the Determination of Information Requirements for Decision Support Systems," *Large Scale Systems,* Vol. 5, No. 2, 1983, pp. 131–167.
40. A. P. Sage, and W. B. Rouse, "Aiding the Human Decision Maker Through the Knowledge-Based Sciences," *IEEE Transactions on Systems, Man and Cybernetics,* Vol. 16, No. 4, July 1986.
41. A. P. Sage, "Knowledge, Skills, and Information Requirements for Systems Design," a chapter in W. B. Rouse, and K. Boff, (Eds.) *The Psychology of System Design,* Elsevier, 1987a.
42. A. P. Sage, (Ed.) *System Design for Human Interaction,* IEEE Press, 1987b.
43. R. C. Schank, *Reading and Understanding,* Lawrence Erlbaum Associates, 1982.

44. B. G. Silverman, "Analogy in Systems Management: An Information Processing View With Implications for Comparison Guiding Aids," *IEEE Transactions on Systems, Man, and Cybernetics,* Vol. 13, No. 6, November/December 1983.
45. R. J. Sternberg, *Intelligence, Information Processing, and Analogical Reasoning: the Componential Analysis of Human Abilities,* Erlbaum, 1977.

'Tin Cans and Strings': The Concept, Design and Construction of an Evolutionary C^2 System

William J. Livsey

The draft of this article stayed on my desk for a long time; I looked at it almost every day. I picked it up time and again but, kept putting it down. I did so for a number of reasons. First, I was not sure I wanted to try to publish something about command and control (C^2) because I learn something new about this business every day and I have been working at it for quite a while. And I just did not know where to start or finish. I have trouble getting my arms around it, because the subject of C^2 means so many things to so many people. It can be as broad or as narrow as you want it to be to fit your purpose.

I kept thinking about it and decided to go back to the manuals as a place to begin. There were too many so I looked at the "head" manual, JCS Publication 1 (1 Apr 84). It says that C^2 is "the exercise of authority and direction by a properly designated commander over assigned forces in the accomplishment of the mission. Command and control functions are performed through an arrangement of personnel, equipment, communications,

This article was written especially for* ***Principles of Command and Control.*

facilities, and procedures employed by a commander in planning, directing, coordinating, and controlling forces and operations in the accomplishment of the mission." Well, that is probably a pretty good definition. However, it is probably broader than some would want and narrower than others would want. It caused me to focus on the word "mission," the job of fighting a war, and that seemed to be the right place to start an article about C^2.

In any analysis of what the military is all about, warfighting takes priority over all else. Just what is it? From a practical standpoint, warfighting is the synchronization of combat power at a time and place of one's choosing. That is a tall order, and easier said than done. It begs the question of how you get to that point where you try to bring your optimum combat power to bear on the enemy so as to cause decisive results in your favor. Much is involved: We have to synchronize the cutting edge of the ground, air and naval forces while integrating the combat support and the combat service support for these forces. In Korea, we also have to do this in a binational environment in genuine coalition warfare.

Some of these requirements are discussed in what is popularly known as the "operational art of war" that leads to campaign planning, intelligence preparation of the battlefield, and all of the other things that must be done before any synchronization can take place. In short, it is all of the actions that staffs and commanders must do to cause things to happen.

Another aspect of this synchronization of combat power is the structure and the equipment that are part of the C^2 process. There is the chain of command, the tactical operations centers (TOCs), communications nodes, circuits, computers, software, message centers, etc. All of this leads to a discussion of "tin cans and strings," the means by which we tie our forces together. Vital elements are the information conduits through which information must flow. Very few times the conduits may be "clean," such as with fiber optics; more often, the "strings" are not so clean, like WD-1 with knots in it. The problem begins when you try to push 12 diameter inches of information through a one inch diameter conduit. If we are going to synchronize combat power effectively, we must have critical information in the proper volume and of the right quality.

Conduits are not without limit, however, so the corollary of how much and how well we push information through is based on deciding what is critical information and how it applies to each echelon. How do you get this information where it ought to be in time to cause the synchronization of combat power?

Let me try to answer these questions practically by charting what we have been attempting to do in Korea to cause critical information, hopefully predictive in nature, to be in the hands and heads of those who must have it when they need it. This is C^2 in Korea as I see it:

We begin by differentiating between data and information. Data, or facts, consist of such things as intelligence on the enemy's order of battle. Such data certainly have value; they tell us what the enemy *can* do. But when what they tell us goes from enemy capability to intent, data become critical information and are much more useful than the mere collection of facts.

There's an important variable in this information process that must not be overlooked. The need for and the value of information vary with time and circumstance. In other words, there is information and there is critical information. And, what is critical information for a theater commander looking ahead 96 hours and beyond is different from that of the division commander on the forward edge of the battle area (FEBA) where the head-knocking is already taking place. We also find that what may be just information today could go critical tomorrow, and vice versa.

Who is responsible for deciding what is critical information? We all are! Who needs this critical information? We all do. And who sets this process in motion? The commander does. He begins by determining what he needs to fight the battle at the level he intends. His staff takes this guidance and refines it. Normally, this exchange takes place at the commander's daily briefing, an activity that usually takes too many people too long to prepare and present. In order to meet the schedule of pre-briefs and dry runs, the cut-off time for new material is several hours. So much for timely, critical information. This problem has characterized most commands that I know about.

In an age of reliable, high speed communications systems, the problem only gets worse. We can be overwhelmed with data and hardpressed to "mill" it into information, much less into critical information. It seemed logical that if technology could contribute to the availability of data, it could and should be used to manage that data. To do just that, I challenged my staff to give me something that permitted the action officer to collect, store, and manipulate a lot of data which would compose the "field" from which facts could be "harvested" and processed into information. As needs changed, "new facts" could be selected and "old facts" returned to the field where they could be updated or recycled when they were again needed.

The staff, meanwhile, needed an accurate and responsive means to refine information into critical information and blend it with critical information from higher, lateral and subordinate elements to ensure no one was operating in splendid isolation. As we all know, operations cannot function in an intelligence vacuum, but many times it tries to. (Are we fighting the right enemy at the right time in the right location?) Similarly, logistics has meaning only in terms of operational requirements. I really never understood "short tons" and when logisticians talk in these terms it is useless to me on an operational level. Logistics critical information must be productive and many times the distribution system becomes much more important than days of supply of ammo and other material. I could give you a lot of similar examples regarding soldiers' replacement and evacuation during war. So each staff element is involved in the process. What I needed was a system that could hold and process a good bit of data and a staff to sort out critical information, make it immediately available to those in the command who really needed it, and to do so in a changing environment without miring me or the staff in countless hours of briefings and associated preparation. Most of all, I needed that capability now.

My requirements were clear and straightforward: the capability to tie together key elements of the command throughout the peninsula in an automated C^2 system that permitted near to real time access to critical information and real time communications using electronic mail. I did not want to waste time or money to "re-invent the wheel." I wanted something that was available, tested, reasonably priced, user-friendly; and also I wanted the "keys to the system," permitting us—not some outsider—to change the system over time, if necessary, and not be beholden to a "provider" or "contractor" whom I could not take to war. In other words, the command would pay for the system and therefore the command should own it, lock, stock and barrel, especially its software.

To come up with this evolutionary system, we knew that it could not be "militarized." It would take too much time and cost too much. Instead, the system should consist of easily available and replaceable parts. Buying "off-the-shelf" would give me reasonable assurance of finding additional equipment for expansion and spares. The software should be as widely applicable as possible. We tried to stay away from as many "theater uniques" as we could. Although there would be Korea-unique requirements, these would represent only a small part of the overall system. In short, my goal was a system that would look and act pretty much the same whether it were located in Europe, Korea, or anywhere else.

The Army has not yet fielded an echelon-above-corps system, and since we were about to create something that might be a prototype, that was all the more reason to look for universality. One way to ensure this evolutionary effort was to buy commercial software where appropriate. Combined Force Command/U.S. Forces Korea (CFC/USFK) had no interest in developing or maintaining an integrated software package (data base management, spreadsheet, word processing, etc) when several very good and inexpensive products could be bought at the Post Exchange, or at the neighborhood computer store. I wanted to limit to the greatest extent possible anything so unique that its use, cost, or maintenance could limit the warfighting capability of this command, for which the system was being built.

The first step toward getting this system was to task the automation/information management people to tell me what capabilities in that target range existed and to point me in the right direction. Fort Lewis, Washington was the place and an experimental system called Distributed Command and Control System (DCCS) was the starting gate. I asked the Department of the Army (Information Management) to support the unfunded requirement and had the DCCS contractor come to Korea to look at our problem and offer solutions. I also took a different approach to program management within the command by having the C/J/G3 (operations) head the effort on behalf of the users. My intent was to involve the users early, continuously and deeply in the evolution of a system that was, after all, primarily for their use. It was a C^2 system and not just another communications system.

I remain convinced that the user-provider-maintainer relationship must start at the beginning of the project and remain through its life cycle. Only then can you be assured that the users are responsible, and are held accountable, for providing the requirements and that the communications/automation people deliver based on those requirements. Far too often we find a major disconnect in project design or development because the primary players were involved too late or not at all. We were not about to re-learn that lesson. A by-product of that policy has produced the most integrated operator-communicator relationship in any organization I know. This relationship was important to the success of the effort because the signal community had to work the "strings" portion of the system. Even the greatest of "tin cans" cannot overcome the limitations on the "strings" and we were dealing in an environment where there are few "strings" to begin with and many of those did not have the quality required for data (in the communications sense) transmission.

Our C^2 communications lash-up now consists of a combination of the original Ethernet coaxial cable Local Area Network (LAN) upgraded to fiber optic in garrison and, where available, to some of the subordinate elements; fixed site, ROK military; leased commercial and; experimentally, tactical systems. We were limited by available capability, cost, and time, but we made it work down to and including the "last 50 feet" of linkage—the proverbial choke point of even the most modern communications networks. I knew that it made no sense to have a "12 inch string capability" if somewhere along the line it connected with a "one inch" line. The experience of learning these Korea-unique lessons first hand has been institutionalized in our design requirements for any follow-on signal effort, including C^2.

From concept to initial operational capabilities (IOC), the new system was developed, installed, and operating on time and within budget in 12 months. It was far from perfect, but our goals were realistic and we were willing to accept limitations. To paraphrase a popular advertisement, neither we nor the Department of Defense have pockets so deep that we can afford to automate every aspect of our business. We had some hard choices to make and we made them. Theater Automated Command and Control System: Korea (TACCS-K) supports many of our most pressing needs. It provides us with a tool to bring together the principal commanders and staffs in the defense of the ROK. It is, by every definition, a modest system; however, it quickly, efficiently, and effectively took us out of the grease pencil and briefing board era (although we continue to retain that "skill" in reserve) and into an age of automation. We quickly became "automation literate" to better know and state our requirements for our final C^2 system, Theater Automated Command and Control Information Management System (TACCIMS).

TACCIMS will be everything TACCS-K is and more. We will design and engineer into TACCIMS our major requirements and all of the valuable lessons that TACCS-K has taught us, to include the same reasonable, logical approach we took with the acquisition and use of TACCS-K. It will be an evolutionary process; we will rely heavily on available technology to keep the project on target, on time, and within budget. We will use hardware and software designed for Service and DOD-wide systems with which it must operate or be compatible; and we will not fund activity for which someone has already paid the "price of admission." We will use common acquisition, equipment, and procedures to minimize costs associated with its operations, maintenance, and sustaining base.

This system is designed to support the user and not the reverse. Among its chief attractions will be its simple design and ease of use. Any unique requirements will be limited to those absolutely essential, such as the bilingual English-Hangul capability we now enjoy in only limited fashion with TACCS-K.

The most unique quality of TACCIMS will be the combined nature of its development, as well as its operations, as evidenced by the ROK/US joint program management team—the first of its kind. I cannot overstress the importance of our defense partnership in the Combined Forces Command. It is the best example of coalition warfare capability anywhere in the world.

Now how do we bring all of this together? How do we integrate everything I have talked about so far into critical information flow? This answer is the same as for any other military skill: training and more training. Training develops competence; competence builds confidence; and confidence is the greatest single determinant of success in battle.

If I have emphasized one military concept more than any other, it is the need to understand the battlefield and what it comprises. I think the greatest obstacle to managing warfighting is the lack of understanding the differences between battle and the battlefield. You have to know what is going on everywhere all the time. Too often, we focus on the FEBA, forward line of troops (FLOT) or whatever acronym we can find to describe where the head knocking takes place, but there are many other equally important battles going on at the same time that will contribute to determining the outcome. They include control of the skies and the seas; they also include the rear area, the battle to protect our freedom of action; and we place great emphasis on the deep battle area where the enemy's forces are marshalled, and from which they are supplied as they prepare for the echelonment that will bring them to the focal point of the conflict. The battles are many, but the battlefield is just one. It consists of all of these battles as well as political, economic, and psychological factors.

In sum, the battlefield must be viewed as the confluence of everything that can affect the conflict, and we must be prepared to meet the enemy on each "front." It is too late if we wait until he has formed his forces for combat at the FEBA, because the initiative is lost. Only when we understand the integration of the battles and accept the totality of the battlefield can we properly prepare ourselves to deter the enemy, or failing that, defeat him.

A basic step is the intelligence preparation of the battlefield, the collection and evaluation of intelligence about the enemy's capability and intent. The entire battlefield—the headknocking area, the deep battle area *and* the rear area must be taken completely apart, analyzed, and put back together. Only if you look at every detail of the terrain, the enemy, his strengths and vulnerabilities can you begin to optimize your combat power. It is not enough to know what he *can* do. We have to know with confidence what he *will* do, and *where* and *when* he will do it. This is a difficult task, it is hard to do, but we must do it. We can pretty well assess the elements over which the enemy has control (size, disposition of forces, quality of training, etc); know those factors which can help or hurt him, but which he cannot control (geography, weather); and understand his goals and motivation (political, economic, cultural, military). From these flow his likely courses of action, our plans to deal with them, and the logistics necessary to support those plans. This is the stuff of which critical information is made. It comes from all of the data available, refined into information by the judgements of those up and down the line and across the staff, reduced to its purest form—critical information—which gives us the enemy's "who, what, when, and where" so we can defeat his plan and plan his defeat. Critical information is not more, but better, information with which we can synchronize our combat power. When distributed quickly to all vital locations (major subordinate commands, main/rear/mobile TOCS), etc., decisions can be made quicker and, ideally, within the enemy's decision cycle.

We have designed our C^2 system to automate critical information flow. TACCS-K uses a series of micro host computers located throughout the Combined Battle Staff in garrison as well as our warfighting headquarters. We have other hosts around the country to support the component and subordinate command elements. All of these hosts are securely linked by a series of interconnecting LANs. Each host contains the same database. The decision to go this way was one of the hard choices, but I believe it was the right one. We recognized that a distributed versus a replicated database is normally a more efficient use of computer storage and processing, but that is not true in combat. The loss of any host with a distributed database can result in the loss of critical information and, thus, combat capability. Theoretically, the loss of all but one replicated host leaves us with an austere, but still functional, automated capability.

Each of the host computers can hold up to seven workstations although we operate with something less than that number to give us a surge capability and to keep from burdening the system. Each workstation, itself a powerful personal computer (PC) with 20 MB of storage space, serves as the point of entry into the LAN through the host or as a stand-alone processor for database management, spreadsheet, and word processing functions. The action officers use the PCs to store data and process information unique to their staff or line functions. Only a small portion of their information will qualify as critical information at any one time and that alone will reside in the host, immediately available to the staff/line principals and decision makers at appropriate levels. As critical information changes, the action officer will update or remove it from the host. Although all users have access to the total database, only those responsible for the individual data elements can change or remove those elements. This keeps us from corrupting or losing the database.

The next major consideration was how we could format and display critical information. TACCS-K and TACCIMS are systems for echelons-above-corps. Since my information needs and those of the component and field commanders are not exactly the same, I asked them to submit their requirements along with my staff. Even if their critical information needs differ from mine, the principle of the process remains the same. As with all users of critical information, they will determine what meets the requirement for their given situation.

One of the early and important judgements was the decision to limit critical information largely to summary material. This is not to say that detail is not important. However, at my level, summary information is more simply displayed, more easily understood, and far more useful than an endless array of mind-boggling numbers or statistics. I am, for example, less concerned about the number of air interdiction sorties available for the next 24 hours than with our ability to meet the requirement. Summary data, color-coded or displayed in bar graph or pie chart, lets me know at a glance if there is a problem. I can then turn to the expert to find out when the problem will be fixed and what impact there is on our operations. Similarly, the number of artillery rounds available to a given organization is critical to its commander, but only incidental to me unless it impacts the overall mission.

Access to too much non-critical information not only clogs the critical information flow arteries, but bogs us down with too much detail. Worse yet, it gets me or my staff in someone else's area of responsibility. It is difficult to avoid micro-management when one has the capability to do so, but we must be cautious in commanding and managing only what is ours and leaving to others what is theirs.

The next step is to display all of the critical information in one setting. This is where the "rubber meets the road." My staff and I have developed some 36 automated summary graphics which show the core of the critical information. They are largely universal in nature and run the gamut from weather to logistics and everything in between. In short, we think they are pretty much what every other theater commander would need. Some scenario-dependent and theater-unique displays reflect the fine tuning and tailoring to meet our specific needs and those of this command. Where our graphics differ is in the summary nature of their contents, the breadth of the line/staff input, the simplicity (formats are flexible, not rigid), the brevity (30–40 minutes) with which they may be presented, and the automation which displays them on a large screen before the staff or on individual monitors. The latter capability lets me or anyone else look at the graphics to update information without having to shut down the staff's planning responsibilities to give that multi-hour briefing I discussed earlier.

One important measure of merit for effective C^2 is to ensure that key commanders and alternate command facilities at remote locations have access to the same critical information and can interact with each other during the critical phases of mission planning and execution. This is particularly true when dealing with activities that lend themselves to graphic displays such as a front line trace. These visual aids provide an enormous advantage in time and accuracy over hard copy messages. We have a limited graphics capability in TACCS-K, but will have a full-up capability in TACCIMS. This will guarantee that my major subordinate commanders and I are "reading from the same sheet of music."

Perhaps the most important quality of the system reflects the combined nature of this command. The English-Hangul keyboard permits bilingual graphics and electronic mail. This capability is a recognition of the cultural and linguistic differences within CFC and the need to deal with them in a partnership of mutual understanding. Until such time as U.S. military members learn to speak Korean as well as the Koreans speak English, the bilingual capability will remain vital to the successful employment of TACCS-K and TACCIMS.

What, then, have we done in Korea that contributes to the command and control mission? First, we hope we see the battlefield as it really is: a complex, interrelated series of battles on land, sea and air and superimposed on and interacting with the mobilization plan and governmental agencies of the ROK. We think this all leads to a more realistic deterrent posture and also prepares us for a wider range of responses should deterrence fail. It gives us greater confidence in the sustaining structure of well-trained and tested people and procedures operating a reliable, redundant, and survivable system of "strings and cans" to help manage the flow of critical information vital to the defense of this republic. We cannot afford to sit back and admire our creation, however; we have to work and train hard so that critical information becomes second nature to us. "An absolute doctrine is impossible," writes Major General J. F. C. Fuller. "For once a doctrine and its articles become a dogma, woe to the army which lies enthralled under its spell."[1] The process is on-going, but, with adequate training and when properly constructed, operated and maintained, C^2 gives the commander the ultimate capability—to put his synchronized combat power when and where he wants it with the greatest effect on the enemy and at the least risk to his own forces.

Footnote

[1]S. L. A. Marshall, COL, AUS; *Men Against Fire,* 1947 reprinted, Peter Smith, Gloucester, Mass., 1978. p. 107.

Army Command and Control Information Systems Requirements Definition

Donald W. Kroening

Abstract—The process of developing Army command and control information systems has been less than totally successful in producing fielded systems. One critical but often neglected aspect of that process, the requirements definition phase, is discussed. It is argued that the dynamic, high-risk, stressful military decision environment has unique characteristics which make it difficult to adapt information systems from, for example, a business environment. It is further argued that, within the military, the Army in particular has special characteristics that complicate the requirements definition process. Finally, the total acquisition process, particularly the time taken by that process, is shown to work against effective requirements definition. A series of solutions to these problems are proposed within the context of the concept based requirements system.

**This article originally appeared in IEEE* Transactions on Systems, Man and Cybernetics, *Vol. SMC-16, No. 6, November/December 1986.*

Introduction

There is little doubt today that Army command and control (C^2) information systems development is recognized as a very high priority requirement within the Army. Some go so far as to say that our doctrine will not work without significant improvements in information collection, processing, and dissemination. This increased awareness of Army C^2 is beginning to show up in all phases of the combat material development process and even, more importantly perhaps, in the Army budget. Why is it then, that despite this increased priority, the Army has failed to type-classify and field the C^2 information systems that we all want and that we know are well within the state-of-the-art of information system technology?

There are a number of possible answers to this question. There are certainly many problems in both the combat developments process and the material developments process which we could point to as the cause of the situation we are in today. It is the intent of this paper, however, to focus on a particular aspect of combat developments—the C^2 information system requirements definition process. Perhaps no other phase of the acquisition process is more important to the successful fielding of C^2 information systems and yet so misunderstood and neglected by the community at large. Andriole (Reference 1) has commented that "systems analysts have for years produced and reproduced evidence which suggests rather conclusively that without an explicit detailed definition of user experience and requirements, the interactive system will go underused, be misused, or ignored altogether." Boar (Reference 2) reports that 20 to 40 percent of all systems problems can be traced to problems in the design process, while 60 to 80 percent can be traced to inaccurate requirements definition. While technology, both for hardware and software, continues to advance at almost an alarming pace, our ability to define the requirements against which the technology is to be applied is progressing at a snail's pace, if at all. We have the situation where defining Army C^2 information systems requirements is perhaps more difficult than building the systems to meet those requirements.

This paper will further develop this hypothesis through discussion of specific problems affecting Army C^2 information system requirements definition which in turn have historically contributed to the more fundamental problem of developing and fielding these systems. Possible solutions to these problems will then be explored within the context of the Army's Concept Based Requirements System.

Specific Problems

The process of defining C^2 information systems requirements has many specific problems associated with it that contribute to its difficulty. Some of these problems are unique to the Army by virtue of the harsh nature of the tactical environment within which these systems must operate. Others are not unique at all, but represent common problems that the other services and commercial industry also face. Specific problems discussed in this paper are categorized for presentation purposes into three general groups: those problems which result from the very nature of the military decision problem being addressed by C^2 information systems and supporting technologies; those problems which result from the uniqueness of the Army's C^2 requirements; and those which result from the particular management approach, procedures, and/or analysis tools currently applied, or in some cases not applied, to Army C^2 information system requirements definition process.

C^2 information systems, by their very nature, are uniquely different from other systems. Both the military and commercial communities have come to this realization over the years. This uniqueness causes problems both for the requirements developers and the systems developers and, in large part, is responsible for the high degree of complexity associated with these systems. But why are these systems unique and complex? What is it that sets them apart from standard payroll systems, accounting systems, corporate decision support systems, or other computer systems? The answer, it seems, is that C^2 information systems exist to support the human decisionmaking process in a time pressured dynamic environment with multiple, often conflicting, goals, and with a high volume of data which nevertheless does not provide an acceptable level of valid data. We do not know all we need to know about human decisionmaking under such circumstances, but we know enough to understand that the decisionmaking process is unique to each individual, that there are different styles of decisionmaking that require varying degrees of information and that there is considerable uncertainty as to what information is even required for tactical decisionmaking.

The uniqueness and complexity of the tactical decision environment causes us to rely heavily upon technology for C^2 information systems that will provide solutions to all our problems. However, while technology development certainly solves many problems, the pace of technology development actually causes other problems. Microelectronic technology has made commonplace today's computer capabilities which we only dreamed about a few short years ago. The speed at which this technology changes is far greater than the speed at which we can effectively apply it. Technology, also, does not come free. It comes with significant overhead consisting of the personnel to apply and maintain it, training programs for all who will use it and procedures and policy to ensure its more acceptable and efficient integration into a larger system. These elements of overhead take time to develop and employ and often are overlooked, particularly by those who are blinded by the bright lights of flashing technological gadgetry. It is becoming more evident, however, that our best technology will never be effectively applied unless we pay attention to this overhead and ensure that our acquisition process is responsive not only to the application of new technology, but also to the overhead that comes with the technology.

An additional problem that results from the very nature of the military decision problem and the user population is the fixation we tend to have with hardware. Hardware is the part of the system that we can see and touch, and it is quite natural to want to focus our attention there. Software, on the other hand, is not understood by most people, and it is not physical in nature. It seems much more acceptable to spend $2 million or $20 million on new hardware "solutions" than to spend $1 million or $10 million generating, debugging and documenting *x* number of lines of program code. Additionally, from a system development standpoint, it is easier, less time consuming and less costly today to develop hardware than it is to develop software. A similar statement can be made about requirements definition. How often have we seen hardware requirements, such as mean time between failures, specified down to eight decimal place accuracy, while the functional software requirements for the same system are not even defined. Should not functional requirements drive hardware development? An equally disturbing occurrence is our propensity to perpetuate old software on new hardware. In some cases this is necessary from an economic standpoint, but in other cases it simply is evidence of our resistance to re-examining how we might perform tasks differently with improved automation capability. It is much easier to do business the same way, albeit perhaps faster with new hardware, than it is to do business in a different and perhaps more efficient way with new software.

Additional problems in C^2 information systems requirement definition are unique to the Army. Both Army and commercial systems are designed to support decision making by providing timely and accurate information. The information is very different, but the functions of the C^2 information systems are remarkably similar. The unique aspects of Army C^2 information systems result from the environment within which these systems must operate, not from any real difference in what the systems are intended to do.

The Army also faces different problems than its sister services do. A well-known contemporary army general officer once said that the Army equips its men while the other services man their equipment. This not-so-subtle distinction says a lot about the difficulties associated with the Army's C^2 information system requirements definition process. The command and control of any army of men in combat is a completely different problem from the command and control of a relatively small group of ships or planes with effective communications, inertial navigation systems, etc. For example, it is extraordinarily difficult to keep track of the location of subordinate units, personnel, and equipment; the personnel frequently are not sure themselves where they are relative to one another or to key battlefield locations. Army C^2 information systems must survive within a much more severe environment than those C^2 information systems of the other services. So the Army is faced with a more difficult problem in fielding C^2 information systems due not only to the relative complexity associated with defining requirements for what and how the system will operate, but also in defining the requirements for operating the system in a harsher environment.

One unique problem facing Army C^2 information system requirements development today is to determine the degree of militarization required. The problem centers on the trade-off which exists between the exorbitant cost of militarized, fully survivable systems on the one hand and much less costly, but less survivable commercial off-the-shelf systems on the other hand. Two aspects of this problem deserve mention. First, we find that the decision to militarize or not militarize is often based upon political and other considerations rather than on the actual requirements of the objective users of the system. In stating his/her requirements for systems in general, the user most often will not state as a requirement that he/she needs a militarized system, but he/she may state that certain functions performed by the system be operational continuously. Stating the requirement in this way provides considerable flexibility to the system developers in meeting the requirement. Continuous operations can be provided, for example, not only by fully militarized systems, but also alternatively by redundant, much less costly commercial off-the-shelf systems or by selective temporary manual backup for commercial systems. Requirements for C^2 information systems, however, are rarely stated in this manner.

A second noteworthy aspect of the C^2 information system militarization decision has to do with the lack of consistency with which these decisions are made across the Army. Why is it that in one location on the battlefield we have two computer systems sitting side by side, one militarized and one commercial off-the-shelf? Is the militarized system performing a more critical function than the commercial system, or is it not more likely that the Army simply does not have a consistent set of criteria applied to the militarization decision? The decision to militarize a C^2 information system must be based on the requirement for such militarization which should be driven by an analysis of the threat and the criticality of the functions to be performed by the system. Ensuring that this requirements analysis is done consistently across each battlefield functional area so that consistently militarization decisions are made is a difficult thing to do and is not happening today.

The unique problems facing the Army in C^2 information systems hardware requirements definition are perhaps only exceeded by problems in software requirements definition. Foremost is the Army's inability to define clearly the software requirements for the system and in particular, the information requirements of the eventual users of the system. A number of studies have been conducted by the Army in recent years to determine the information requirements of the tactical commander and his staff. Although each of these studies has produced a list of specific information requirements, there is a commonly stated caveat that information requirements are actually situational and cannot be determined with confidence without consideration of the specific tactical situation. Add to the variability of the battlefield, the variability of the commander, and you have a situation that makes C^2 information systems requirements definition next to impossible.

A problem very much related to defining information requirements is the quantification of such requirements. Quantification not only is required in order to size the C^2 information system but also, perhaps more commonly, to size the supporting communications systems. Although quantification of information requirements is required, it must be recognized that the same variability of the tactical situation and the commander that caused problems in defining information requirements, causes similar problems in quantifying those requirements. It does no good to quantify in erlangs for the communicators the amount of information required to flow between C^2 information systems, when the basic content of the information flow is suspect.

As a final problem in C^2 information systems requirements definition that is somewhat unique to the Army, consider what is commonly referred to as "gold-plating." When one does not know the true requirement and cannot quantify it, one usually overstates the requirement to ensure that the requirement will be satisfied. This philosophy is perpetuated through various levels of staffing of requirements documents until finally the requirement reads as if a gold-plated C^2 information system is required. Developing systems to meet these overstated requirements may ensure that the real requirement is met, but it also means that we have spent more money and time than was necessary to get the needed capability to the field. Certainly a similar problem exists within the other Services, but this problem is more pronounced in the Army because of the previously discussed environment within which the Army operates.

As a final category of problems associated with C^2 information systems requirements definition, let us turn our attention to those associated with the process itself and the Army's management of that process. The problems associated with the Army's Material Acquisition Process immediately come to mind and are well documented within the community. The most common argument made is that the process is too long to permit fielding of new technology before that technology becomes obsolete. Certainly this is true for computer systems technology given the rate that technology is evolving. However, the fact that evolving technology is seen as a problem is itself symptomatic of our failure to adequately define requirements. If we were able to say "we want our C^2 system to perform the following functions . . . ," then we would need to concern ourselves only with the quality of the solution, not the age of the technology. Since we do not know what we want, we constantly choose new technological solutions to our undefined C^2 problems. Efforts have been sponsored recently at the highest level to modify the Material Acquisition Process to make it more responsive to the pace of computer systems development. These efforts address the symptoms (i.e., "we are fielding obsolete technology") rather than the cause (i.e., "we haven't defined how we will use technology to address our needs").

The Army's requirements development community has long been criticized for changing its requirements for C^2 information systems before they can be implemented. A change in requirements can be very disruptive to a project manager who is charged with meeting these requirements. Some such changes may be due to a chase after the chimera of new technology. However, requirements *are* continually changing because of changes in the threat, in our doctrine, and in our force structures. Once again, it is the length of the Material Acquisition Cycle that the requirements development community views as the culprit. Using this system, the Army simply cannot develop and field a system to meet requirements before those requirements change.

Another problem associated with the Material Acquisition Process is that concerning the extent to which requirements are validated by the so-called "user." Who is the user in this context? Is it the actual user in the field or is it a surrogate user established within the Training and Doctrine Command (TRADOC)? Army regulations are clear on this matter and point out that TRADOC represents the user in these matters. This relationship was established so that the common good of the Army is the foremost consideration in requirements development, rather than the individual good of a particular Army unit. Why is it then that the Army had individual units defining requirements and developing their own C^2 information systems? Once again, the common argument given is that the Material Acquisition Process does not permit timely fielding of C^2 information systems to meet the immediate needs of the field user, so the field user defines his own requirements and typically purchases off-the-shelf systems to meet his requirements. While this may solve an immediate peacetime need, it may actually create or exacerbate other problems during wartime. Will these systems survive on the tactical battlefield? If not, do manual backup procedures exist? Who will maintain these systems? Will they interoperate with adjacent, national, joint, and allied systems? What about the logistics tail? These questions must be addressed. The Material Acquisition Process does address them, although it is the deliberate addressal of these questions that contribute to the unresponsiveness of the process.

As a final problem to be addressed, let us consider the role of analysis within the material acquisition process. The Material Acquisition Process either permits, or in some cases explicitly requires, analysis to support decision points within the process. These analyses are necessary, in most cases, to help justify or otherwise contribute to the development of the best product. Over the years since World War II, the operations research community has developed and/or improved operations research techniques, methodologies, and analysis tools to improve the quality of their trade. Unfortunately, this work has concentrated on the analysis of firepower systems. Until recently, command and control was not thought to be important enough to the outcome of the battle and, perhaps more evident, was thought to be too difficult to address analytically. This attitude contributed for years to the lack of development and fielding of C^2 information systems, even though commercial industry had developed the basic technology and had shown it to be effective by demonstrated use.

With the advent of AirLand Battle doctrine, the role of C^2, and in a larger sense, C^3I has been magnified. This magnification has begun to generate enthusiasm within the operations research community. With this enthusiasm, however, comes a realization that a lot more work is required and specific problems require resolution before responsive quality analysis of C^2 information systems can take place. For example, as a basic underlying requirement, the operations research community must develop the analytic methodologies and tools that will permit the evaluation of the effectiveness of C^2 systems. This evaluation cannot be measured by using standard system performance measures, like speed and accuracy, but rather by using the standard effectiveness measures used to evaluate firepower systems. How many destroyed enemy tanks will result from the use of C^2 information system XYZ? The ability to answer this question analytically will contribute immeasurably to answering the questions which now face the Army concerning C^2 information systems fielding.

Proposed Solutions

It would not be fair to present such an account of problems contributing to C^2 information systems requirements development as described in this paper without recommending a set of proposed solutions. While the case has been made perhaps that, collectively, these problems make C^2 information system requirements definition more difficult than the development of systems designed to meet these requirements; this is not to say that these problems are insurmountable. Each problem has a solution, although some solutions proposed herein are difficult to carry out.

Central to any discussion of the Army requirements development process is an understanding of the Army's concept-based requirements system (CBRS). The CBRS is a process, still evolving, designed properly to focus the development of doctrine training force design and material systems to meet the Army's most pressing needs. The overriding objective of the CBRS is to ensure that Army requirements (in this case, C^2 information systems requirements) are based upon approved concepts developed through proper consideration of Army doctrine, missions, history, threat, and technological forecasts. It is within the context of the CBRS, documented by TRADOC Regulation 11-15 (Draft), that the following C^2 information systems requirements development solutions are proposed.

Solution 1: The Army must recognize and accept the variability of the commander's decision making process and the variability of the tactical situation. Perhaps no other proposed solution discussed herein is easier to state, but more difficult to implement in all phases of C^2 information systems requirements development. History shows us that good tactical decisions can be made with very little information and that poor tactical decisions can be made with virtually complete information. Although we should continue to research why this is so, we should not let our lack of understanding of the human decision making process keep us from moving ahead. For example, instead of trying to define every element of information required for every possible tactical decision situation, simply provide the commander with a flexible system for obtaining and maintaining the information that he wants for any particular situation and provide some simple decision aids to assit him in his own unique decision making process. This is well within today's state-of-the-art technology. This need for built-in flexibility is what caused the movement during the 1970s from rigid management information systems to flexible data base management systems with accompanying ad hoc query capabilities. The Army's Maneuver Control System (MCS), to be fielded during the next two years, will provide this capability. Unfortunately, this capability comes a full ten years after the requirements for a data base management system with ad hoc query capability was first defined for the system.

Solution 2: The Army must plan for the battle against technological obsolescence. Although recent attempts have been made, some of which were partially successful, to change the Material Acquisition Process, it is unlikely that any radical changes will occur in the near timeframe to make the process more responsive to computer technological advancements. It also is unlikely that the pace of technological development will slow down. What this means is that the Army must learn to work within the current system to ensure that technological obsolescence is recognized and that the necessary planning has taken place to reduce its effects. This can best be done through preplanned product improvement (P^3I) programs for C^2 information systems. These programs are permitted by the Material Acquisition Process and are an effective way to get a core capability to the field more quickly and to add technological advancements later as they develop. All that is required is a reasonable understanding of what is coming and a concerted planning effort.

Solution 3: The Army must not allow itself to be blinded continually by those bright lights of flashing technological gadgetry. We must learn to recognize what is behind them that we cannot see. Failure to address personnel, maintenance, logistics, and training considerations can render useless the most advanced technology. Additionally, the Army must learn to focus its attention on the development of functional requirements for C^2 information systems rather than upon the development of hardware. The Army's move toward standard hardware will facilitate this. In the development of functional requirements, we must let the user community state precisely what they want the system to do without specifying the hardware or software design. Wherever possible, these requirements should be developed without bias toward the previous way of doing business. Let us take full advantage of new technology and not unnecessarily constrain its application by simply automating the same manual processes that have been perpetuated since World War II.

Solution 4: The Army must develop a rational, consistent policy concerning the militarization of C^2 information systems. While it is true that the concerns of C^2 survivability are paramount, the Army simply cannot afford to permit these concerns to result in overstating the true requirement. The requirement for militarization should not only be based upon the threat, but also upon the criticality of the function performed. These two criteria, if applied consistently across the full range of battlefield functions, are sufficient to form the necessary policy governing militarization of C^2 information systems. A possible reduction in the number of system development jobs, or the parochialism of individual functional proponents cannot be allowed to unduly influence the militarization decision.

Solution 5: The Army must learn to hold C^2 information system requirements constant for a reasonable period of time to permit satisfaction of the requirements. Rarely does the threat, doctrine, or force structure change radically in such a short period of time that would prohibit the holding constant of requirements. Is not an 80 percent solution better than no solution at all? A constantly changing requirement can cause the solution to the requirements never to materialize. The combined effect of holding requirements constant and developing P^3I programs can result, in most cases, in an acceptable 80 percent solution.

Solution 6: The Army must recognize and respect the preeminent position of the surrogate user of C^2 information systems in the Material Acquisition Process. The field user must learn to contribute to the solutions rather than to the problems associated with nonstandard, nonsurvivable and nonmaintainable C^2 information systems. Valuable contributions would include the identification of new approaches to doing business, the identification of specific requirements for information and decision aids, the evaluation of prototype systems and the strict adherence to existing policies calling for working through the surrogate user community to ensure the development and fielding of the C^2 information systems that are in the best interest of the entire Army.

Solution 7: The Army must develop the C^2 analysis tools and methodologies that permit evaluation of the contribution of C^2 systems, doctrine, and procedures to the effectiveness of the force. The effectiveness of a combat force typically is measured by comparing the relative results of the application of combat power upon both enemy and friendly forces. For a long time, C^2 has been thought to be a multiplier of combat power, but little analytical substantiation of this supposition has been generated. Only through the development and application of new models and other analysis tools, or through improvements to existing models and analysis tools, will this situation improve.

Solution 8: As a final solution, the Army finally must recognize the difficulty and importance of requirements definition and properly allocate resources for the C^2 information systems requirements process at a level commensurate with the resourcing of the C^2 information system development process. A comparison of the full realm of resources, to include not only dollar expenditures, but also number and grade level of personnel applied, travel budgets, and other resources, will show a definite slant toward system development rather than systems requirement definition. Hopefully a case has been made herein for the recognition of the complexity and importance of C^2 information systems requirements definition process. This recognition, and perhaps in a large sense, a whole new philosophy or attitude toward system development is required by the entire community before the necessary resources will be appropriately applied.

Summary

Many problems have been addressed concerning both requirements definitions and systems development. There are, of course, many more problems and, hopefully, many more solutions which could have been addressed. This paper, however, make a sufficient case for the need for an expanded awareness and increased attention to be given to the complex and critical job of C^2 information system requirements definition. Without this increased attention particularly embodied in the proposed solutions described previously, the process of C^2 information system requirements definition will continue to be what it is today—the least understood, most complex, and important step in the material acquisition process.

References

1. Andriole, S. J., *Interactive Computer-Based Systems: Design and Development*. New York: Petrocelli, 1983.
2. Boar, B., *Application Prototyping: A Requirements Definition Strategy for the 80's*. New York: Wiley Interscience, 1984.

Air Force Command and Control: Assessment Criteria for Computer Based Decision Aiding Systems

Gerald W. Hopple

Command and Control

The approach of this effort is to descend systematically from the domain of all logistics capability assessment models to the level of sortie generation models and then to bound the scope by identifying relevant command and control characteristics (general attributes of C^2 systems) and components (substantive building blocks of sortie-generation C^2 models). This chapter outlines the generic military and Air Force-specific C^2 process and structure, with an emphasis on *characteristics* that must be taken into account as the

This article was written especially for **Principles of Command and Control.*

alternative models are described and compared. Then, the *components* of C^2 relevant to this effort are delineated.

The command and control based requirements for (and characteristics of) sortie generation capability assessment models revolve around the concepts of *quick response, efficiency* and *effectiveness* of C^2 decisions. The specific prerequisites for such models are in turn closely tied to the decision command level and the two criteria of time available and degree of information aggregation, with both maximized at the higher force management levels (theater and component) compared to the lower levels (force and unit); at the lower levels, then, sortie generation models must be compatible with very constrained time parameters and produce performance and other output measures that are both disaggregated and detailed.

The core concepts of command and control are *missions, assets* and *levels of assessment*. C^2 decision making may concern any one or more of multiple alternative missions; assets available constitute a key input to the commander's decision calculus; the level(s) of assessment treated by the model are also critical. These three core components and two others (the level of indenture and the functional category addressed by the model) emerge as the generic components, which are profiled in general terms and then linked to the four force management levels (theater, force, component and unit).

Characteristics of Command and Control

The most general of the characteristics of command and control flows from the emphasis in doctrine on C^2 systems that function quickly and work efficiently. An extensive knowledge base on tactical planning emphasizes the assessment of the quality of both friendly and opposing force command and control systems in generating courses of action and plans (Hopple, 1986a). Central to this assessment is the determination that C^2 processes work quickly and efficiently. The efficiently of the system is measured by the extent to which the commander's intentions are carried out and the ability to cope quickly and effectively with changes in the situation. A quick response capability and efficiency of operation therefore emerge as preconditions for a candidate sortie-generation capability assessment model in a C^2 context.

Speed and efficiency, however, are necessary but not sufficient prerequisites; in addition, a C^2 system must be *effective*.[3] A command and control process, for example, could operate very efficiently but yield plans and decisions which were totally ineffective or even disastrous in their consequences.

What does it mean for a plan or decision to be effective? Hopple (1986b) briefly discusses one viable approach to answering this question in a survey of decision aiding systems relevant to Air Force C^3 countermeasures problems. The Headquarters Effectiveness Assessment Tool (HEAT), under development for the Defense Communications Agency (DCA), provides a generic framework for the evaluation of military plans. Initially designed as a methodology for measuring the effectiveness of theater level headquarters, HEAT assumes that as long as a plan remains valid, fewer steps must be performed in the control process; therefore, a useful measure of effectiveness at the aggregate level is the length of time a plan is in force.

HEAT measures the *quality* of the decision process (from monitor and understand to plan and dirct as phases in the C^2 flow) by tracking a headquarters' interactions with its environment and its ability to convey and integrate information quickly and correctly. The headquarters is held responsible for "plans" defined as sets of missions, assets, and schedules developed for subordinate commands. Effectiveness is tapped by the ability to develop and implement such plans while adjusting them for this information and assets available. Shortcomings are detected when too much is attempted by too few forces, too quickly, or in the absence of adequate information.

Poor or inadequate plans are ones which require the development of a whole new understanding of the military situation, a completely new plan and a new set of directions. Thus, the percentage of time that a plan is in force is the fundamental measure of C^2 effectiveness. This basic measure is supported by a number of measures of process performance; examples include:

- For monitor: percent of units where HQ data on unit location, status, plans, or operations are outside the desired accuracy window;
- For understand: average error in predictions of the time at which system operability changes because of weather;
- For direct: average number of queries received per directive.

For the sortie generation capability models, outputs become information inputs into the C^2 decision process. Therefore, their potential contribution is in terms of their accuracy and validity and the role they play in effective decision making. Model validity and forecasting accuracy criteria thus emerge as key criteria.

The C^2 measurement criteria thus cluster into two broad categories of characteristics:

- Speed and efficiency (of the C^2 decision *or* aiding process); and
- Effectiveness (of the ultimate decision outputs *or* the information inputs to the process).

In addition, decision command levels also affect the necessary quality of model results. Specifically, both time constraints and information needs vary rather dramatically as one ascends or descends the tactical C^2 force management decision hierarchy.

Figure 1 illustrates this. Higher level decisions take more time to execute or adjust because of their dependence on materiel, maintenance and personnel scheduling and movement. Although time is always a precious commodity, it is even scarcer at the lower levels (force and unit), where preflight and inflight C^2 activities must occur against the backdrop of very compressed time parameters. Also, lower level decisions require more informational detail, while decisions at the higher levels of the C^2 hierarchy tolerate less detail and more aggregation.

DOMAIN	DECISION CATEGORY	DECISION TYPE	TIME AVAILABLE FOR DECISION	DEGREE OF INFORMATION AGGREGATION REQUIRED
PREFLIGHT	PLANNING AND COMMITMENT	APPORTIONMENT ALLOCATION TASKING DETAILED PLANNING	(MORE TIME) ↑	(MORE INFORMATION, LESS DETAIL) ↑
INFLIGHT	CONTROL AND COORDINATION	RESOURCE REALLOCATION TARGET REASSIGNMENT TARGET UPDATE/ INTERCEPT CONTROL THREAT WARNING	↓ (LESS TIME)	↓ (LESS INFORMATION, MORE DETAIL)

Figure 1. Summary of critical tactical command and control force management decisions (Wohl, 1981).

Aids and information systems thus have different criteria depending on the level. The higher the level, the more time is not a critical concern (in a relative sense) and the more that information can be aggregated. Performance and other measures must be most detailed and disaggregated and generated most quickly the lower the level.

In addition to characteristics from the C^2 realm, the literature on the evaluation of forecasting models specifically (e.g., Armstrong, 1978) or decision aids/support systems more generally (e.g., Adelman and Donnell, 1986) features a number of frameworks for assessing such models or systems. Armstrong emphasizes the dual attributes of model *inputs* (particularly assumptions used) and *outputs* (forecasts)—what goes into the model and what comes out. He identifies four central evaluative questions, revolving around:

- The relationship of the real world to the assumptions (whether or not the latter are comprehensive and reasonable);
- The relationship of the assumptions to the model (the logical structure of the model);
- The relationship of the model to its outputs (the replicability of the results); and
- The relationship of the outputs to the real world (ultimately, a question of the model's benefits and costs).

Assumptions, model structure, outputs and costs/benefits represent four very broad gauged criteria for the comparative assessment of alternative capability/sortie generation models.

In addition, a characteristic orthogonal to the model's quality or technical performance is its acceptability or user approval. This ranges from the per-

		ACCEPTABILITY			
		HIGH		LOW	
		High Relevance	Low Relevance	High Relevance	Low Relevance
QUALITY	High	A	B	C	D
	Low	E	F	G	H

A = best combination
H = worst combination

Figure 2. Universe of model characteristic rankings: acceptability/relevance/quality (KSC/Perceptronics, 1985).

ceived quality of the system/user interface to other subjective reactions and more objective measures such as frequency and nature of use.

Ideally, both attributes are necessary conditions. Conceivably, however, a model could be very strong technically (high in quality) but unpopular with users for any of a variety of reasons (low in acceptability). Alternatively, a model might be "popular" but technically weak or even deficient. When a third critical characteristic—relevance to the analytical context, which here is command and control—is introduced, there are eight distinct high-low combinations (as displayed in Figure 2.

Figure 2 isolates the three primary characteristics gleaned from the C^2 and decision aid methodology literature streams:

- Acceptability:
  —User friendly/quality of user-system interface;
  —Ease of operating (inputs and outputs);
- Relevance (= efficiency):
  —Quick response;
  —Real time (or near real time) data;
  —Flexible (capable of evaluating various logistics/C^2 options/scenarios);
  —Useful at one or more force management levels;
- Quality or Performance (= effectiveness):
  —Accuracy of outputs (forecasts/estimates);
  —Validity of inputs (especially assumptions);
  —Logical structure of model; and
  —Replicability of results.

The preferred cell in Figure 2, of course, is A, where high acceptability, high efficiency or C^2 relevance, and high effectiveness or quality all intersect. The worst case is cell H, where all three characteristics are scored low. The eight cells bound the universe of C^2-relevant sortie generation models, underlining the importance of a model evaluation template which satisfies user criteria, relevance/efficiency standards, and quality or effectiveness desiderata as well.

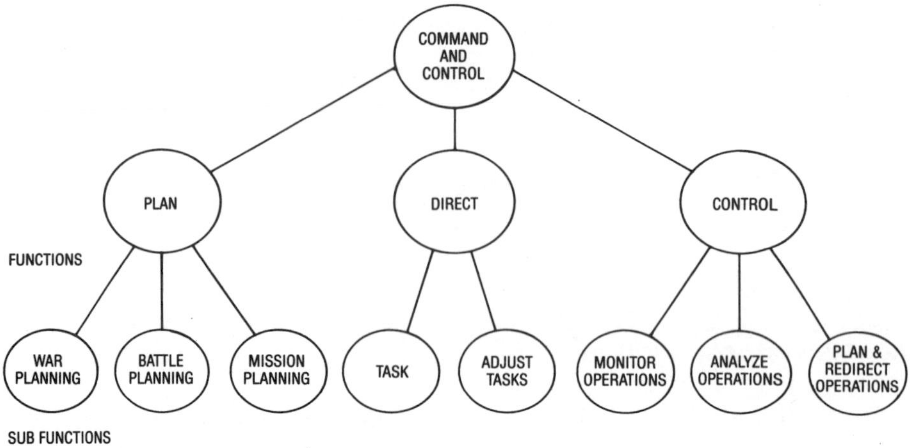

Figure 3. C^2 functions and subfunctions.

Air Force Tactical Command and Control

The standard definition of C^3 (command, control and communications—according to TAC Pamphlet 55-19/TRADOC Pamphlet 525-7)—is:

> The process of and means for exercising authority and direction by a commander over assigned forces or operational controlled forces. C^3 functions are performed by: personnel, equipment and communications facilities and procedures to plan, direct, coordinate and control forces and operations. C^3 capabilities include resources used to:
>
> - Obtain, process and disseminate information to support command planning and decision making;
> - Formulate orders to subordinates;
> - Make decisions; and
> - Monitor the results of actions and the status of forces.

Wohl (1981) distinguishes between command (*preflight* oriented) and control (*inflight* oriented) stages. The command/preflight tasks focus on planning and commitment; the inflight decision processes are oriented around control and coordination.

Command and Control Functions

Various descriptions have been employed to identify principal functions within the Air Force tactical environment. The terms *planning, directing* and *controlling* serve as useful high level descriptors of the central C^2 functions. Figure 3 depicts the C^2 functions and corresponding subfunctions.

> *Planning* is the function of preparing plans for future use. It begins with the decision to conduct a military operation or mission and

includes adjustments to plans before the affected operations period. Planning is subdivided into three subfunctions: war planning or campaign planning; battle planning; and mission planning.

(1) *War Planning or Campaign Planning*. This planning subfunction is concerned with long range planning and occurs at the Joint Task Force (JTF) level. The products of this functional activity include war plans or campaign plans that may cover a large area and an extended period of time. The scope of such planning is broad in nature, focusing on theater-wide objectives. This function embraces a series of related functional activities that begins with an understanding of the mission(s), an assessment of the situation, developing and evaluating options, and determining the best course of action.

(2) *Battle Planning*. This is the planning subfunction that focuses on semi-near-term, short-ranged plans concerned with the employment of the force. It normally has a time window for expected execution. Battle planning begins with the assignment of immediate battle objectives, assessment of the situation, development and evaluation of options, selection of the best course of action and assignment of tasks or missions.

(3) *Mission Planning*. This subfunction takes place at the tactical unit level and focuses on the detailed development of mission composition, schedules, and flight plans in accordance with the directives emanating from the battle planning subfunction.

Directing is the function of issuing orders to plan and execute operations or missions. It includes the direction of the commander's policy and guidance for planning and execution of air operations, the approval and issuance of plans, specific assignment to tasks, and orders for execution; this functional activity also involves the approval and direction of adjustments to previously issued plans/orders.

Controlling is the function of monitoring execution of planning operations, and dynamic planning, directing, and controlling of forces to adjust to demands of the current situation. Controlling takes place during the execution phase of military operations. This function features three subfunctions:

1.) *Monitor Operations*. This functional activity involves those tasks associated with collecting and observing information that will provide the feedback for evaluation of the execution of directed operations.

2.) *Analyze Operations*. This functional activity is an assessment and evaluative process that either determines that current operations require no change or discovers that some change is required.

3.) *Plan and Redirect Operations*. This is the next step in the process which involves all dynamic planning and direction necessary to perform selected adjustments to current operations.

These C² functions are accomplished regardless of the theater in which the Tactical Air Forces (TAF) is employed. The organizations tasked with responsibility for the function may change, but the function remains constant.

Tactical Air Forces: Organization

Tactical air forces are organized, equipped, trained and prepared to conduct sustained air operations aimed at the destruction or neutralization of enemy forces. These forces may be employed as a component of a combined command, unified command, subordinate unified command, a joint task force (JTF) or they may be employed independently. They also must be organized, assigned and employed in a manner that will permit concentration of effort in accordance with the priorities of the commander.

Tactical air forces, then, normally will form the Air Force components of unified and sub-unified commands and joint task forces. These U.S. joint organizations are often further integrated into multinational combined (Allied) command structures.

Four management levels have been identified and accepted as relatively standard within the unified and combined command structure. The C² management levels are: Theater, Component, Force, and Unit.

Theater Level. The theater level exercises operational command over all forces assigned and is principally concerned with establishing broad strategy, objectives and priorities and apportioning forces to meet theater requirements. This level affords the assumption of a broader perspective on current operations than that maintained at lower management levels; however, selected operations of significance that could require responsive theater level actions are monitored in detail. Overall, planning is longer term and entails changes less frequently than is the case at lower levels.

Component Level. The component level exercises operational control and authority over the forces assigned as delegated by the theater level. The air component is concerned with the air strategy, concept of operations, and force allocation in consonance with theater guidance and direction. Planning is both long and short term. An air operations order may go unchanged for a considerable period of time, but a daily Air Task Order is normal. During the execution phase, the component level monitors the progress of ongoing air operations, which is under the tactical control of the force level. The component level is primarily concerned with those significant aspects of current operations that may require major adjustments (such as, change to the air allocation). Additionally, the feedback of the results of current operations contributes to component planning for the next operational period.

Force Level. The force level is delegated authority by the component level for the tactical control of tactical air forces. The principal activity is concerned with the execution of the plans and orders issued by the component

MANAGEMENT LEVEL	AUTHORITY	BROAD FUNCTIONAL ACTIVITY
Theater	Operational Command	Theater Joint Strategy Objectives Priorities, Apportionment
Component	Operational Control	Force Strategy and Allocation Priorities, Employment Concepts, Detailed Planning
Force	Tactical Control	Resource Planning for Mission Execution, Current Operations, Dynamic Adjustments to Current Operations
Unit	Execution	Detailed Mission Planning Execution

Figure 4. Basic division of responsibilities in the generic command and control structure (KSC/Perceptronics, 1985).

level. Included at the force level are those functions relating to the launch, control and execution of all tactical air missions. During the planning phase, the force level is engaged primarily in the preparation of all force level control centers and system elements for the subsequent execution of plans/orders of the component level. Since the force level exercises tactical control of current and ongoing operations, accurate information is required in real time and in detail. The force level must be prepared to respond immediately to the dynamic requirements of ongoing combat operations within the confines of the authority delegated by the component level.

Unit Level. The Unit Level consists of the tactical air wings and squadrons. This level is concerned primarily with detailed mission planning and execution of tactical air missions as tasked by the air component. During current operations the force level may task the unit level within the authority delegated to it by the air component. The most current and precise information on enemy activities and capabilities is required relative to missions to be executed.

Figure 4 provides an overview of the authorities and functional activities of the C^2 management levels. Although the management levels are generic, it is easy to associate current regional TAF structures to the levels (as shown in Figure 5).

The process of planning, directing, and controlling air operations is iterative at all C^2 levels. The process requires continuous internal coordination and feedback between levels. Figure 6 portrays the basic functional interrelationships at the theater, component, force and unit levels and their decision time horizons. What is difficult to depict graphically is the dynamic layering of the functions within a time and space context and the fact that all three functions are analytically distinct but in reality occur simultaneously.

The command and control functions discussed above are performed within C^2 organizational structures. A Tactical Air Force performs its mission and tasks as the air component in a Joint Task Force (JTF), unified command, or combined command and is organized and configured to meet the requirements

LEVEL	CONTINGENCY	CENTRAL REGION	KOREA
THEATER	CENTCOM	AFCENT AAFCE	UNC CFC/USK
COMPONENT	TAF TACC/COMBAT PLANS	AAFCE ATAF ATOC SOC	ACC/AFK TACC/CURRENT PLANS
FORCE	TACC/COMBAT OPS CRC ASOC	ATAF ATOC SOC CRC ASOC	TACC/CURRENT OPS CRC ASOC
UNIT	MISSION AIRCRAFT WOC	MISSION AIRCRAFT WOC	MISSION AIRCRAFT WOC

Figure 5. Command and control element equivalency (KSC/Perceptronics, 1985).

of the theater of operations. The air component commander (ACC) directs the activities of his headquarters and exercises operational control over assigned resources in accordance with the overall plan of action set forth by the theater commander.

Overall responsibility for the conduct of combat air operations rests with the ACC. Normally, the ACC authorizes subordinate agencies of the Tactical Air Control System (TACS) to perform certain tactical control actions. The ACC, through the TAF deputy for operations, establishes the allocation of the air effort planned to support each assigned air task. The allocation is based on the apportionment guidance of the theater commander and the requirements for air support from lateral component commanders.

The ACC's headquarters includes the personnel and resources needed to support the commander. A TAF headquarters and its echelons develop con-

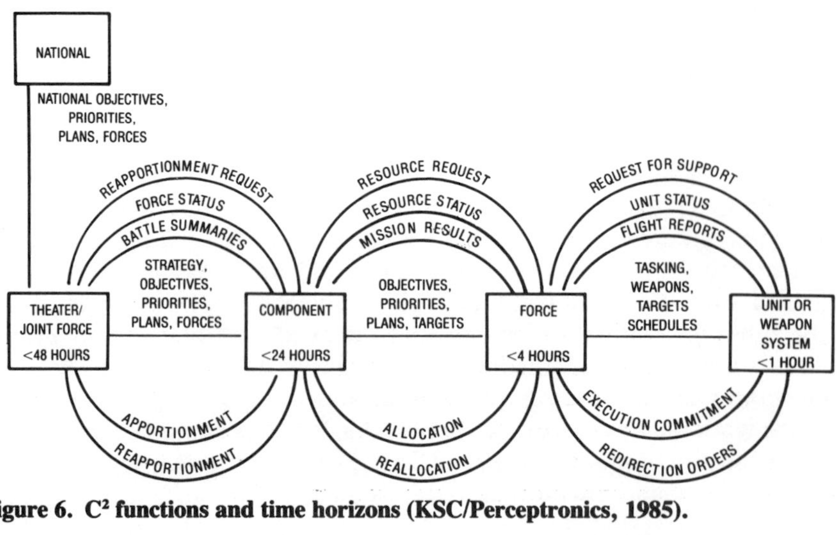

Figure 6. C² functions and time horizons (KSC/Perceptronics, 1985).

cept plans, operation plans and orders, fragmentary operation orders, air directives, air operations orders, air tasking orders and other guidance necessary for the centralized control of resources that conduct combat and supporting operations. The primary ACC senior battle staff members include: deputy commanders for communications and electronics, intelligence, logistics and operations.

The deputy for logistics (LG) assists the commander in providing logistical support for the TAF. The LG exercises staff supervision over subordinate logistics elements and has ultimate responsibility for logistical planning, supply, maintenance, transportation and procurement. The LG staff works closely with the deputy for operations (DO) and the Tactical Air Control Center (TACC) in requesting replacement equipment and monitoring the movement of supplies essential to the operation of the TACS.

The TACS is an integral part of the ACC's organizational structure and consists of units specifically trained and equipped to support the command and control process; it centrally plans, directs and controls tactical air operations and integrates air operations with the operations of other Service components. It is the system that is used to complete combat planning, sortie allocation, force tasking and control. The missions of counter air, air interdiction, close air support, electronic combat, tactical air reconnaissance, tactical airlift, special operations and air combat support operations are coordinated and integrated through the TACS. A TACS also provides the facilities required for air defense and airspace control in an area of operations.

The TACC is the focal point of the TACS. It is the operational facility in which the ACC, DO, and the deputy for intelligence (IN) have centralized the functions of planning, direction, control and execution over tactical air resources. Figure 7 portrays the organizational structure of the TACC.

The Director of the TACC is responsible for the centralized planning, direction, control and execution of assigned and attached air assets. The director is responsible to the TAF DO for the allocation and tasking of all tactical air operations. He directs and supervises the combat plans and operations divisions, translates air operations direction and guidance into specific allocation of air assets, and ensures that the Air Task Order (ATO) incorporates the ACC's priorities and objectives. The director guides the monitoring, evaluation and adjustments needed to meet changing tactical situations. He advises the commander and staff of problems or factors affecting the achievement of assigned objectives to include recommended resource allocation, use of specialized assets and relocation of assigned forces or units.

The central planning activities associated with the daily planning for tactical air operations involve force apportionment, force allocation, force employment and tasking and mission execution. War objectives emanate from National Command Authority/Joint Chiefs of Staff (NCA/JCS). The JTF produces apportionment recommendations, which are sent to the next level, where allocations are generated (the ACC); finally, the allocation is translated into tasking (TACC DO) and then specific orders to the units involved. The process looks like this:

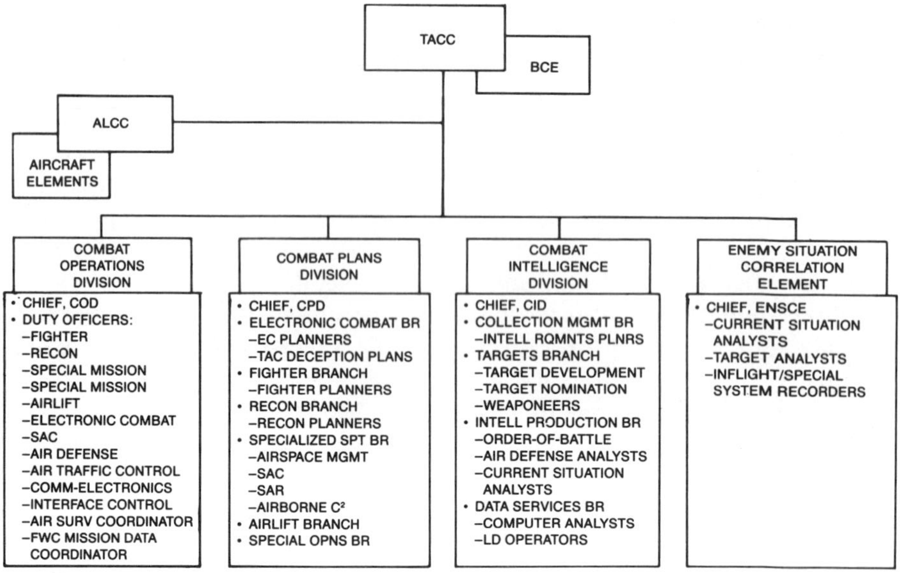

Figure 7. TACC organizational structure (KSC/Perceptronics, 1985).

Decision		**Command Level**
Apportionment	→	JTF Commander
Allocation	→	ACC (ATAF in NATO)
Tasking	→	TACC (ATOC)
Detailed Unit Level Planning	→	Tactical Air Base Unit Level

The major key decisions of the command and control elements center on the apportionment, allotment, allocation and tasking of air resources:

- Apportionment: This is the assignment of the total expected effort by percentage or priority to various tactical air functions for given periods of time. Apportionment is a responsibility of a theater commander in satisfying theater as well as component requirements for tactical air. The air commander will recommend apportionment decisions to satisfy the theater commander's requirements.
- Allotment: This is the temporary change of assignment of tactical air force resources between subordinate commands (in NATO, for example, from the Fourth Air Tactical Air Force [4ATAF] to the Second Air Tactical Air Force [2ATAF]). The power to allot is vested in the commander who has operational command (for example, the Joint Task Commander).
- Allocation: This decision translates the apportionment decision into a total number of sorties by aircraft type for each function; this is the responsibility of the Air Force Component (AFC) commander and is a function of the TACC.

- Tasking: This translates the allocation into specific orders to the units involved. This process assumes the form of an air task order (ATO) from the TACC.

All of these decisions are made within the bounds of the theater commander's decisions on objectives. These key decisions are executed through the apportionment directive, the air operations order, and the air task order. There are numerous decisions required to support the issuance and execution of these directives. Moreover, there are other decisions not directly related to these periodic directives which support current operations—the control function—and which are required for strategic planning and direction.

Components of Air Force Command and Control

Core Concepts

The core concepts here are *missions, assets* and *levels of assessment.* C^2 decision making may concern one or more of multiple alternative missions; assets available represent a fundamental input into (and constraint on) the decision calculus; the level(s) of assessment covered by the model are equally crucial to a determination of its uses and potential relevance. In addition, the level of indenture and the specific functional category addressed by the model are noteworthy features and have discernible impact on an assessment of the model's utility in a C^2 environment.

The generic categories and sets of illustrative and mutually exclusive (but not necessarily exhaustive) examples include:

- Missions:
  - —Tactical air force;
  - —Strategic bombers;
  - —Strategic air defense;
  - —Strategic airlift;
  - —Tactical airlift;
  - —Air refueling tankers;
  - —Medical evacuation;
  - —Reconnaissance;
  - —Search and rescue;
  - —Special operations forces;
  - —Tactical airborne command and control systems; and
  - —Airborne command posts.
- Assets/Commodities and Resources:
  - —Primary weapon systems
    - —Aircraft
    - —Missiles
    - —Other
  - —Engines and spares (line, shop and sub-shop replaceable units [LRUs], [SRUs], [SSRUs]);
  - —Munitions;

—Ground support equipment;
—Petroleum, Oil and Lubricants (POL);
—Facilities;
—Tanks, racks, adaptors and pylons (TRAP); and
—Other.

- Levels of Assessment:
  —Air Force wide;
  —Theater;
  —Base; and
  —Unit.
- Levels of Indenture:
  —Subsystem level (two digit work unit code);
  —LRUs;
  —SRUs; and
  —SSRUs.
- Functional Category:
  —Requirements determination;
  —Distribution;
  —Base maintenance;
  —Depot maintenance;
  —Intermediate maintenance;
  —Organizational maintenance and weapon system servicing;
  —Base supply/resupply;
  —Facility management;
  —Warehousing;
  —Inventory control; and
  —Shipping.

Force Management Levels

It also is advisable to consider the pertinent components on the basis of force management levels. The outline below arrays the various elements of each level, showing the similarities across levels (for example, decision making at each level involves the factoring in of intelligence data) and simultaneously highlighting the differences (such as, the fact that the components at the unit level are both more focused—constraints, for example, are not explicitly analyzed—and involve more detailed data elements). The components identified at each level reflect the fact that the focus shifts from very broadly defined objectives, priorities and plans at the national and headquarters levels to apportionment and operational command at the theater level, allocation and operational control at the component level, execution and tactical control at the force level, and tactical execution at the unit/weapon system level.

I. Headquarters, U.S. Air Force
   A. Forces

- B. Logistics
- C. Intelligence
- D. Constraints

II. Theater
- A. Resource Status
  1. Airfields
  2. Units
  3. Special assets
- B. Logistics
  1. Munitions
  2. POL
  3. Spares
  4. Aircraft
  5. Other
- C. Intelligence
  1. Air order of battle (AOB)
  2. Ground order of battle (GOB)
  3. Targets
  4. Enemy intentions
- D. Constraints
  1. Political
  2. Environmental
  3. Mission support
- E. Requests
  1. Army
  2. Allied Tactical Air Forces (ATAFs)
  3. Allied Forces Command

III. Component
- A. Resource Status
  1. Airfields
  2. Units
  3. Special assets
- B. Logistics
  1. Munitions
  2. POL
  3. Spares
  4. Aircraft
  5. Other
- C. Intelligence
  1. AOB
  2. GOB
  3. Targets
- D. Constraints
  1. Environmental
  2. Mission support

- E. Requests
  1. Army
  2. Air Tactical Operations Centers (ATOCs)
  3. Sector Operations Center (SOCs)
  4. Territorial commanders
- F. Direction and Guidance
  1. Apportionment
     a. Areas of responsibility
     b. Forces
     c. Priorities
  2. Task
     a. Targets
     b. Missions

IV. Force
- A. Logistics
  1. Munitions
  2. POL
  3. Spares
  4. Aircraft
  5. Other
- B. Intelligence
  1. AOB
  2. GOB
  3. Targets
- C. Constraints
  1. Environmental
  2. Mission support
- D. Requests
  1. Army Corps
  2. ATOCs
  3. SOCs
- E. Sorties (from units)
- F. Direction and Guidance
  1. Allocations
     a. Airfields
     b. Units
  2. Tasks
     a. Airfield Interdiction (AI)
     b. Offensive Counterair (OCA)
     c. Defensive Counterair (DCA)
     d. Alert
     e. Battlefield Air Interdiction (BAI)
     f. Suppression of enemy air defense (SEAD)

V. Unit
- A. Resource Status

1. Aircraft
   a. Fully mission capable (FMC)
   b. Partially mission capable (PMC)
2. Aircrew
   a. Qualifications
   b. Duty day
3. Airfield
   a. Runway
   b. Taxiways
   c. Arresting gear
   d. Weather
      (1) Launch/recovery
      (2) En route
      (3) Threat area
   e. Airspace control
   f. Launch/recovery pattern

B. Logistics
   1. Munitions
      a. Air-to-air (AA) preferred
      b. AA alternate
      c. Air-to-ground (AG) preferred
      d. AG alternate
      e. TRAP
   2. POL
      a. Storage (Consumption/Resupply)
      b. Hydrants
      c. Trucks
      d. Pipeline
   3. Maintenance
      a. Launch/recover
      b. Load/refuel
      c. Fix airplanes
         (1) Spares
         (2) Skills
      d. Aerospace ground equipment (AGE)

C. Intelligence
   1. AOB
   2. GOB
   3. Targets

D. Task
   1. Range/duration
   2. Routing
   3. Targeting
   4. Electronic combat
   5. Tactics

Tactical Air Force Command and Control Analysis/Decision Aids

Quite a few analysis and decision aids have been targeted at various tasks and phases within the TAF C^2 process. Tactical C^2 decisions within the U.S. Air Force frequently involve both information input uncertainty and consequences of action uncertainty, the decision context that is most challenging to the unaided decision maker and most in need of supporting information, analysis and decision aiding systems.

Currently, many aspects of command and control and battle management generally are time consuming, manpower intensive manual processes supported by diverse automated intelligence data handling systems (IDHS) and force management systems. Included here are Display Control/Storage and Retrieval (DC/SR) and Computer Assisted Force Management System (CAFMS) and Central Air Force Command (CENTAF), Korean Air Intelligence System (KAIS) and Constant Watch Phase III by the Korean Tactical Air Control System (TACS), and U.S. Air Force Europe Tactical Air Intelligence System (UTAIS) and EIFEL by the European TACS.

The primary function of a computer-based force management system, such as CAFMS, is to assist in the combat planning and combat operations of the TACC. Distribution of terminals within the TACC and other TACS facilities depends upon the operational situation and available communications. CAFMS provides the TACC with an improved capability to:

- Manage large volumes of data such as:
  —Airfield/facility status;
  —Aircraft/aircrew status;
  —Munitions status;
  —Weather status; and
  —Aircraft losses;
- Prepare, review and disseminate an air task order (ATO) in:
  —JINTACCS or internal Air Force format;
  —Mission schedule generation and update;
  —Strike package display;
  —Unit sortie recap; and
  —Mission sortie recap;
- Transmit data rapidly to other TACS elements using:
  —Digital data via voice link to remote terminals; or
  —Offline Automatic Digital Network (AUTODIN), Tactical Network (TACNET), and Teletype (TTY).

Constant Watch is a multiphase program designed to improve the air component commander's ability to effectively execute an air war within the Korean area of air operations; the objective of the Automated Data Processing (ADP) projects within the Constant Watch program is to develop, implement, and manage automated systems. Unique to the effort is the marriage of an intelligence ADP system and an operations ADP system to provide automated war fighting capabilities to the tactical forces. Included in Increment One of Phase III is an automated means for producing and disseminating the air task order (the ATO preparation will aid the combat plans staff in estimating available sorties, building the interdiction part of the ATO package from the target recommendation list and available forces) tasking those forces in either smooth flow or surge operations, preparing other parts of the ATO using preformatted shells and disseminating the ATO.

The EIFEL system is an automated tactical C^2 system that is used to plan, task and monitor offensive air operations. It consists of data processing facilities, remote terminals at subordinate units and a communications network. The overall EIFEL program provides automation to air tasking, resource status keeping and intersite coordination functions throughout the tactical C^3 environment. EIFEL 1 will automate three basic functions for users: air tasking; resource status keeping; and intersite coordination.

The performance of the air employment planning function uses four major categories of input data:

- Direction and guidance data;
- Friendly force and situation data;
- Enemy force and situation data; and
- Environment data.

The data are received within the TACC via record, data, and voice communications media. For example, direction and guidance and friendly force data are retrievable via a force management terminal connected to CAFMS, EIFEL, or Constant Watch Phase III computer systems.

Friendly force and situation data include:

- Current and Forecast Force Resource Data: by units, facilities, weapon systems, POL, communications, special equipment, aircrew, load, and maintenance crews. Particular elements include: type, unit designator, time, status, location, number, mission/role, in-commission, abort, and sortie rates, and limiting factors. These data are provided by the TACS elements, the Deputy for Communications and Electronics (DC) and logistics (LG) representatives.
- Friendly System Operating Capabilities: This information includes data on the operating capabilities and weaknesses of assigned and attached weapon systems and associated support systems to include: type of system/equipment, internal and external stores configurations, range, radius, speed, and parameters for mission profiles, and employment doctrine strategy and tactics.
- Combat Operations Friendly Situation Data: This includes: performance of the current ATO, limiting factors occurring as a result of the execution of the ATO, system losses, and mission results. These data are provided by Combat Operations, DC, LG, and TACS elements.

Finally, the role of friendly force and situation or logistical data should be specified. If option generation (the identification of alternatives) and choice (the actual decision) constitute the "heart" of the command and control process, logistics information functions in the calculus as a resource. In addition to resources, option generation involves the concepts of objectives, options and constraints.

An *objective* is defined as a goal that is desired; there can be more than one objective and objectives can be specified at different levels of abstraction. An *option* is defined as a decision alternative or course of action that, if adopted, can be expected to lead to the achievement of the objective with a given degree of probability. A *constraint* is defined as a fixed "rule" or precondition that could limit the appropriateness of pursuing one or more options; rules of engagement exemplify this variable. A *resource* is defined as a specific item that commanders have available that facilitates the implementation of one or more options toward achieving one or more objectives; resources include fighters, bombers, tanks, ammunition, troop strength, etc.

The military option generation environment is anchored in a tactical military command structure (as described above) and options are delineated as part of a C^2 decision process. In such a process, only the commander or his designated representative is authorized to make command decisions. Because of the complexity of the situations that arise, most commanders must rely on their staff members for recommended courses of action or solutions to problems. The commander's staff is composed of experts in the various functions that make up operations and primary support. Most commanders require their staffs to offer multiple options (two to four), to recommend a single option from this short list, and to furnish rationale for the selection. The planning staff consists of specialists in intelligence, operations, resource management and weather; the resource manager deals primarily with limiting factors, such as munitions, aircraft or fuel—that is, consumables. These will represent constraints to be imposed on the options.

Option generation is the creative component of command and control decision making. Technically, the option space for a decision is the union of all *possible* courses of action available to a decision maker; in the absence of constraints, the space is unbounded. The option space is pruned as constraints are imposed or imperfect or flawed option generation processes occur. The option space, then, is bounded by environmental and psychological constraints. Psychological or process related constraints concern such features as the inherent difficulty of a task (such as surprise, novel threat, deception versus the relative "simplicity" when a trained pilot implements standard operating procedures in an emergency), human cognitive and information processing limitations and similar characteristics. In contrast, environmental or context related constraints, the focus of this project, include resources available as well as a host of other "objective" constraints (including goals or objectives, rules of engagement [ROEs], institutional biases, etc.).

The parametric option space is the set of options produced by an ideal option generator. This space is upper bounded by the exhaustive (that is, totally unconstrained) option space and lower bounded by the option space produced by an unaided human. The environmental option space is that portion of the parametric option space associated with environmental constraints; as such constraints are relaxed (for example, as sortie generation capability increases), this space tends to expand.

Notes

[1]The literature on command and control is voluminous. In preparing this chapter, the following sources were relied on: overviews of C^2 are presented in Harris (1979) and Wohl (1981); the decision making perspective or human dimension of C^2 is discussed in Andriole and Hopple (1982) and Teates (1982); C^2 decision aiding methodologies are the subject of Hopple (1986c), Workshop (1985) and Andriole (1985); Air Force C^2 is profiled in Wohl (1981), Courand et al. (1984), Hopple (1986b), KSC/Perceptronics (1985) and Perceptronics et al. (1985).

[2]C^2 decision aiding taxonomies are developed for the Navy in Miller et al. (1979), for the Marines in Crolotte and Saleh (1979) and for the Air Force in Wohl (1981).

[3]Logically, survivability is a third criterion for evaluating C^2 systems.

[4]In the context of this report, C^2 and C^3 are assumed to be equivalent.

References

Adelman, L. and M. L. Donnell. 1986. "Evaluating Decision Support Systems: A General Framework and Case Study," pp. 285–310 in S. J. Andriole (ed.), *Microcomputer Decision Support Systems: Design, Implementation, and Evaluation*. Wellesley, Massachusetts: QED Information Sciences.

Andriole, S. J. 1985. *Decision Aids for Command and Control (C^2)*. Burke, Virginia: Armed Forces Communications and Electronics Association/Professional Development Center.

Andriole, S. J. and G. W. Hopple. 1982. "They're Only Human: Decision-Makers in Command and Control." *Signal* (March): 45–49.

Armstrong, J. S. 1978. *Long-Range Forecasting: From Crystal Ball to Computer*. New York: Wiley-Interscience.

Courand, G., C. O'Reilly, and J. R. Payne. 1984. "OCA (Offensive Counter Air) Mission Planning." Prepared for the Rome Air Development Center. Mountain View, California: Advanced Information and Decision Systems.

Crolotte, A. and J. Saleh. 1979. "Development of a Decision Taxonomy for the Marine Command and Control Environment." Prepared for the Office of Naval Research. Woodland Hills, California: Perceptronics.

Harris, D. R. 1979. "Final Report: OSD Sponsored Colloquies on Command and Control." Arlington, Virginia: ESL, Inc.

Hopple, G. W. 1986a. "Alternative Defensive Plan Generation and Evaluations: Object-Attribute-Value Knowledge Base." Prepared for the U.S. Army, Communications and Electronics Command. Marshall, Virginia: International Information Systems.

Hopple, G. W. 1986b. "The C^3CM Cumulative Aiding Backdrop: Ten Relevant Aids and Aiding Systems." Marshall, Virginia: International Information Systems.

Hopple, G. W. 1986c. "Decision Aiding Dangers: The Law of the Hammer, and Other Maxims." *IEEE Transactions on Systems, Man, and Cybernetics,* Special Issue on Information Technology and Command and Control.

Knowledge-Systems-Concepts/Perceptronics. 1985. "Automated C^3CM Battle Management Decision Aid: Functional Description." Rome, New York: KSC.

Miller, A. C., III, T. R. Rice and M. R. Metcalf. 1979. "An Analytical Characterization of Navy Command and Control Decisions." Menlo Park, California: Applied Decision Analysis.

Perceptronics/Martin-Marietta/Knowledge-Systems-Concepts/Ardmore. 1985. "Functional Description: Option Generation Aid." Prepared for the Rome Air Development Center. Woodland Hills, California: Perceptronics.

Teates, H. B. 1982. "The Role of Decision Support Systems in Command and Control." *Signal* (September): 45–49.

Wohl, J. G. 1981. "Force Management Decision Requirements for Air Force Tactical Command and Control." *IEEE Transactions on Systems, Man and Cybernetics,* Vol. SMC-11 (September): 618–639.

Workshop. 1985. "Third Annual Workshop on Command Decision Aiding Technology for Military Command and Control." November 5–7. Jointly sponsored by the Space and Naval Warfare Systems Command and the Office of Naval Research in cooperation with the Joint Services Working Group on Decision Augmentation.

A Modular Building Block Architecture

Donald C. Latham and David R. Israel

For the past few years, a team led by the Defense Communications Agency (DCA) has been developing a Modular Building Block (MBB) architecture for the design and implementation of a wide range of command, control, communications and intelligence (C^3I) systems. Recognizing that the impact of such an architecture and such an approach would extend beyond the normal responsibilities of DCA, this effort has been supported strongly by the Assistant Secretary of Defense (ASD) for C^3I.

The Problem: Implementation and Upgrading of C^3I Systems

The MBB approach addresses the largely unsatisfactory experience of the past several decades in the acquisition of C^3I systems. Most acquisitions are extremely lengthy and very expensive; it is not unusual for acquisitions, from concept to an initial operational capability, to exceed 10 years. However, the threat, the technology and the requirements change much faster. By the time a system becomes operational, both the system supplier and the user normally find that requirements have evolved due to:

- changes in threat;
- inclusion of new missions;
- availability of additional technical capabilities;
- desire for new modes of operation; or

This article originally appeared in the May 1986 issue of* ***SIGNAL.*

- the need for other changes and capabilities.

These acquisition problems are compounded further by the fact that C³I systems normally are designed and built without adequate attention to provisions or capabilities for facilitating those changes required before delivery; permitting the systems to evolve over a period of time; or implementing major changes or upgrades later in the system life cycle.

Fixed and mobile systems are characterized by significant amounts of cabling and wiring that interconnect the various system requirements. This cabling is expensive, heavy and not conducive to implementing changes. Retrofit and modification of these C³I systems becomes prohibitive at worst, and very time-consuming and expensive at best.

Many of these problems were recognized in the AFCEA Command and Control System Acquisition Study of August 1982. Of direct pertinence is Conclusion 5: "A potential for chaos exists if C² system acquisition proceeds without an architectural framework, including flexibility to facilitate growth."

The characteristics of the MBB approach meet the requirements of the evolutionary acquisition approach endorsed in the AFCEA report. Specifically, the MBB approach directly satisfied one of the report's three major recommendations: "Assure the use of system architectures and development techniques that can accommodate growth, change, continuous user and developer learning, and the insertion of new technology with a minimum of redesign and impact on existing systems."

There are other C³I developmental and operational problems that either formed the basis of the MBB effort or may provide long-needed solutions. These include providing high altitude electromagnetic pulse (HEMP) protection at a reasonable cost; minimizing the number of operators required to operate communications security (COMSEC) and communications equipment and minimize their exposure to fallout; and providing a short-term flexibility to "tailor" systems rapidly for specific missions.

Mobility/transportability and shelters have caused problems. As systems have grown in size and complexity, the tendency has been to develop custom shelters and larger enclosures. However, large moving van sized systems cannot be deployed easily, and certainly not by a C-130 or a helicopter.

Efforts to standardize types of shelters across the services have not been very successful. They have not included defense agency shelters, which may be fielded with the services, and they have not halted customizing standardized shelters with respect to entry panels, antenna mounting and other features. There is major concern that customizing standardized shelters has facilitated the enemy's task of using the unique optical signatures of shelters to identify units and operational functions.

Finally, major problems are experienced in the longer life cycle aspects of C³I systems. Beyond the retrofit shortcomings cited above, it is not possible to establish effective and efficient logistics capabilities while supporting so many specialized systems and equipment. Standardization and minimization of life cycle costs normally do not receive priority attention by busy and harassed program managers.

The desired attributes of early availability, flexibility, transportability, standardization and interoperability of C^3I systems are goals that largely have defied the best of intentions and design efforts. It is in the context of these troublesome C^3I experiences and problems that the MBB architecture was developed and now shows considerable potential for improving the implementation and evolution of C^3I.

A Solution: The Modular Building Block Concept

In summary, MBB is essentially a standardization and packaging approach that is independent of specific systems or mission equipment. The primary objective is to be able to implement, and later to retrofit, a wide range of C^3I systems directly, easily and economically from a compatible set of standardized functional elements or modules (shelters, work stations, cabinets, etc.). These elements are interconnected and communicate over a versatile data bus, which accommodates a wide variety of signal and control information, and hence can be used to implement remote monitoring or control of the modules. Computer assistance is available for these processes. Modules are designed to be independent of the shelter or fixed facility in order to facilitate system upgrading and retrofit and to permit equipment to be moved from shelters to fixed facilities, or the reverse, and to be reconnected quickly where the concept of operation calls for such a mode.

Packaging standardization is emphasized primarily at the individual equipment cabinet level with the cabinet being the basic modular building block. A shelter (or fixed facility) is implemented by installing and interconnecting a series of cabinets. Cabinets are independent of the shelter, permitting cabinets to be filled, assembled and checked out by multiple sources and allowing integration and checkout outside the confined space of a shelter. The isolation and standardization of the cabinets facilitates the assembly and checkout and the later replacement of entire cabinets without a major effort.

In the beginning of the design stage, it was agreed that over 100 db of shielding was necessary for HEMP protection (in addition to the appropriate filtering of all power and signal penetrations of the shielding enclosures). To obtain all of this shielding from the shelter would be expensive. There were questions as to whether it could be done, and if it could, whether it would be maintained after the shelter had been in rugged use. The decision was made to design for zonal HEMP protection, with roughly half the shielding obtained from the cabinet and the rest from the shelter.

It is important to emphasize that the MBB concept does not attempt to standardize specific types of electronic equipment, communication equipment, automatic data processing (ADP), terminals, etc. The concept relates only to the packaging and interconnection of these items in a way that focuses on interchangeability by virtue of standard physical and electrical interfaces.

It is interesting to note that this MBB approach to the packaging of C³I hardware has certain similarities to that currently employed in the business and personal user software market, where a wide variety of standard software packages for the most common and frequently needed capabilities are readily available. This standardization and availability of basic software packages has a clear application to C³I systems; a start has been made in this direction, but much more remains to be done.

The MBB Design

An MBB design, intended to meet the needs of a wide range of fixed and transportable C³I systems at the strategic and theater levels (echelons-above-corps), has been achieved. Beyond the various requirements and objectives cited above, the design was to be air-transportable and possess a limited off-road capability, just short of traveling over rough terrain, for general-purpose, ground mobile applications. To provide fallout protection, communications, COMSEC and ADP equipment had to be capable of being operated remotely, without the need for personnel to be located in each shelter or with major pieces of electronic equipment. Another requirement was being able to use commercially available telecommunications and ADP developments and equipment.

Racking and Stacking: The Cabinet

The standard MBB cabinet (Figure 1) is an enclosed equipment rack that provides 60 db of the desired HEMP shielding. To accommodate commercially available equipment, the design uses a 24 inch wide cabinet that incorporates a standard Electronic Industries Association (EIA) 19-inch rack. The extra width provides space to assure adequate air flow. The cabinet depth was selected at 30 inches which permitted two parallel rows of cabinets in the MBB shelter separated by a 30 inch aisle.

The cabinet sits on a base that provides a section of the air conditioning duct. Lighting and power junction boxes also are part of the cabinets, making it possible for the cabinets to be largely independent of either the building or the shelter.

The Basic Enclosure: The Shelter

Current MBB shelters are eight feet by eight feet in cross section and available in lengths of 13 and a half feet and 20 feet (Figure 2). Road mobility and air transportation were taken into consideration when these dimensions were selected. Eight feet by eight feet is the cross section of the standard International Organization for Standardization (ISO) shipping container. (The

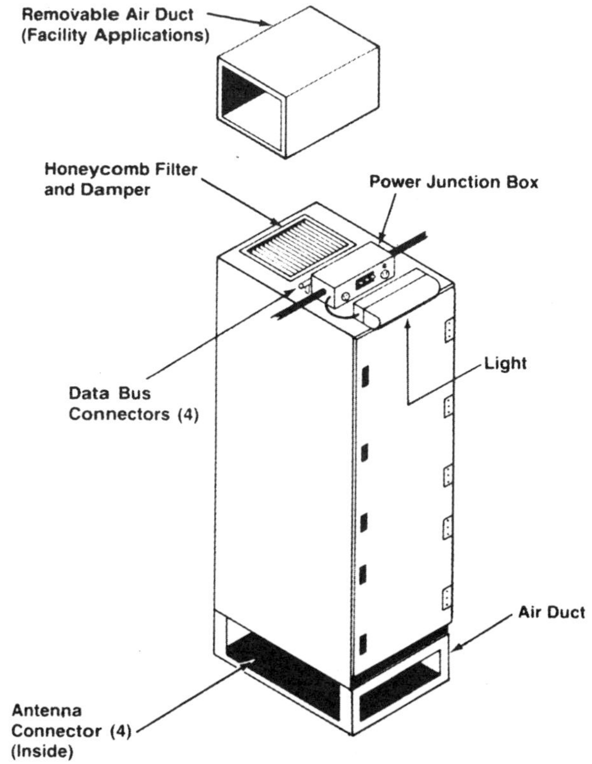

Figure 1. The standard MBB cabinet has exterior dimensions of 24 inches wide, 30 inches deep and 82 inches high. The interior height is 66 and a half inches. The front door is latched and the rear panel is removable. Two versions are available: a full-door version and a split-door version with an interior Tempest shield forming two compartments for use with COMSEC equipment.

shorter MBB shelter is slightly larger than an S-280 shelter; the 20 foot shelter is similar in size to an ISO commercial shipping container.) Three 13 and a half foot or two 20 foot shelters can be carried on a C-130.

Conventional shelter construction techniques are used. The shelter provides 80 db of the desired HEMP protection. To maximize the integrity of this protection, no penetrations of the sidewalls are permitted; rather, penetrations are restricted to the end sections, which are fully removable. One end section is for personnel entry and provides air flow for environmental control units located just inside that end section. The other end section contains the penetrations for power, signal and antennas, together with the necessary electromagnetic pulse (EMP) and Tempest filters and power distribution capabilities; it can be customized, if required, for specific systems. Removal of the full end sections facilitates loading of the cabinets and also permits end-to-end coupling of shelters to obtain longer lengths as required (Figure 3).

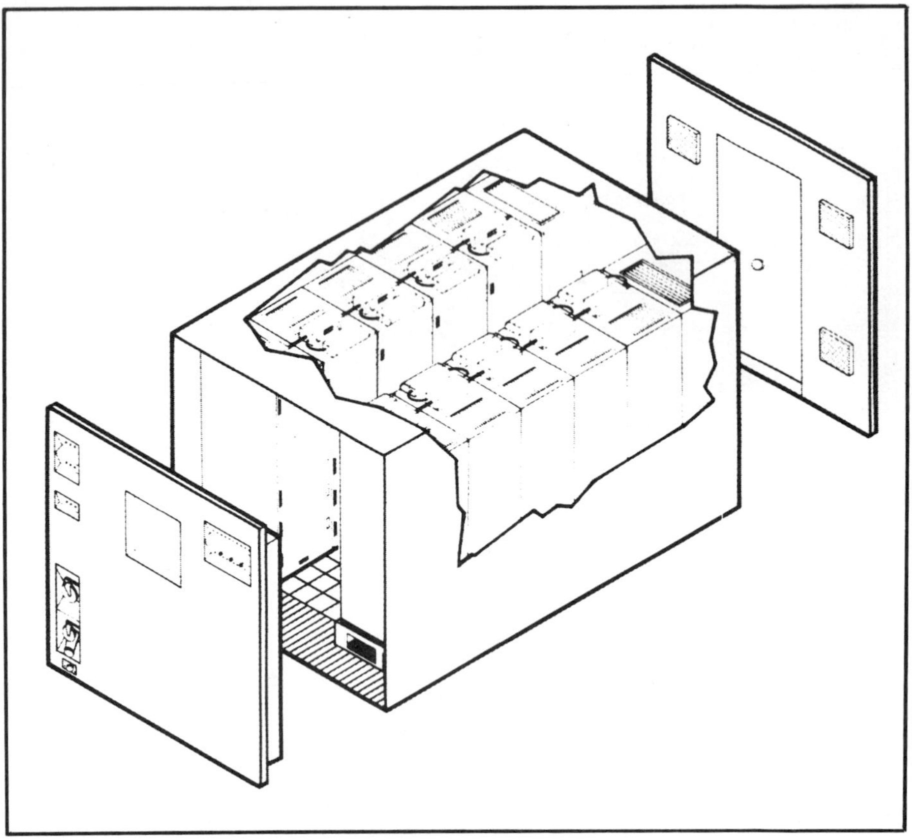

Figure 2. The basic MBB shelter is eight feet by eight feet in cross section and is available in two lengths: 13 and a half feet and 20 feet. The end sections are fully removable. ECUs are located adjacent to the entry end section. With 30 inch deep cabinets and 30 inch aisles, it is possible to open a cabinet door and slide out a drawer for servicing, but it is not possible to remove a cabinet from the shelter via the aisle.

Equipment cabinets are installed on either side of a 30 inch aisle. Environmental control units (ECUs) are packaged with the same 30 inch depth as equipment cabinets, but they come in two widths: 48 inch wide, five ton units and 24 inch wide, three ton units. These units were specified for low noise and vibration; the internal ECUs are fully compatible with internal work stations and personnel operating in the shelters.

Power for the electronic equipment in either shelters or fixed facilities can be provided by commercial sources or towed power generation units. The RF cabling from receivers and transmitters to external antennas runs under the false floor of the shelter to the electronic end section. Standard antennas would be used; antennas and their cabling can be carried in the center aisle of the shelter during transport.

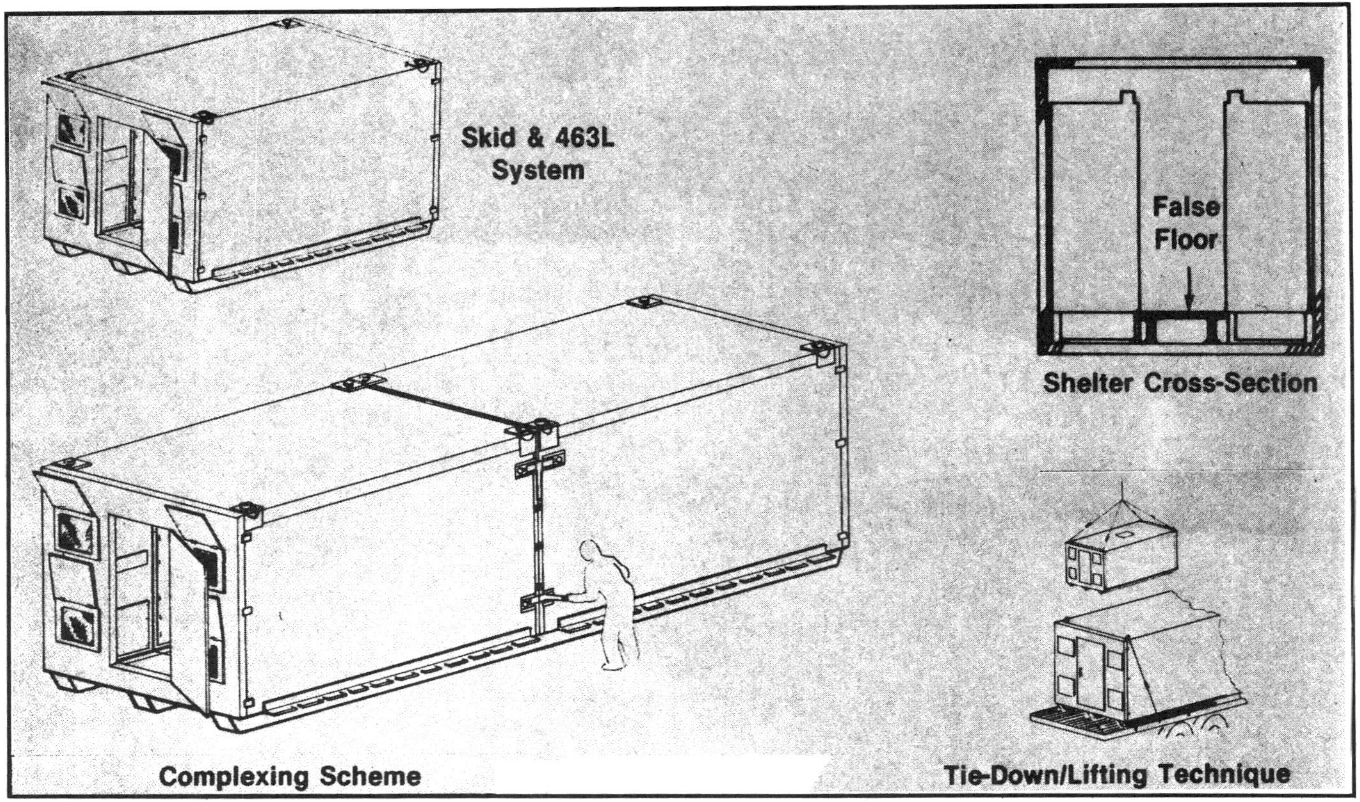

Figure 3. MBB shelters are manufactured with skids and are compatible with the 463L aircraft loading system. Tie-downs and lift-rings are available for truck mobility and helicopter movement. By virtue of the removable end sections with latches and special gaskets, two shelters can be connected to achieve longer enclosures of 27 feet (13 and a half + 13 and a half feet), 33 and a half feet (13 and a half + 20 feet), or 40 feet (20 + 20 feet). The connection or complexing normally would be done at a depot.

Intermodule Communications: The MBB Data Bus

A key component of the MBB architecture is an information or data bus that minimizes the intermodule and intershelter wiring by multiplexing all data control signals onto a single bus.

The selection of the MBB data bus was governed by two principal objectives. The first objective was to provide sufficient capability for as many applications as possible. The second objective was to accommodate the maximum number of different signals and formats to include audio (both voice and modem tones), synchronous digital data, asynchronous digital data (as with COMSEC equipment), video and various vendor local area networks (LANs) (e.g., ETHERNET, WANGNET, etc.). Another objective was not to develop a new data bus. After a careful evaluation of available broadband and baseband systems, the broadband ICS-SCAN architecture pioneered by Bill Carpenter at the Naval Ocean Systems Center was selected and recently adopted as a U.S. Navy standard for ships, submarines and aircraft.

In this design, a number of channels or circuits are created by frequency division multiplexing. The bus is implemented on a coaxial cable and many channels can be created in the 300 to 400 MHz available; the number of signal channels depends upon the bandwidth assigned to each. Time division multiplexing then may be used within a channel to provide packet service channels or further signal multiplexing. Bus Interface Units (BIUs) are used to route signals on and off the data bus. Some of these BIUs had to be developed (such as the frequency agile synchronous digital data unit), but a large number of them are available commercially and may be used without modification (other than repackaging).

Remote control and monitoring of devices on the data bus is accomplished with Device Adapter Modules (DAMs), microprocessor based units that provide the correct electrical interfaces. Generally, it has been necessary to design a DAM for each type of device placed on the bus, but more and more equipment now comes with this type of interface capability built in.

The bus dramatically minimizes the wiring in a shelter. A typical MBB application might use a red bus and a black bus to separate classified and unclassified information. In such a case, radios and other external communications generally are connected to the black bus, while local communications and ADP are connected to the red bus. Located between the two buses is the computer, which supports the facility control position as well as COMSEC units. These units are controlled by the facility control position and are switched to the appropriate communication channels through the data bus rather than being hardwired. The data bus also services all work stations and shelters.

At the facility control position, an operator can, with appropriate computer assistance:

- operate, tune and control communications equipment;

- establish configuration of equipment (ADP, COMSEC and communications);
- configure circuits;
- continuously monitor and check the status of system modules and equipment;
- monitor communications link performance;
- perform limited diagnosis of faults; and
- accumulate performance data.

The software supporting the facility control position is highly modular and specifically reflects the modularity of the hardware. A system independent kernel comprises 50 to 60 percent of the code and represents that portion of the code that does not change no matter what kind of devices are in the system or what kind of man-machine interface terminal is used. Smaller device control programs contain the code that implements control and status monitoring of the individual devices. These are replaced easily in line with the overall philosophy of simple retrofits.

Operator Interfaces: Work Stations

To serve the facility control position or for use as operator positions for command posts or operations centers, the standard work station shown in Figure 4 was designed. By turning the front of the work station 45 degrees, operators are not located back-to-back and are less cramped for space. The MBB approach does not stipulate the various work station accessories (i.e., terminals, printers, telephones, etc.) they are selected according to the specific application.

There are some situations in which more work area is required after a transportable system has been deployed and circumstances allow work stations to be moved into a fixed facility. In these cases, it is desirable to have a transportable work station in which the components of the work station are packaged into boxes with the door on one of the boxes serving as the connecting shelf. As indicated in Figure 5. The boxes may be carried in the center aisle of the shelter during the transport mode.

Operational Utilization

As indicated in Figure 6, with the interchangeability of MBB modules, shelters can be configured with combinations of work stations and equipment cabinets.

Beyond the data bus and basic packaging modules, use of the MBB approach envisions building up standard configurations of equipment within cabinets for functional purposes such as communications, COMSEC or ADP. The term "assembly" has been adopted to describe a cabinet filled with specific

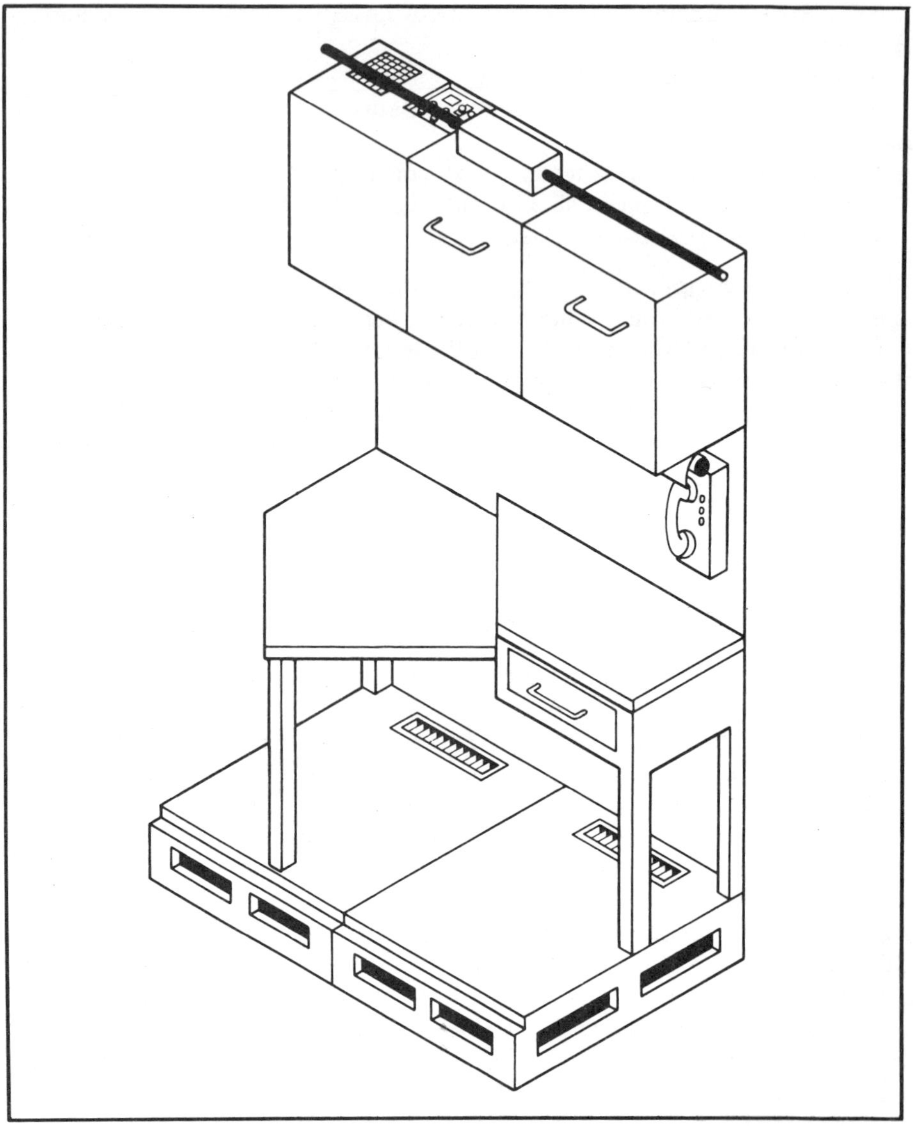

Figure 4. The MBB work station is 30 inches deep and 48 inches wide and was designed for physical and electrical compatibility with MBB cabinets. Work stations can accommodate a wide variety of telephone, terminals, keyboards, displays, printers, etc.

equipment. A set of representative standard assemblies, for a large number of users, might include a data bus control assembly, red and black private branch exchanges (PBXs), specific COMSEC assembly, high frequency (HF) assembly, ultrahigh frequency/very high frequency (UHF/VHF) assembly, etc. In these cases, a complete assembly, with necessary interfaces to the data bus, might be available. Note again the similarity to recent trends in available business and personal software packages.

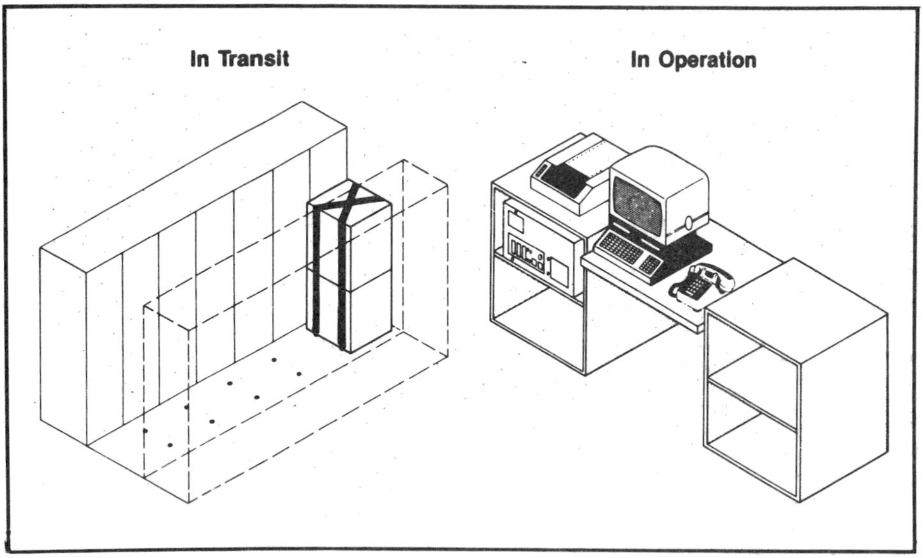

Figure 5. Transportable work stations, which can be deployed away from shelters but are connected to them by data bus, are shown in the transit mode and operational mode.

The modular building block approach naturally evolves into one of having a broad selection of standard units or items that can be used to build all or a major part of any given system. (Any specific system undoubtedly would have some unique electronics; this would have to be packaged into the standard MBB cabinet for compatibility.) An MBB "catalog" then might list the available shelters, end sections, ECUs, power generators, work stations and the standard assemblies (Figure 7).

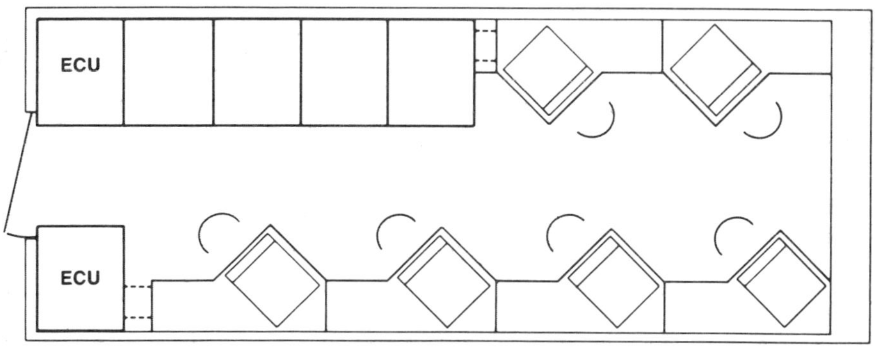

Figure 6. Shelters can be implemented with various combinations of equipment cabinets and work stations. ECUs must be located adjacent to the entry end section for air inlet and outlet. Other than considerations of weight distribution, there are no other restrictions on the locations of cabinets and work stations within a shelter.

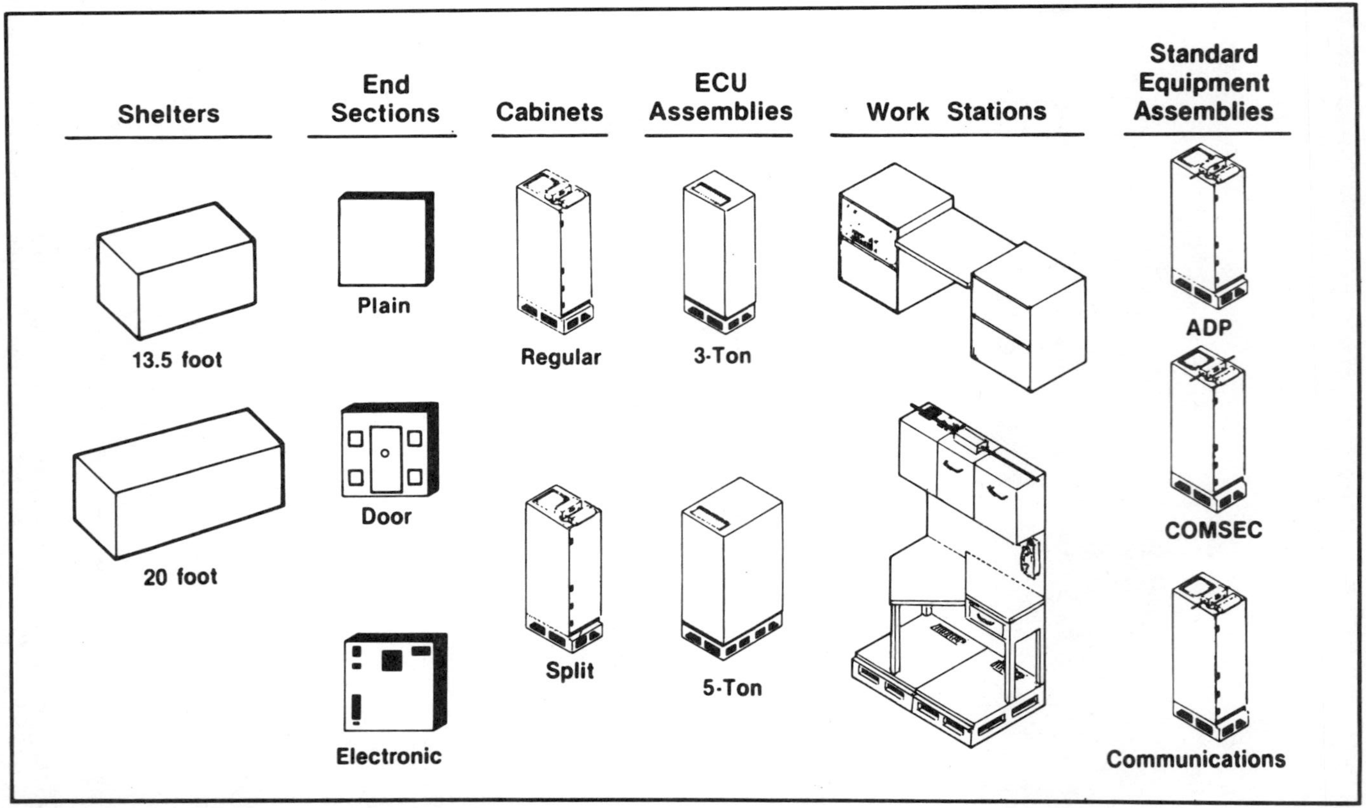

Figure 7. The current selection of MBB modules—shelters, end sections, cabinets, ECUs and work stations—is shown. As systems are built, various functional assemblies—cabinets plus electronics—become available. The ADP, COMSEC and communication assemblies depicted are representative of a large class of possibilities.

To assemble a system, standard items first could be selected from the catalog. Unique equipment cabinets plus standard assemblies then would be assembled on the floor at a depot or other staging location (Figure 8). When the cabinets are placed against one another with appropriate connections to the data and power buses, it is possible to check out that group of cabinets or assemblies outside the confined working space of the shelter. After this is done, the row of cabinets can be moved into the shelter and secured to the wall and the floor. Upon reconnection of power, data bus and antennas, the system is ready to operate.

Thus in the MBB approach, the equipment is checked out and integrated before being transferred into the shelter; this differs from the more conventional approach in which one begins with a shelter and installs equipment and racks or cabinets in it (with major problems in "running" all the cabling) and then does all the integration within the shelter.

Applications of MBB

MBB shelter transport options are very broad: mobilizers, trucks, trailers, aircraft, helicopters, rail, barges or ships.

A wide range of applications for the MBB concept and design are foreseen at the strategic and theater levels (echelons-above-corps). Both fixed or mobile/transportable type C^3I systems are amenable to this method of packaging and construction. With the ability to change or replace individual cabinets in a matter of hours, shelters can be configured or tailored to meet urgent ad hoc mission requirements. Systems also can be built in a transportable configuration and then can be converted to fixed installations by moving the cabinets/assemblies into buildings. The transition problem of replacing older, fixed ADP systems with new computers can be facilitated by packaging the new equipment in MBB cabinets and shelters, then deploying the new shelters (with HEMP protection) outside the existing facility, but connected to it by data bus. When the older ADP equipment is removed finally and space is available, the MBB cabinets can be moved into that space.

The primary benefits and advantages of the MBB architecture are flexibility and adaptability. (The former refers to rapid configuration changes and tailoring to meet changing mission requirements; the latter is the longer term evolutionary modification of systems.) There is flexibility and adaptability at the equipment cabinet assembly level: the cabinets/assemblies can be used in shelters and fixed facilities, and they can be removed and replaced readily as the equipment suites change to meet new needs. A second level of flexibility is at the shelter level: the 13 and a half foot and 20 foot shelters can be carried in cargo aircraft, lifted by helicopters or transported by truck, train, barge or ship.

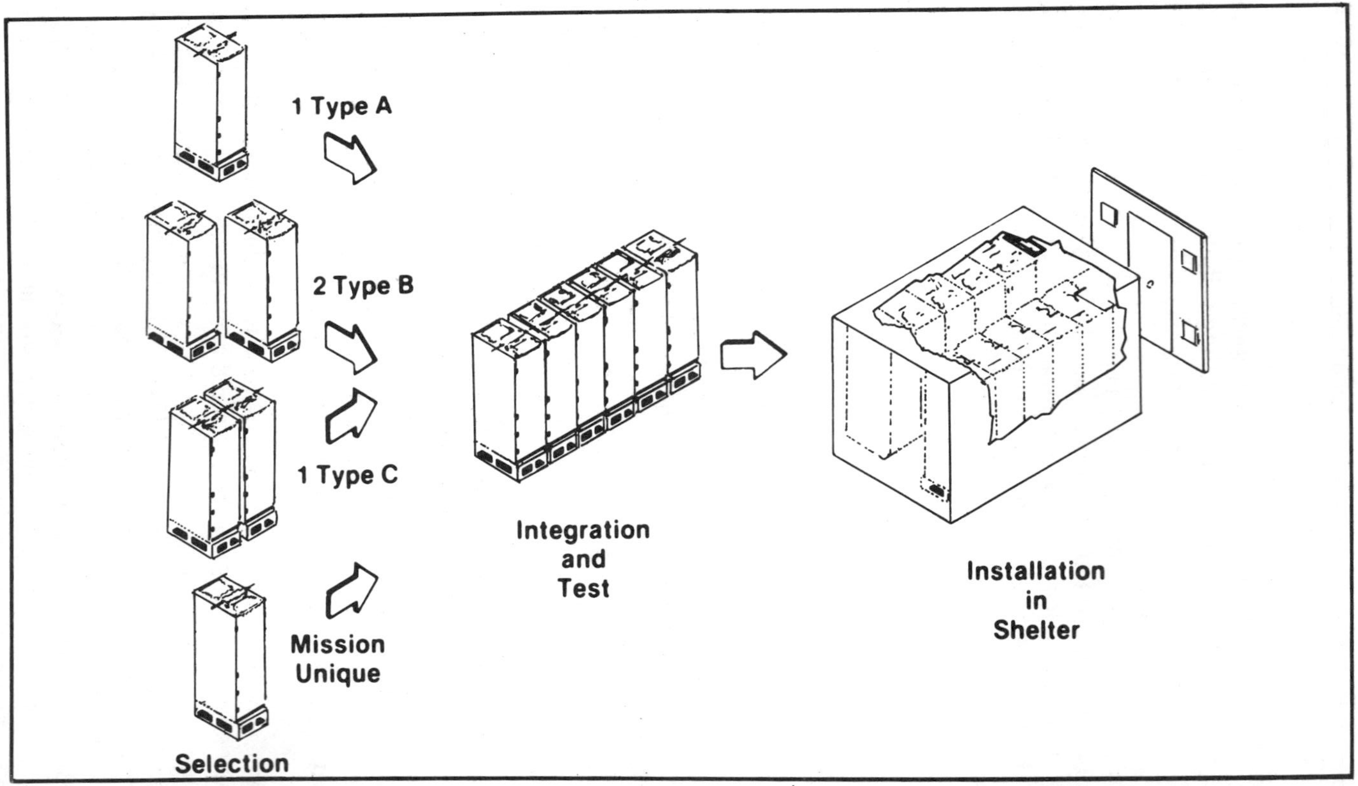

Figure 8. A major focus of the MBB approach is the integration and test of cabinet/assemblies, work stations and ECUs outside the confines of a shelter. Both standard modules and mission unique cabinets would be involved. The MBB design, including shelter end removal, power and data bus implementation, etc., facilitates the subsequent relocation of a checked-out system to a shelter or to a fixed facility.

An equally important feature of the MBB concept is the high potential for significant savings in cost and time when acquiring or retrofitting systems. These savings result from such items as:

- use of preengineered, preassembled, pretested modules that reduce design, engineering, implementation and test efforts;
- modularization, with carefully defined interfaces, which permits independent parallel work efforts, and enhances problem localization and correction;
- assembly, integration and test activities that are not inhibited by shelter constraints;
- ease of module removal and replacement, simplifying retrofit;
- retrofit testing that can focus on affected modules; and
- the strong electromagnetic interference (EMI)/EMP protection of the MBB cabinets and shelters that makes it easier to use commercial equipment in the cabinets.

Other benefits are anticipated. Manpower savings (quantitative and qualitative) should be realized from the implementation of remote control of equipment and from the benefits of the automated facility control. If a family of systems uses the MBB concept, logistic savings could be significant. Security advantages result from the basically unclassified nature of individual modules and assemblies: the classification of a system may be deferred until it is assembled completely and software added.

Lastly, modularization would appear to offer certain benefits of increased competition in the procurement and retrofit of systems. Also, it would seem to be a natural for production sharing and multination involvement for NATO C^3I systems.

Status

The modules described above for strategic and theater applications (echelons-above-corps) have been built and tested. MBB is more than a concept; it is a reality. Design details are available for the shelters, cabinets, work stations, data bus and most assemblies. DCA's goal is to transfer this technology and design data to the services and the defense agencies for application and institutionalization.

As a part of the technology transfer process, the ASD (C^3I) has encouraged DCA to implement demonstration or proof-of-concept systems. The major effort is for the Strategic Air Command (SAC) and is termed POC/ET (proof-of-concept/experimental test bed). This is a joint venture of DCA, the Defense Nuclear Agency (DNA) and SAC and will result in a communications-electronics capability packaged in three 40 foot shelters (three pairs of 20 foot shelters) and three 20 foot shelters. The POC/ET design began in December 1984; system integration and testing is to be completed by spring of 1987.

While the original MBB concept and design has been oriented to strategic and theater applications where HEMP and adaptability were key considerations, recent design activity has been oriented to crisis/contingency applications where operational flexibility and minimal airlift are the key requirements, and to mobile tactical applications for battlefield use.

For tactical utilization where the principal means of equipment transport will be 5/4 ton vehicles (high mobility multi-purpose wheeled vehicle or commercial utility cargo vehicle), the shelters and cabinets described above are too large and heavy. A series of "downsized" modules is under development for these applications. These modules, which might include lightweight transit cases, would still be kept independent of the shelter. Modules and shelters again would be interconnected by the data bus.

The impact of such a downsized design on tactical operations could be significant and might include:

- interchangeable standard equipment modules, with the flexibility, retrofit and logistic advantages described earlier;
- the data bus would reduce significantly intra- and inter-shelter wiring/cabling;
- a remoted facility control would minimize operators;
- computer assisted facility control should reduce required operator/maintainer skill level;
- built-in testing and exercising capabilities could be added; and
- improved airlift.

Arrangements currently are being made to implement an appropriate POC/ET of the downsized version of MBB in association with the U.S. Army. DCA also is continuing to work with the services and defense agencies to identify additional applications of the MBB technology and to consider broader standardization of modules and the data bus.

Summary

The MBB approach offers a low-cost, rapid and flexible approach to the implementation and upgrading of a wide range of C³I systems. A specific design has been implemented for systems at the strategic and theater levels; downsizing of this design for the widely different requirements of tactical applications is now in progress.

Evolutionary Development of Command and Control Systems: The Fort Lewis Experience

David J. Welch

Introduction

Evolutionary development (ED) has been a popular "buzz word" of the early 1980s. It seems likely that with shrinking budgets and steady or increasing demand, the emphasis on ED will continue. In fact, the Packard Commission's interim report on defense management notes that:

> All too often, requirements for new weapon systems have been overstated. This has led to overstated specifications, which has led to higher cost equipment.

It further states:

> Developmental and operational testing have been too divorced, the latter has been undertaken too late in the cycle and prototypes have been used and tested far too little.

**This article originally appeared in the November/December 1986 issue of IEEE Transactions on Systems, Man and Cybernetics.*

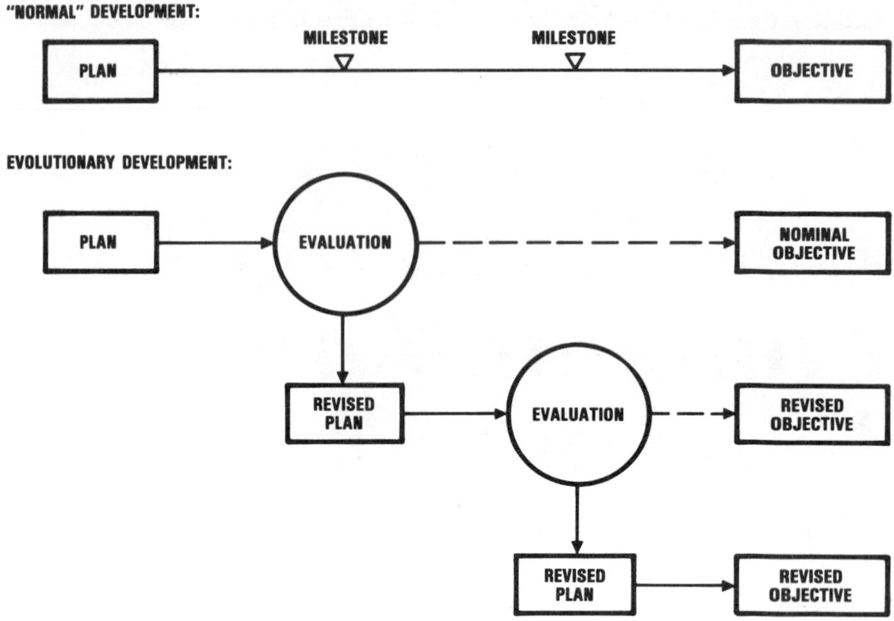

Figure 1. Evolutionary development.

These comments and the Commission's recommendations reflect the philosophy of evolutionary development. However, effective utilization of evolutionary development is not without challenges to include the need for judicious application.

Application requires understanding, but often the term ED means very different things to different developers and suppliers. At one extreme the standard development process is modified little, but recognition exists of the need for increased feedback during the development process. On the other extreme is a substantially different approach, one in which little of the current research and development structure remains. The following is a discussion of the latter process as exemplified by the distributed command and control system (DCCS) program for the Ninth Infantry Division (MTZ) at Fort Lewis, Washington. Figure 1 is a comparison of the normal and evolutionary development processes. This chart implies similar overall time lines, but in fact ED can reduce development times significantly, especially when coupled with nondevelopmental items (NDI). More importantly, the ED process assures that the ultimate objective system meets user requirements when fielded and allows technology insertion during the development process.

The Fort Lewis effort was led by the Army Development and Employment Agency (ADEA) at Fort Lewis and the Army's Communications and Electronics Command (CECOM). The DCCS was renamed MCS 2.0 when it became part of the Army's mainline tactical command and control system, the Maneuver Control System (MCS).

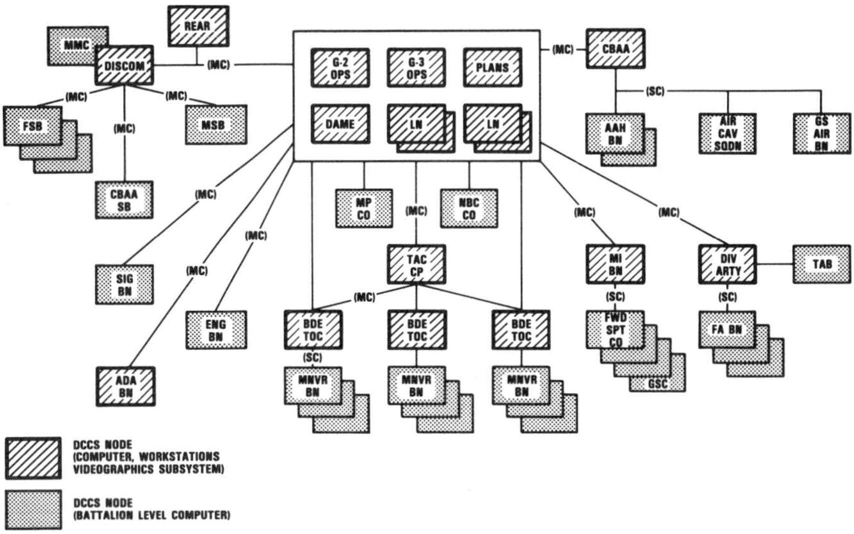

Figure 2. Ninth Infantry Division configuration.

The MCS 2.0 System

The MCS 2.0 provides division level tactical command and control automation assistance at various locations throughout the battlefield. The MCS 2.0 equipment is deployed at the division's main, tactical (TAC) and rear command posts (CPs) and major subordinate commands and battalions (See Figure 2). The system uses the existing single channel and multichannel communications systems with the addition of packet switches within the upper echelon. In parallel with the introduction of computer equipment to the division, a modular command post, the Integrated Command Post System, (ICPS) has been developed and deployed. This system is another example of successful evolutionary development, perhaps even more demanding than the development of a command and control system. The ICPS, illustrated in Figures 3 and 4, provides a small, flexible and more mobile CP for the division while also affording some protection for the commercial equipment used in MCS 2.0. The ED processes discussed in this paper as applied to computer systems were also used in the development of the ICPS.

The MCS 2.0 system provides a wide range of support functions to the division commander and his staff. In addition to normal office automation functions such as word processing, spread sheets, electronic mail and database management, the system provides automatic communications interfaces, map overlays, graphics and automatic status reporting. The status reporting capability is generated by information flow and rollup from the battalion level to brigade and division. Custom displays permit the commander or his staff to obtain a quick overview of unit status down to the battalion level (See Figure 5).

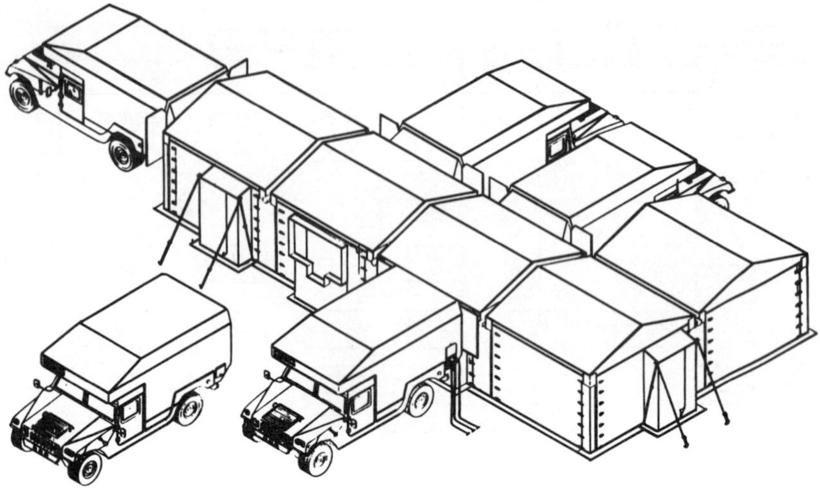

Figure 3. Integrated command post (exterior).

The equipment used in the MCS 2.0 command and control system consists of the commercial computer, terminals with graphics capability, a dot-matrix printer and a videographics subsystem (VGS). The VGS consists of a 19-inch monitor, a graphics overlay unit and a video disk player. The latter displays standard Army maps which have been transferred to video disks under another Army program.

The final piece of equipment is the communications interface unit (CIU). This device, which permits interface to a variety of standard Army communications media, is the only developmental item included in the MCS 2.0 system.

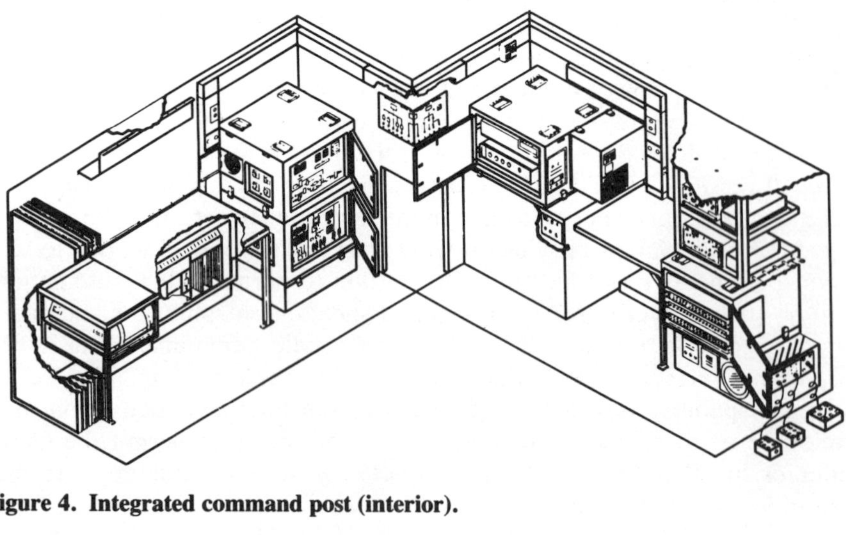

Figure 4. Integrated command post (interior).

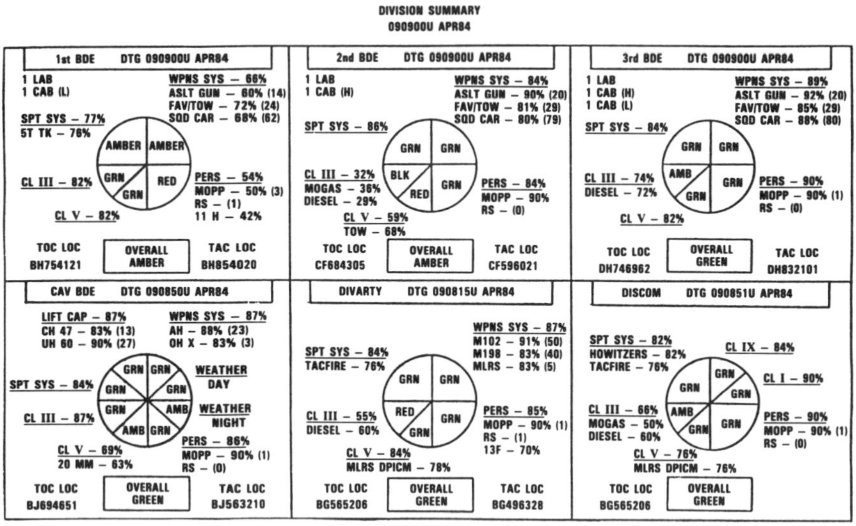

Figure 5. Graphics display division summary.

The software for MCS 2.0 has been developed by integrating commercial software into a new executive program. Software packages have been written for the communications, videographics and database processing functions. The system utilizes the UNIX operating system and new code is written in the C language. Future plans include evolution to Ada as required.

Evolutionary Development of MCS 2.0

The remarkable aspect of MCS 2.0 is not the aforementioned equipment. It was intended from the outset that the hardware comprise commercial standard off-the-shelf components. The notable point is that the Ninth Infantry Division now has an *operational* tactical automated command and control system that extends from battalion level to division headquarters and associated elements. No other system of similar magnitude and capabilities exists in the U.S. Army. Furthermore, this capability was achieved in about two and one-half years from program start.

During that time, the system has been to the field for eight major exercises, including field tests at the Yakima Firing Center, Fort Bliss, and Korea. It has been subjected to temperatures above 120 degrees Fahrenheit and below freezing. It has been operated in high and low humidity and has been exposed to the extremely fine-grain ash fallout from Mt. St. Helens. During this period, the system evolved to meet new requirements based upon new understandings by the soldier user and the collocated developer.

The evolutionary process is perhaps best understood by examining the planned program and the actual program as it developed over the past three years. Table 1 presents a summary of the planned schedule and the actual

DATE	ORIGINAL PLAN	ACTUAL ACTIVITY
October 1982	Program Start	
March 1983		Program Start
May 1983	Demonstrate hardware with commercial packages but no new software (OCI)	As planned plus first executive and electronic mail
August 1983		Field deployment to Yakima
November 1983	First custom software (OCII)	First rudimentary communications. First videographics. Expanded requirements for videographics and database defined. Equipment installed in vehicles
February 1984		CPX at Ft. Lewis (Caber Base)
		Alternate communications approach tested. New user software with graphics
August 1984	Test in the field at the Yakima Firing Center (Laser Strike) (OCIII)	Packet switch approach introduced. Continued evolution of database and displays. First integrated CPs deployed
October 1984		CPX at Ft. Lewis (Bold Venture)
January 1985	OCIV	
March 1985	OCV	Deployment of system to Ft. Bliss for operations and test. (Border Star)
August 1985		Caber Flash CPX at Ft. Lewis
September 1985	Final field tests (OCVI) before acceptance. Acceptance by 9th Infantry Division	Final acceptance postponed until December to accommodate improved communications
November 1985		Enhanced communication approach fully tested in CPX (Cascade Peak) at Ft. Lewis. Continued enhancement of database and videographics
December 1985		System declared operational by 9th Infantry Division
March 1986		Deployment to Korea for operational use during Team Spirit

Table 1. Evolutionary Development of MCS 2.0.

activities. The plan was never assumed to be fixed; it was always considered to be a baseline for expected variations. The major variations have been 1) a compressed schedule with more activities, 2) more emphasis on the graphics features, 3) a change to a packet switched communications architecture, and 4) introduction of the integrated command post system.

The change to more emphasis on graphics, map displays and the database functions illustrates the evolutionary development process well. These efforts began with poorly defined requirements and have evolved to become the heart of the system. Through soldier use, requirements were defined and refined. While the initial requirement definitions were poor, they did provide the basis for a hands-on definition of the final product. It was learned that it is difficult to describe a good system, but it is easy to recognize one.

The database system in particular proved difficult to explain to the user until he had hands-on experience with the system. Explanation of files and records is difficult, but users can evaluate displays that quickly and accurately present rolled-up summaries of division-wide resources.

The communications approach evolved in a similar manner. The only factor important to the soldier/operator is that communications operate easily and effectively. This is a challenge since the MCS 2.0 system relys upon the exchange of digital data over existing tactical communications. After several false starts, a packet switched architecture, using SUN computers and C-30 switches at the communications nodes, proved successful. This approach—derived not by in-depth engineering analysis but through the application of existing technology and practical field experience—was necessitated by the uncertainties of using existing communications equipment (a requirement) which was not designed to support a high volume of digital traffic.

The Evolutionary Development Process—How It Works

The accomplishments of MCS 2.0 obviously were not achieved through a ''business-as-usual'' approach by either the developer or the contractor. How was the successful result achieved?

A primary requirement is contract flexibility by both the developer (ADEA/ CECOM) and the contractor (in this case R and D Associates and others). In the course of the contracted activities not one engineering change order has been generated. And yet the requirements, system design, and schedule changed drastically many times in the 2-year + period. Clearly, this is not the way to build an airplane or a tank, but technologies exist for which this approach is appropriate—those technologies that are evolving rapidly and that are being driven by the commercial sector.

A second consideration is flexibility in the system architecture to permit changes in hardware and software as required. For example, the MCS 2.0 equipment (although less than 3 years old) is now approaching obsolescence. This was anticipated; thus an architecture that would easily accommodate changes was selected to include off-the-shelf equipment, modular software and the UNIX operating system. The payoff for this approach was the deployment in 1985 of a derivative command and control system that was developed for the Combined Field Command's war headquarters, CP Tango. New workstations, a local area network (LAN) and new displays were introduced without major redesign of the system. An initial operating capability (IOC) was achieved in less than six months.

The use of nondevelopmental items also can be a key to successful evolutionary development since it reduces deployment cycle times. While NDI availability makes evolutionary development easier, it is not a prerequisite. When appropriate technology is available, NDI greatly reduces the overall acquisition cycle, perhaps reducing what often takes 10 years or more to less than 3 years. NDI is particularly appropriate for command and control (C^2) applications since commercial technology has provided the basic tools for a number of years.

A significant improvement in automated C^2 can be achieved by providing a commander with the same tools and database capabilities available for years to analysts and corporate officers. One also might postulate that development of a command and control system *demands* evolutionary development and NDI applications since the typical 10- or 15-year R & D cycle provides little hope of deploying a system with near state-of-the-art capabilities.

Lessons Learned

First and foremost, the evolutionary process works, given the right application. The user can define and communicate his requirements in an effective manner. To maximize progress, the time between successive field tests should be short (6 months or less) to maintain user involvement and keep pace with technology. For applications where change is anticipated (such as C^2), the result using ED is far better than that achieved by an *a priori* estimation of the desired product. User support of the MCS 2.0 and CP Tango systems substantiates this claim.

Second, it is reasonable to assume that, generally, a new user does not know what he wants on the *first* definition. It is only after use that he begins to understand the technology and system capabilities to the point that he can better define his needs. In the normal developmental process his first estimate is fixed by a specification, and contractual terms make changes expensive and time consuming to implement. It is no wonder that the products of that process are so often unsatisfactory.

Another lesson of the evolutionary development process is that significant user time is required. Collocation with an operational unit (such as the Ninth Infantry Division at Fort Lewis) maximizes information exchange; however, the process results in an additional workload on participants. This should not be viewed as a drawback since the additional work is directly related to the unit's wartime missions and in many cases sharpens procedures, operations and training.

Obviously, the ED-NDI process is not without pitfalls. The process is, by its nature, at odds with many quality control and configuration management practices. Because of its rapid pace and many changes, it also requires new approaches to management, support and training. However, it is possible to develop new procedures that will permit continued exploitation of the evolutionary development process without compromising quality.

Summary

Evolutionary development is not the old game under a new name. Successful application requires movement away from the fixed program plan approach to a series of phases, each of which is subject to major changes depending upon the lessons learned from the previous phase. Both the developer and the contractor must be flexible and tolerant. A basic trust must exist since a rigorous paper mill supporting each decision would make the process impossible. The orientation of all participants must be toward satisfying the user's requirements and achieving the best possible product, not on corporate or bureaucratic considerations.

Systems Engineering and Integration in the U.S. Navy

Keith Collard

The ultimate goal of command and control (C^2) in any combat scenario is to place the right ordnance on the right target at the right time (although there are many intermediate objectives). It is equally important not to put ordnance (right or wrong) on the wrong target at the wrong time. The latter is particularly critical in a cold war crisis management or limited conflict situation where ineffective C^2 could result in an unwanted escalation. To accomplish this goal, an increasingly complex set of command, control, communications and intelligence (C^3I) and combat support systems, ashore and afloat, are being developed that support Navy decision makers. The systems must be engineered and integrated to provide an unbroken chain between the decision maker and the person who has a finger on the trigger. The design of these systems and their interfaces must be sufficiently flexible to accommodate any situation wherein either the commanding officer of a tactical unit has been delegated the decision making authority (a hot war situation) or the higher echelons ashore, including the National Command Authority (NCA), have retained the decision making authority on weapons employment. NCA may be in direct contact with the on-scene commander, bypassing intermediate echelons in the normal chain of command, as was the case in incidents such as Mayagüez and the Cuban blockade.

The systems engineering and integration challenge has been and will continue to be heavily influenced by the application of computer technology and

**This article originally appeared in the January 1985 issue of SIGNAL.*

the manner in which the military has introduced that technology. Historically, the application of computer technology has been a decentralized process. In the Navy, the first afloat digital C^2 system, the Naval Tactical Data System (NTDS), was introduced in the early 1960s and was initially oriented toward Antiair Warfare (AAW). NTDS interfaced with analog weapons computers, which were state of the art at the time. Ashore, the Navy antisubmarine warfare (ASW) maritime patrol community was the first to introduce tactical digital computers. As the advantages of computerized systems became apparent, more and more individual sponsors in the office of the Chief of Naval Operations (OPNAV) and Principal Development Agencies (PDAs) in the Naval Material Command (NAVMAT) climbed onto the computer bandwagon. Unfortunately, the decentralized management and procurement procedures, adequate for person to person interfaces, were no longer adequate for complex computer to computer interfaces. No guiding computer oriented system architecture assisted in the definition of the functional relationships, interfaces and communications connectivities among the many automated systems that were being developed independently. Requirements for electrical interfaces between systems were often not identified until late in the design process, occasionally after the first systems were deployed. New interfaces frequently could not be accommodated without an increase in the core or computational speed of the original system. Funding constraints often mitigated against upgrading systems. In addition, the interoperability of the collection of subsystems in a battle group (BG) versus single unit environment was often not thoroughly examined.

The net result of this growth process has been the development of a collection of systems, ashore and afloat, that have not been initially designed and integrated as a complete macro system. In the future, more and more specialized systems will be added to the existing collection, thus, adding further complexity to the overall systems design.

For semantics clarity, the remainder of this article uses the term "System" to refer to the macro system and subsystem to refer to any component of the System. The System is thus an integrated collection of subsystems, ashore and afloat, that must provide the support required by all levels of command. It includes C^2, C^2 support (including intelligence), communications, sensor and weapons subsystems.

The Navy and the military in general have only in the last few years come to appreciate the necessity and challenge of systems engineering and integration. Efforts are now underway to treat parts of the System collectively. However, the efforts are generally inadequately funded, and only a very limited number of bona fide systems engineers are experienced in the total system approach. Furthermore, it is difficult to overcome the bureaucratic inertia of doing things "the old way" and to resolve issues that arise in the process (i.e. finding the person in charge who can make critical design decisions). Nevertheless, progress, albeit slow, is being made and will be discussed later in the article.

The current system engineers and integrators remain captive to the past and therefore must be very pragmatic. They often must accept the System design as it is, improve on it where possible and ensure that the future subsystems added to it are engineered properly and integrated into the existing System. The engineers also must accept the vagaries of a budget process that funds subsystems individually. They must take into account the possible impact on the System design if slippages or cancellations of major subsystems occur.

There is not enough money in the Federal Treasury to start with a clean sheet of design paper. Real world constraints apply equally to the system analysts who tackle the whole System. Too many theoretical, top-down design proposals do not meet the litmus test of reality; System enhancement is more typically a bottom-up evolutionary process, rather than a top-down revolutionary process.

The challenge for Navy systems engineers and integrators is to minimize the cost and to optimize the operational effectiveness of a very complex System composed of many subsystems. Meeting the challenge will enable the Navy to perform its forseeable and unforseeable missions and, when directed by higher authority, to deliver the right ordnance on the right target at the right time.

The Challenge

In illustrating the challenge, some limits must be applied. The number of subsystems that constitute the System, stretching from NCA to the crew of an F-14, is too broad a spectrum to cover in one article. For this reason, the Navy interfaces with the Worldwide Military Command and Control System (WWMCCS) are not included; however, this does not imply that they are not important. To limit the discussion further, only the following key engineering and integration challenge is addressed:

"Operational centers, U.S. and Allied, can be linked together so that each will see the same graphic symbology. Each seeing the same picture can ensure the most efficient utilization of friendly resources,"[1] which is a statement made by MGen. Vaughn Lang, USA, in the February 1984 issue of *SIGNAL*.

The systems engineering and integration ramifications associated with providing the above capability in Navy ashore, afloat and airborne (e.g. E2C) command spaces clearly illustrate the challenge and the procedural steps that need to be taken to meet the requirement. The steps include:

- have a definite statement of the requirement;
- access the requirement;
- breakdown new requirements into functions;
- allocate the functions into subsystems;
- define interfaces and communications requirements; and
- size the system.

Definitive Statement of Requirement

The quote by MGen. Lang was obviously not intended to be a formal statement of a requirement. In the Navy, the Chief of Naval Operations (CNO) OPNAV is responsible for stating requirements to which engineers design and build. Drafting a requirements statement to satisfy the requirement would not be without perhaps disabling controversy. The first item to be defined would be the meaning of operational centers. If the centers were to include shore centers starting with the National Military Command Center, (NMCC), down through the Fleet Command Center (FCC) to the afloat Tactical Flag Command Center (TFCC) and then to an individual ship's Combat Information Center (CIC), there may be reluctance to provide the same picture. Senior naval officers with battle group command experience have legitimate concerns about being micromanaged and receiving rudder orders from decision makers ashore at any level who can never have the complete picture seen by the Officer in Tactical Command (OTC). Such factors as rapidly changing on-scene tactical developments, weapons availability, fuel states or weather may be too small grained to watch in the big picture, but could dominate the local scene. In a hot war situation, remote commanders, for example, Commander in Chief, Pacific Fleet (CINPACFLT), should not or could not become involved in directing the specific tactical activities of deployed battle groups. This potential and legitimate concern of an OTC, however, must be balanced against the requirements of crisis management situations. If key decisions are to be made ashore, the counter argument might be that the ashore decision makers should have the capability to display the same picture so that their decisions are based on the best information available. This requires rapid two way exchange of information between ashore and afloat centers, which is not always possible today. However, it would not be technically difficult, for example, to design an FCC for CINCPACFLT that is flexible enough to look at the broad picture commensurate with the fleet commander's overall responsibilities and yet that is capable of focusing on a specific crisis management situation. The decision to design this falls into the category of an electropolitical decision, which must be made before systems engineers and integrators can proceed.

If the decision is affirmative, the requirements statement must provide a definition of what the same picture means quantitatively. If, for example, all centers are to have a common display of tactically significant tracks in the area of interest, what timeliness constraints are involved? How much can the ashore picture lag the afloat picture—30 seconds, two minutes or two hours? Conversely, if OTC is relying on surveillance data from a center ashore, what are the timeliness, accuracy and update criteria the product must meet? If it is too late to have a positive effect on the outcome of the battle, it is of little use.

The above discussion illustrates the need to start with a valid set of requirements. For discussion purposes only, the remainder of these steps assumes that the requirement to provide the same picture to Navy command centers (ashore, afloat and in the air) is valid.

Assess the Requirement

Once the stated requirement is provided to the systems engineers and integrators, they must review the current status of the System to determine where it falls short of meeting the new requirement. Then they must examine funded enhancements to the current base line System to see the projected impact on the new requirement. Next, engineers must survey new subsystems that are in the development cycle in order to assess their impact on meeting the requirement. This assessment must also include consideration of the missions and responsibilities of the organizations and people that are supported by the various command centers. In the Navy, quite a variance exists, which may not be as true in the other services. For example, a short based Submarine Operations Control Center (SOCC) has a limited interest in the air picture. It will probably be found that all centers do not require identical data bases, but where they are common, the data must be the same so that all decision makers operate with the same set of facts. Once this assessment process is complete, the systems engineers and integrators will have a good grasp of where the System design is headed, and what needs to be done to meet the new requirement.

Breakdown the New Requirement

The traditional engineering approach is to examine a new requirement and tier it downward into functions, subfunctions and tasks that the System must perform. Again, this cannot be done in a vacuum. The new requirement represents the latest in a series of requirements that led to the current System design. All of today's Navy command centers provide displays of the tactical picture. The degree to which they are the same varies considerably between the collection of subsystems. For example, NTDS (now often referred to as CDS—the Combat Direction System) on an aircraft carrier (CV) and guided missile cruiser (CG) have the same data to display because they are tied together via a real time digital data link (Link 11). The FCC ashore may have similar data; however, it may be time late due to the current FCC design and its communications connectivities. Therefore, it is not the same. As was the case in assessing the new requirement, the functional breakdown must consider how closely related current functions are performed and must often use them as a base.

Providing a common tactical picture between command centers involves the exchange of tactical messages in a variety of forms, character oriented (COM) or bit oriented (BOM). They are either directive (e.g., mission assignments, rules of engagement, engagement orders) or informational in nature (e.g., surveillance reports, intelligence assessments, readiness and logistics data, battle reports). Surveillance reports are by far the most numerous. Existing and planned sensors that are land based, space based, airborne and organic to a battle group are and will become dramatically more capable of providing a deluge of data. The data must be collected, filtered, evaluated, correlated, disseminated and displayed in the various command centers as timely, relevant and unambiguous information. Maintaining this information, common between command centers, is a data base management and communications challenge of the first magnitude. In the functional breakdown of the new requirement, data base management would be a high priority concern.

Allocate the Functions to Subsystems

Once the System requirements and functions have been identified, the next step is to allocate the functions to subsystems. It should be noted that most Navy command centers are spaces that contain a number of supporting subsystems. For example, TFCC aboard flagships will contain, as a minimum, a Flag Data Display System (FDDS) console and an NTDS console, either of which can drive a large screen display. These two systems in turn interface with many other subsystems. The net result is that many subsystems directly or indirectly support the information displayed in TFCC.

The allocation of picture building functions, including surveillance data collection, filtering, multisource correlation and dissemination, requires a coordinated design and great deal of interactive communications between subsystems. Discussion of the functions can best be addressed by temporarily dividing the System into two parts: ashore and afloat. The ashore side is commonly referred to as the Navy Command and Control System (NCCS) Ashore. Programmatically, NCCS has an afloat node (TFCC); however, the term C^3I and Combat System will be used because it bridges the historical programmatic gap between C^3I and combat, weapons and sensor subsystems that exists because of organizational demarcations in OPNAV and the Naval Material Command.

Figure 1 illustrates NCCS ashore command centers in the Pacific Fleet. Table 1 defines the acronyms used in Figures 1 and 2. Although the number of ashore command centers is remaining constant, all are being upgraded. The major growth ashore will be in sensor and related subsystems that provide inputs. The sensors depicted are illustrative only and do not represent the entire gamut for security reasons. This collection of subsystems must be designed to provide, where common, the same information to all ashore subsystems at the same time. It also must provide a single, unambiguous, timely, tailored surveillance product to the afloat forces. It must accept and correlate feedback from on-scene forces and disseminate the integrated picture to all users, including selected ashore subsystems. Systems engineers and integrators must examine the various architectural options for the man-

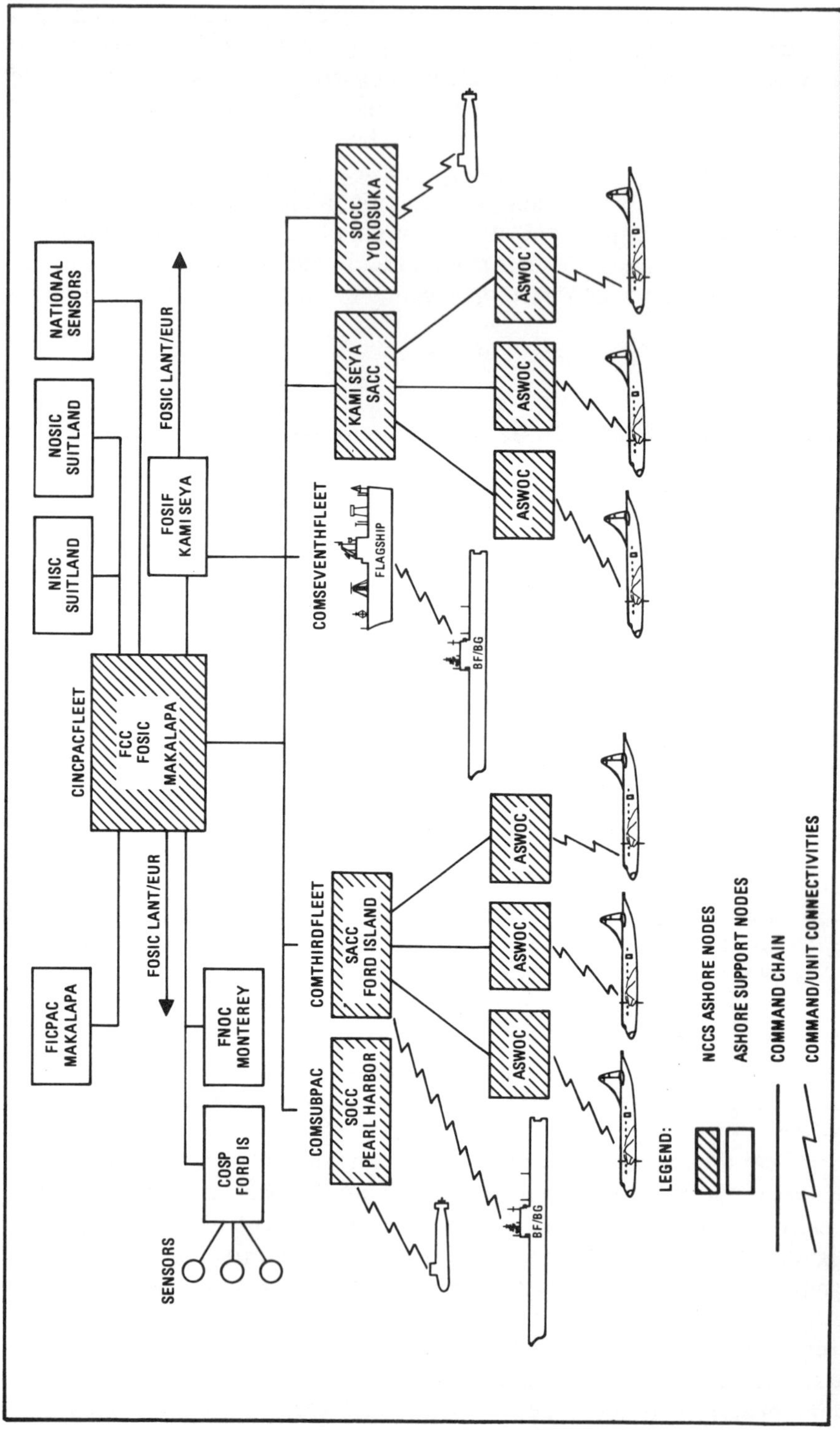

Figure 1. Navy command and control system ashore.

ACS	–	AFLOAT CORRELATION SYSTEM
ASWOC	–	ANTISUBMARINE WARFARE OPERATIONS CENTER
BF	–	BATTLE FORCE
BG	–	BATTLE GROUP
CDS	–	COMBAT DIRECTION SYSTEM
COSP	–	COMMANDER OCEAN SYSTEMS PACIFIC
C^2P	–	COMMAND AND CONTROL PROCESSOR
FICPAC	–	FLEET INTELLIGENCE CENTER PACIFIC
FNOC	–	FLEET NUMERICAL OCEANOGRAPHIC CENTER
FOSIC	–	FLEET OCEAN SURVEILLANCE INFORMATION CENTER
FOSIF	–	FLEET OCEAN SURVEILLANCE INFORMATION FACILITY
JTIDS	–	JOINT TACTICAL INFORMATION DISTRIBUTION SYSTEM
NISC	–	NAVAL INTELLIGENCE SUPPORT CENTER
NOSIC	–	NAVAL OCEAN SURVEILLANCE INFORMATION CENTER
SACC	–	SHORE ASW COMMAND CENTER
SOCC	–	SUBMARINE OPERATIONS CONTROL CENTER

Table 1. Acronym list.

agement of the surveillance data. Several options are available, i.e. a centralized approach, a decentralized approach or combinations in between. The assignment of multisource correlation functions to a particular subsystem is of key importance because of the necessity to meld multisource data from diverse sensors, which may be reporting the same contacts, into an integrated, timely, coherent product.

Three factors are driving the increase in the number of computerized subsystems afloat.

- The development of specialized C^2 support and communications systems;
- The automation of sensors and weapons; and
- The response to the Navy's decentralized approach to conducting warfare.

Under the Composite Warfare Commander (CWC) doctrine, warfare is divided into AAW, ASW, Antisurface Warfare (ASUW) and Electronic Warfare (EW). The trend is toward producing dedicated warfare area subsystems for force level commanders and coordinators as well as producing ship level subsystems for warfare area coordinators. The result is a very distributed or federated architecture and is dramatic with respect to numbers. For example, in the aircraft carrier of 1973 were only five warfare oriented computerized subsystems with four interfaces among them. In 1993, projections are for 31 subsystems with 42 interfaces among them.

Most Navy units—air, surface or subsurface—are multiwarfare capable. Multiwarfare coordination between decision makers in the use of these assets in the real time execution phase of warfare requires that the many subsystems

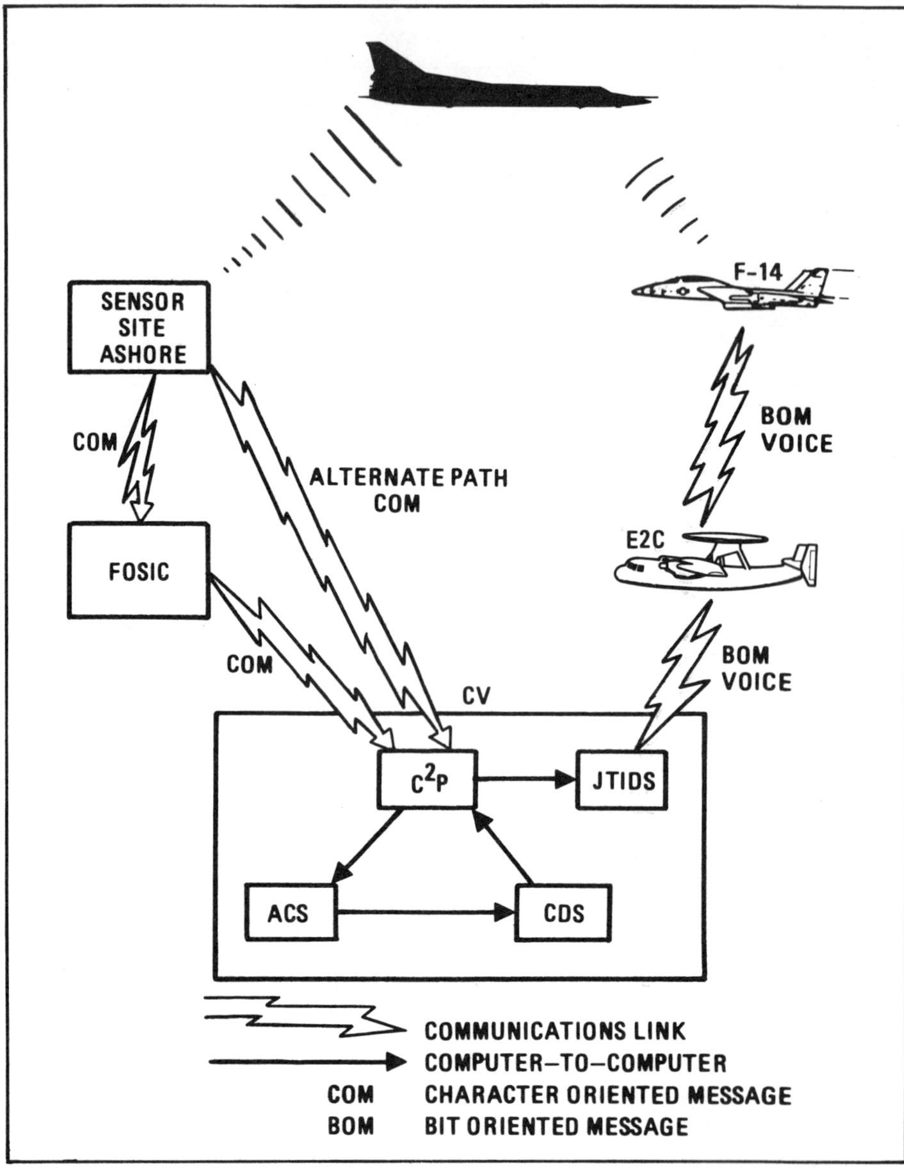

Figure 2. Surveillance data path.

involved have the same picture at the same time. This creates significant new communications and intelligence management issues, which have to be resolved.

The management of surveillance information afloat is equally as complex as that ashore and must be managed at the BG level so that all units have the same commonly held information. Maximizing compatibility and interoperability internally aboard units via interfaces and externally among units via digital links is extremely important and can, in a hot war situation, result in the destruction of more targets with less ordnance. The correlation of organic (BG generated) and nonorganic (remotely generated) tactically significant information demonstrates the point. Many ships and submarines in a BG will be capable of receiving and processing both the nonorganic and organic information. Regardless of how sophisticated the correlation algorithms in the computers are, a person will be included in the loop to resolve ambiguities. In a BG, one cannot have two people resolving the same ambiguity; they may derive different solutions and muddle the picture. Handling this data base management problem requires the development of procedures among operators at the BG level and internally aboard a given ship between operators supported by different subsystems and the associated communications connectivities.

The above example is just one aspect of interoperability. Systems engineers and integrators must keep a close eye on other interoperability impacts and the ability to provide the same picture.

Define Requirements; Size the System

Once functional relationships, allocations and boundaries between subsystems are defined, their resultant interdependencies define the need for information exchange. The exchange can take place via point to point computer interfaces, local area networks (LANs), the Defense Data Network (DDN), digital or voice links, or other means. The key is to define the "what" before defining the "how."

With the System functions initially allocated and subsystem interfaces defined, it is necessary to ensure that the System is properly sized with respect to capacities and response times. The importance of doing this is illustrated in Figure 2, which traces the path of a single contact report. Only the major computer systems are depicted. Not all communications processors are included. Because BOM and COM messages have different formats, a conversion process is necessary.

If tactical time constraints dictate that the contact report be delivered within a maximum delay of three minutes from the time of the observation to receipt by the ultimate user (the F-14), a System time budget would be established. The nine subsystem computer processes and four transmission links operating serially would have to meet the time budget. As a System, the architectures and speeds of all the intervening computers and the transmission speeds of all the links would have to be examined. Bottlenecks must be identified and corrected. Any real world constraints to meeting the requirement (e.g., funds to upgrade a key computer or develop a high data rate communications device) might force examination of different options. The systems engineers and integrators would next impose a projected heavy load on the System to check for capacity. A simulation tool that supports this would be very useful. Any shortfalls would be reported to superiors with recommended resolutions.

How the Challenge is Being Met

While progress in meeting the challenge is being made, no sustained Navy systems engineering and integration effort currently covers the total System as described in this article. The Navy C^2 Plan produced by CNO (OP 94) is an excellent start and contains a projected architecture that extends across much of the spectrum. It is, however, a high level document that is not intended to address the level of detail required by systems engineers and integrators.

Other ongoing activities are oriented toward either ashore or afloat activities. A recently established NCCS Systems Engineering and Integration (NCCS SE&I) program is aimed primarily at the ashore portion of the system. On the afloat side, the recognition of the challenge has led to the establishment of a joint working group within the Naval Material Command that has produced a C^3I and Combat System Interface Requirements Document (IRD). The C^3I and Combat System is defined as: ". . . . that assembly of systems installed aboard naval warships which provides for the collection, processing, evaluation and dissemination of information necessary to support the planning, direction, evaluation and reporting of naval combat operations. This assembly has connectivity with combat system sensors for data input and weapons control systems for weapons direction."[2]

IRD identifies eight major functional areas that the collection of subsystems must perform: navigation and time; correlation; data handling and display; communications; performance monitoring; environmental support; electronic warfare and intelligence support; and command support. A table for each functional area specifies what subsystems will support the area and allocates subfunctions to the subsystems. This process defines the functional relationships between subsystems and the information flow required to meet their interdependencies. Each functional area table has a corresponding figure that depicts in high level terms the information flowing among the interfaces. IRD treats subsystems generically because many are used on a variety of ship types; however, it also illustrates the configurations for specific ship types.

IRD is an excellent start at meeting the challenge for the afloat portion of the System. It has generated the needed communications among the many PDAs. IRD is also assisting in the development of a small cadre of people who are becoming more knowledgeable in the C^3I and Combat System. IRD also highlights the complexity of the system design as currently programmed and provides the basis for examining ways of simplifying the design and reducing the overall cost while optimizing operational effectiveness.

What More Can Be Done?

The following comments and recommendations are offered for constructive improvement of the systems engineering and integration effort:

- design requirements;
- develop and promulgate system documentation;
- establish system management procedures; and
- develop a cadre of pragmatic system engineers and integrators.

The tendency is to treat new requirements in isolation. This often results in the vertical development of new subsystems to meet each new requirement without understanding what is concurrently happening across the spectrum of ongoing actions, which may impact the design. This approach has contributed to the proliferation of subsystems. The previously mentioned development of individual subsystems to support both force level and ship level commanders and coordinators are examples. While this may be very desirable in theory, if we could afford the ultimate System, pragmatic reservations exist as to whether it could be done within the shipboard space, weight, berthing, logistics and other current constraints. In all good conscience, PDAs will overdesign systems without recognizing these constraints. The well-known quote attributed to Adm. Gorshkov applies—"Better is the mortal enemy of good enough."

In the requirements definition process it is recommended that:

- an assessment be made to determine whether existing programs will satisfy current requirements;
- all new requirements be consolidated and addressed collectively to determine their relationships;
- an assessment be made of how the current System is meeting requirements that relate to the new requirements;
- an examination of all funded enhancements to the current System be made to identify their impact on new requirements;
- all options be examined for meeting the new requirements; enhance current subsystems; build a new subsystem if necessary; or a combination of the two.
- if a new subsystem is required, fit it into the current system mosaic and identify all of its interdependencies, interfaces and communications connectivities (if required) while it is still in the conceptual design stage. Think interoperability from the start.

Successfully maintaining control over an entity as complex as the System requires an authoritative System documentation hierarchy that has teeth in it and directs the design of the System. The hierarchy should include documents in the research and development (R&D) phase of a subsystem design. The hierarchy should lead to the incorporation of the new subsystem into the System and the development of a MILSTD 490 subsystem specification. A hierarchy of this type would do much toward eliminating many documents that do not contribute to the System design and that only gather dust on shelves or clutter up safes. Paperwork is expensive. It is recommended that OPNAV develop and promulgate a System documentation hierarchy that is complete yet cost effective.

The subsystems that comprise the System have sponsors throughout OPNAV and PDAs throughout the Material Command. Maintaining control over the System design is a management challenge equal to or greater than the technical challenge. It requires a tremendous amount of coordination. Recent positive moves have been made to enhance the process. The designation of a C^3I System Engineer in NAVELEX, the Combat Systems Engineers in NAVSEA, the NCCS SE&I program and the C^3I and Combat System IRD working group are examples. Two additional System management procedural recommendations are offered.

- Formally establish a small system requirements group in OPNAV that provides coordinated System requirements to the Material Command and responds to any issues involving interpretation of the requirements. Subsystems sponsors should be members.
- Formally establish and provide sustained funding for a small, responsive Systems Engineering and Integration group in the Material Command that crosses Systems Command boundaries. Funding should be separate from any subsystem funding so the group can act as the honest broker for the System. Its responsibilities would include the production of System documents that respond to OPNAV requirements. The group would interact on a continuing basis with the subsystem PDAs and provide them direction. It would flag issues that require OPNAV resolution. Keep it small or it could turn into an unresponsive bureaucratic stumbling block that could hinder not help the production of the System. In the interim, close coordination between the NCCS SE&I program and the C^3I and Combat System effort would assist in defining the connectivities required between shore centers and afloat forces.

Designing the System requires officers with operational experience in C^2, communications, intelligence, sensors and weapons to provide the "what." Engineering Duty officers to assist in the "how," and highly qualified civil service personnel to provide continuity are also required. The group should be innovative in looking at new technologies that could be applied to the System, yet pragmatic enough to discern what is achievable based on real world constraints. Currently, the number of personnel with backgrounds in the systems approach is very limited. The base must be expanded and the field made career enhancing.

Summary

The systems engineering and integration challenge is here today and will expand in the near future. Current efforts to meet the challenge are a step in the right direction; however, more can be done. The recommendations presented represent a greater front loading of the System definition, procurement and design cycle. It has a price tag in people and dollars; however the potential return on the investment is great. Every dollar spent in effective front end efforts will possibly save $100 in the long run. It would perhaps reduce the number of false starts that result from good ideas that run afoul in the real world. It offers the opportunity to simplify the design of a very complex system, to ease the burden on the sailors who must operate and maintain the system and to ensure that the focus is on interoperability, which leads to combat effectiveness.

Footnotes

[1] *"Interoperability—The Key to C³ Systems in Support of USPACOM," MGen. Vaughn O. Lang, USA, SIGNAL,* February 1984.

[2] "C³I and Combat System Interface Requirements Document," August 17, 1984.

The Role of Manned Simulator Test Beds in Evolutionary Acquisition of C³I Systems

Peter G. Freck

This article is based heavily on work the MITRE Corporation has performed for the U.S. Department of Defense. In particular, it draws liberally from information that I, and others, developed for a workshop that MITRE conducted on the use of manned simulator test beds in the counter-air battle. This workshop was briefed by the developers of 23 test beds. It was held at MITRE's Washington Center in June of 1982.

This article will cover the following topics: First, I will discuss the basic principles that the U.S. Department of Defense feels should guide the acquisition of C³I systems. Next, I will discuss mission oriented evaluation, which provides the framework in which I feel it is most appropriate to evaluate C³I systems throughout their life cycle.

Then I will discuss the types of tools required to conduct mission oriented evaluations. My focus will be on manned simulator test beds and the attributes they offer. Finally, I will present a status report of where I think we

**This article was originally a speech presented at the Armed Forces Communications and Electronics Association's Fourth Annual NATO Symposium and Exhibition on October 26–28, 1983 in Brussels, Belgium.*

- **The C^3I-weapons system mix must be viewed as a totality**
- **The C^3I system is a lucrative target for potential adversaries — efforts must be undertaken to ensure that it endures**
- **Our services must work as an integrated force, frequently in coalition with our allies — mandating adequate interoperability**
- **People are critical elements of the C^3I-weapons system mix — they must train as we expect them to fight**
- **The C^3I system gradually evolves in time — mandating tailored acquisition approaches**

Mr. Donald C. Latham
Deputy Under Secretary of Defense (C^3I)
22 June 1982

Figure 1. C^3I Principles

stand in the U.S. with regard to the conduct of mission oriented evaluation, and with regard to the use and management of manned simulator test beds.

Basic Principles

At the June 1982 MITRE workshop, Mr. Donald Latham, then Deputy Undersecretary of Defense for C^3I, stated five basic principles that underly the current C^3I program in the U.S. (See Figure 1.) I hope that after reading this article you will be convinced that manned simulator test beds offer the potential—and I emphasize that word—to be a key ingredient in an integrated approach to evaluation that will allow us to achieve and to do business by most if not all of these principles.

First, I assert that such test beds help us to examine and evaluate the C^3I/weapons system mix as a totality.

Second, I also believe that they allow us to look at extremely severe conditions that cause our C^3I systems to operate in degraded modes. In most cases, these conditions would be impossible to duplicate in live tests because of economic or safety constraints. Data developed under these severe conditions make it possible to estimate whether a system will endure, when it fails, and how "gracefully" it fails.

Third, manned simulator test beds provide a means for examining problems regarding the interoperability of systems and subsystems—and by that I mean not only the C^3I/weapons systems of a single service, but also those of an integrated force.

Fourth, it is clear that manned simulator test beds involve people as a central element and allow us (for both training and system acquisition purposes) to examine the decisions they make.

Finally, these test beds enable us to examine the evolving capabilities of large systems and their individual subsystems. To be most useful, such examinations must involve users, system architects, developers, testers and trainers, and thereby must provide a common focus for these diverse groups to discuss problems and exchange ideas, throughout the course of system acquisition.

Mission Oriented Evaluations

An important aspect of the C^3I principles stated by Mr. Latham is the concept of mission oriented evaluation, in which the mission effectiveness of a C^3I/weapons system combination is examined in the context of a realistic, self-consistent set of assumptions. Note that this concept goes well beyond technical performance measures such as signal-to-noise ratio, and evaluates systems on the basis of their operational utility. Within any mission oriented evaluation (in our workshop, we examined just one particular mission area—the counter-air battle) it is possible to establish benchmarks of performance for the existing (or "baseline") system and to determine its "weak links," as well as to assess any potential enhancements in capability that might be achieved by introduction of improved or new systems and subsystems.

Many think such mission oriented evaluations should be conducted during the entire lifetime of a C^3I/weapons system, including the requirements definition, system development, test and exercise and training phases.

Figure 2 is an example of how one might begin a mission oriented evaluation. During the course of our work we examined the counter-air battle, its associated sub-mission areas and each function that must be performed within these sub-mission areas. For example, this figure presents the functions that are part of the defensive counter-air and the air defense sub-mission areas. Defensive counter-air (a U.S. Air-Force mission) is shown on the right and the air defense (a U.S. Army mission) on the left.

We made an overall qualitative assessment—based on the most current literature— of the ability of NATO to perform each function as well as the major specific deficiencies in performance that existed in 1982.

I should emphasize that these assessments, as well as the specific deficiencies in each functional area, were viewed only as hypotheses to test using the tools that are available to conduct mission oriented evaluation.

The first is straightforward analysis that may involve the use of fast-time, pure computer simulations (for example, the COMO III model which is well known in NATO, particularly at the SHAPE Technical Center).

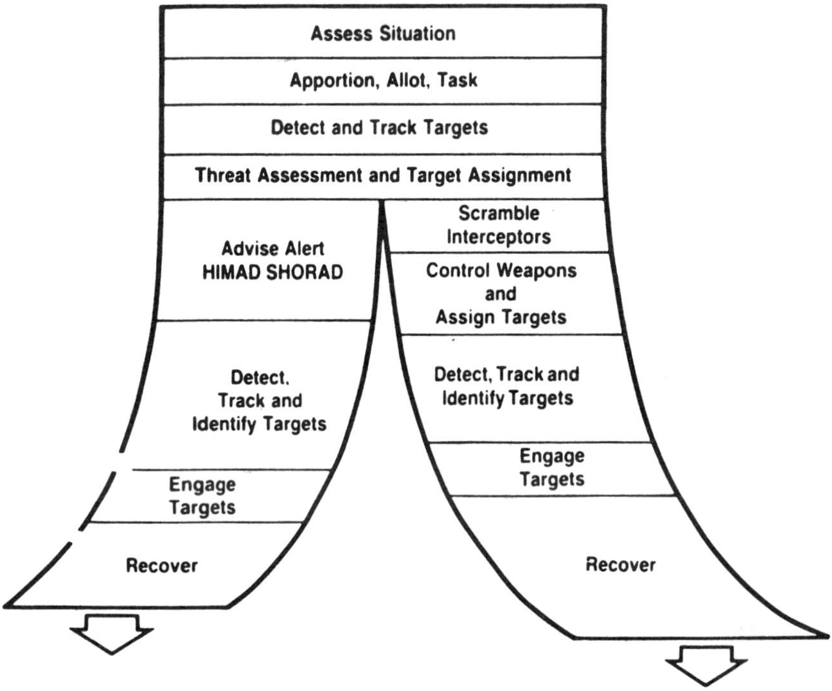

Figure 2. Defensive Counter-Air (Aircraft) and Air Defense—(Surface-to-Air Weapons)—1982

At the other extreme are field tests and exercises, where generally one can collect only gross data on how well forces are doing, including some basic time information about how long it takes for things to happen.

Between these two extremes is the subject of this article: manned simulator test beds, which are central to any mission oriented evaluation of combined C^3I/weapons systems.

Manned Simulator Test Beds

What are manned simulator test beds? Well, they involve facilities, hardware, software, procedures and people. Foremost among the elements of manned simulator test beds are the people: decision makers who are placed in situations where they must make decisions very much like, if not identical to, those they would make under combat conditions.

The second component is the simulated total system environment. This involves not only the external environment such as terrain and weather, but also the physical and electronic threat, and an "adequate" fidelity simulation of both friendly C^3I and weapon systems.

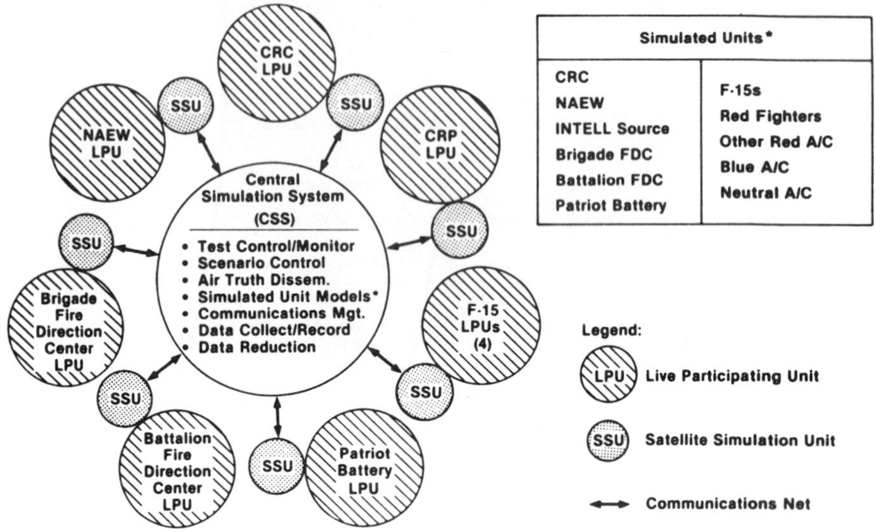

Figure 3. IFFN Test Bed Architecture

The final and very important components are the data collection, processing and evaluation facilities that are necessary to develop the measures of performance and effectiveness that are key to mission oriented evaluations.

An example of a manned simulator test bed is shown in Figure 3. This test bed, now under development by the Identification Friend Foe Neutral (IFFN) Joint Test Force at Kirtland Air Force Base, Albuquerque, New Mexico, is a major U.S. joint service initiative.

In 1985, major tests began using existing and newly developed test bed facilities throughout the United States that were linked together and integrated with a central simulation and evaluation facility located at Albuquerque. This is really a "super test bed" or a test bed of test beds.

To give you an idea of the geographical distribution of this IFFN test bed, the NATO Airborne Early Warning (NAEW) facility is at Boeing in Seattle, Washington, and is the simulator now being used for engineering development there. The Patriot simulator is at Fort Bliss, Texas. The Control and Reporting Post (CRP) node will be the 407L CRP at Hurlburt Field, Florida, that normally is used to support testing there.

All of the systems shown on the chart (Figure 3) are tied together by the appropriate communications, and driven by the central facility in Albuquerque, which provides common scenario and threat data. In turn, performance data for each subsystem is collected and transmitted back to Albuquerque, where an assessment is made of the effectiveness of the overall integrated air defense system in determining the identification of friend or foe.

Manned	• **Human component is critical to C^3I** • **C^3I is "command personality" sensitive** • **Little is understood about** — high level decision making — group decision making
Simulator	• **Save time and money** • **Examine severe conditions**
Testbeds	• **Gain understanding before full scale development** — user — developer • **Accommodate changes** — internal system performance — environment — interfaces

Figure 4. Why Manned Simulator Test Beds?

All of the "flying," if you will, of airplanes and missiles is done by simulated means; however, there are people at each of the "live participating units" indicated on this chart who are making the kinds of decisions that pilots, weapon controllers and force allocators must make during wartime.

The test philosophy is that initial trials are relatively simple, involving individual systems, and they will work up to more complex, combined systems as the tests progress.

Why do we need manned simulator test beds? (See Figure 4.) First, we want the system to be *manned* because of the criticality of human decisions to C^3I functions. Also, to a large extent, C^3I systems are sensitive to the personality of the commander, and we want to be able to identify the degree of sensitivity to differences in the commanders using the system.

In addition, little is understood about high level and group decision making. Therefore, it is important (until we have a better understanding) to include people in the system so that we can develop some data about how decisions are made, and thereby help to develop valid computer models of decision makers.

Second, we want to have a *simulator* to save both time and money. Large scale field exercises are extremely resource intensive and time consuming. As indicated earlier, simulators also enable us to examine extremely severe conditions that cannot be duplicated in exercises or tests.

Finally, we want a *test bed* so that we can gain a better understanding of systems and their requirements before we go into full scale development. Test beds also accommodate changes in the various aspects of the systems problem shown in Figure 4—and therefore can be of great benefit to evolutionary acquisition of C^3I systems.

When do we use manned simulator test beds? As you can see in Figure 5, test beds can be used during virtually all phases of a system life cycle but are

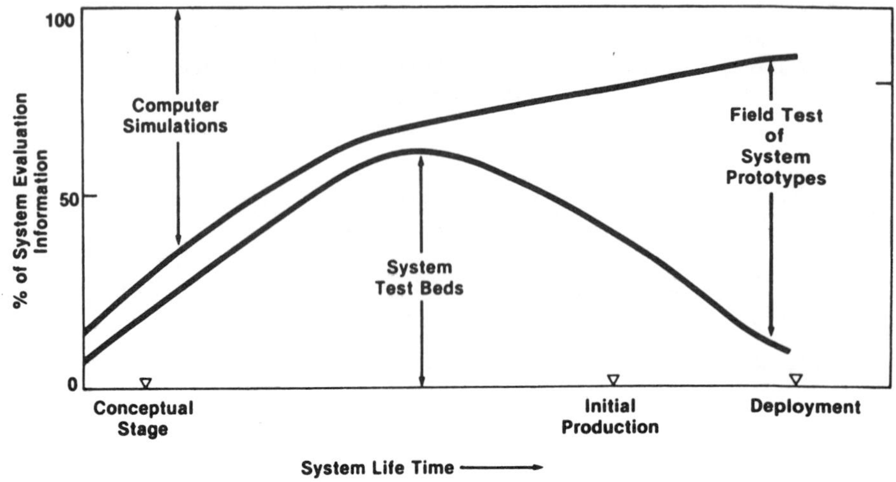

Figure 5. Manned Simulator Test Beds—When?

most heavily used somewhere after the conceptual stage and before initial production.

On the other hand, pure computer simulations are used most heavily in the conceptual stage, when we have very little information on a system and must run many cases to see under what set of conditions the system might prove to be effective.

We then bring the man "into the loop" through the heavy use of test beds during the middle stages of development. Finally, we run field tests of system prototypes when those become available in the final stages of development of a system.

In summary, it is important to orchestrate the use of these tools. I think it also is important to remember that manned simulator test beds represent only one of several tools available for evaluation. However, they are of primary importance when dealing with C^3I systems.

We can use analyses and pure computer simulations to model some of the inputs and to hone them, so that when the manned simulator itself is used, we have reasonably valid inputs. We can use field tests and exercises to validate the results of manned simulators and enhance their credibility with the user.

If the results of the simulator are at least grossly in agreement with those of field tests and exercises, then the manned simulator test beds will have much greater credibility, particularly in the operational community.

We would begin the test progression with a single autonomous weapon system and evolve to a fully integrated C^3I weapon system configuration. We would run rather extensive tests of the former to make sure that we well understand the capabilities of individual weapons and C^3I systems, before we try to pull the integrated system together.

As I said earlier, the philosophy is to establish benchmarks of performance and then to evaluate changes in performance due to system enhancements.

Components	Sub-System Performance	Functional Performance	Force Effectiveness
Elements	• Sensors • Computer/I-O Devices • Communications • Navigation	• C^3 Nodes and Weapons Systems • System architecture • Procedures	• Weapon System Characteristics • Operational Concept
Measures Performance/ Effectiveness	• Technical Factors (e.g. Range, Capacity) for Alternative Environments	• Probability of Correctly Performing Function (e.g. Identification, Classification) • Associated Statistics — Temporal — Spatial	• Engagement — Probability of Successfully Engaging a Target • Encounter — Loss Exchange Ratios — Targets Killed/Sortie • Battle — Casualties — Weapons Attrition — Movement Rates
Tools	• Analyses/Fast-time Simulations • Limited Field Tests	• Manned Simulator Testbeds • Field Tests/Exercises	• Manned Simulator Testbeds • Field Tests/Exercises • Fast-time Simulations

Dr. Stuart H. Starr
Director, Long Range Planning & Systems Evaluation
ODUSDRE (C^3I)
22 June 1982

Figure 6. Components of Mission Oriented Evaluation

Components of Mission Oriented Evaluation

Figure 6 shows the various components of mission oriented evaluation, the elements within each component, the measures of performance and effectiveness that might be used and the tools that are appropriate.

When we look at the various levels of analysis that could be performed in a mission oriented evaluation, we see that for performance of sub-systems, such as sensors or communications, the primary measures deal with technical factors, such as range or capacity, and—although they were inadvertently left off the chart—with programmatic issues, such as schedule and cost. The tools that are most useful here are analyses and computer simulations, along with some limited field tests.

When we examine functional performance, such as how well a particular C^2 node performs, we get into measures of performance like the probability of correct identification. Manned simulator test beds perform a critical role here. Some field tests and exercises are useful to validate simulator results.

When we move to force effectiveness, we have to include the weapon system characteristics and operational concepts, as well as the things that were part of the functional performance area. Our measures of performance and effectiveness dealing with engagements, encounters and battles become increasingly complex.

Manned simulator test beds play an important role in many of these evaluations, but as we progress to battle assessment, it is obvious that we must make heavy use of computer simulations because manned simulators really cannot include all the factors necessary to understand what is going on in a large battle area.

On the other hand, we can put estimates of the necessary factors into computer simulations, particularly if we have made judicious use of manned simulator test beds to understand the pieces of the C^3I process.

Status Report

Where do we currently stand with regard to mission oriented evaluation and manned simulator test beds in general, at least in the United States?

First, we found that there are many very useful existing and planned tools for mission oriented evaluation, including many manned simulator test beds. Second, most of these tools cannot be used to help with near term acquisitions. On the other hand, they should be available to help in future acquisition decisions.

A major problem in their utility is that they are not being used in any coherent manner, nor, in the vast majority of cases, is there any plan to do so. Also, of particular importance is that, in the United States, the use of test beds is not coupled to the acquisition process. Furthermore, for the reasons shown, they are not producing very credible results.

Overcoming these problems requires centralized coordination and decentralized execution of operations, not only of the manned simulator test beds, but also of mission oriented evaluation in general. We feel that centralized coordination is important to gain some coherence to the programs and to ensure that their use is focused sharply on the acquisition process. It is highly unlikely that individual manned simulator test beds will (or should) be taken away from their developers and given to some central "evaluation or test bed czar." However, it is necessary to centralize the use and scheduling of test beds, particularly when dealing with problems that are broader than those of the single service or organization. Centralized coordination also will help limit unnecessary duplication of effort and focus new developments on critical "gaps" in the capabilities of our evaluation tools.

With regard to the resolution of specific issues, we feel that if the available tools are used properly they will result in some very valuable and comprehensive, albeit very expensive (tens if not hundreds of millions of dollars), time phase evaluation programs. In the near term, we recommend that a plan be developed by mission area to integrate the use of these tools. In the longer term, we not only need to implement such a plan, but also to foster development of such things as common data bases and means for certifying and validating test beds and simulations.

As these findings indicate, we are really just beginning to go down the road of conducting coherent, comprehensive mission oriented evaluations, but I am of the opinion that such evaluations, particularly if they make proper use of manned simulator test beds, can provide the information necessary to allow decision makers to conduct rational, evolutionary acquisitions of C^3I systems.

Navy Battle Force Command and Control—A Tactical Coordination and Tactical Communications Management Perspective

R. E. Ward and William J. Brennan

Not since the battle of Okinawa has the U.S. Navy and its ability to execute its primary mission of power projection been so challenged by a single threat other than that posed by nuclear weapons. At that time, the threat was the kamikaze for "The Fleet That Came To Stay." Now the threat is the antisurface missile, described by RAdm. Wayne Meyer, USN, the airborne early warning ground environment integration system (AEGIS) Antiair Warfare (AAW) combat system leader, as the "unmanned kamikaze."

The numbers, tactics and capabilities of the "Devine Wind" strategy of kamikaze aircraft tactics threatened to make the price of the Okinawa oper-

**This article originally appeared in the November 1985 issue of SIGNAL.*

ation excessive. The Navy counter was the accelerated development of Combat Information Center (CIC) concepts, tactics and equipment to provide tactically coordinated defense in depth. The VHF radio equipped radar picket destroyers controlling Combat Air Patrol (CAP) fighter aircraft were stretched to their operational limits to provide defense in depth that gave fleet antiair gunners with proximity fused ammunition acceptable engagement odds so that the lands could continue. Tactical command dimensions and structures ushered in a new era of long-range tactical coordination of sensors and weapons.

Today, the anti-surface missile threat provides a strikingly similar, although more complex, operational challenge to the naval operator. This is true of its various configurations whether air, surface or subsurface launched, and whether from the perspective of the present day gunner or tactical commander. The resulting requirements for defense in depth while enroute to or carrying out offensive operations introduces major information exchange and interdependence demands on both combat and command and control (C^2) systems.

The compression of the time available for defense reaction and the need for decentralized decision making have been well documented in a number of studies, including the Outer Air Battle. The multidimensional, multi-threat operational environment also has been addressed in C^2 doctrine, and in the evolving Composite Warfare Commander doctrine. These procedures and concepts are detailed in Naval Warfare Publications eight and 10. The Navy Command and Control Plan guidance and architecture and the supporting Navy Command and Control System—System Engineering & Integration (NCCS SE&I) effort deal with fielding the required responses from this C^2 perspective. The Navy AEGIS program and Advanced Combat Direction System project are fielding combat system oriented responses to meet these same threats.

The combat and C^2 system approaches are different, but they are being harmonized through an evolving Battle Force System Engineering initiative. Understanding these different perspectives is important when trying to understand the Battle Force tactical coordination requirement realities. Figure 1 compares the operational scenario battle phase breakdown as defined in C^2 and combat system development concepts.

The C^2 perspective is essentially a commander's top-down perspective, oriented toward supporting the command structure in carrying out the full range of Navy missions and tasks. It addresses nonorganic sensors, the supporting command structure and intelligence functions, as well as the overall Battle Force. The combat system perspective is more of a bottom up gunner's approach. It starts by putting a weapon on the threat to include the necessary control and detection capabilities required to make this possible. The focus is on the individual tactical unit, its sensors and weapons and the support for the commanding officer and his tactical team. A recognized challenge is to integrate this C^2 oriented sensor to shooter system, formerly called the Integrated Tactical Surveillance System (ITSS) (now called NCCS SE&I), with both tactical unit and Battle Force levels.

C^2I	COMBAT SYSTEM
• INDICATIONS AND WARNING	• DETECT
• EARLY WARNING	• CONTROL
• THREAT EVALUATION	• ENGAGE
• TARGETING	

Figure 1. Operational scenario battle phase comparison.

The requirements for the Battle Force Tactical Information System to support C^2 and combat system functions are understood best if we have a clear picture of the tactical command structure, representative Battle Force geometry, required timeliness criteria and resultant information flow challenges.

Tactical Command Structure

Starting with the command structure shown in Figure 2, we see how NCCS is structured. Its fleet commander in chief (CINC) interfaces with the World Wide Military Command and Control System (WWMCCS), and its tactical unit interfaces with combat systems. This command structure supports and directs the various Battle Force and Battle Group (BG) commanders and the Officer in Tactical Command/Composite Warfare Commander (OTC/CWC) in the execution of both strategic and general-purpose (tactical) missions.

Here the differing requirements of the strategic and general purpose forces should be emphasized. The command structure and supporting information network for the strategic forces are governed by specific constraints that impart characteristics of rigidity and direction. Exceptional reliability requirements, a stable and redundant command structure and well documented operational concepts and procedures characterize this mission for Navy operators. General purpose forces, on the other hand, are challenged by a broad range of operational tasks, widely variable force composition and potential multiple theater commitments. The high degree of flexibility required by the command structure and supporting tactical information systems places difficult demands on C^2 and combat system developers. Add to this the fact that the multi-mission nature of many Navy units requires systems that can support either strategic or tactical missions and the dimensions of the challenge emerge further.

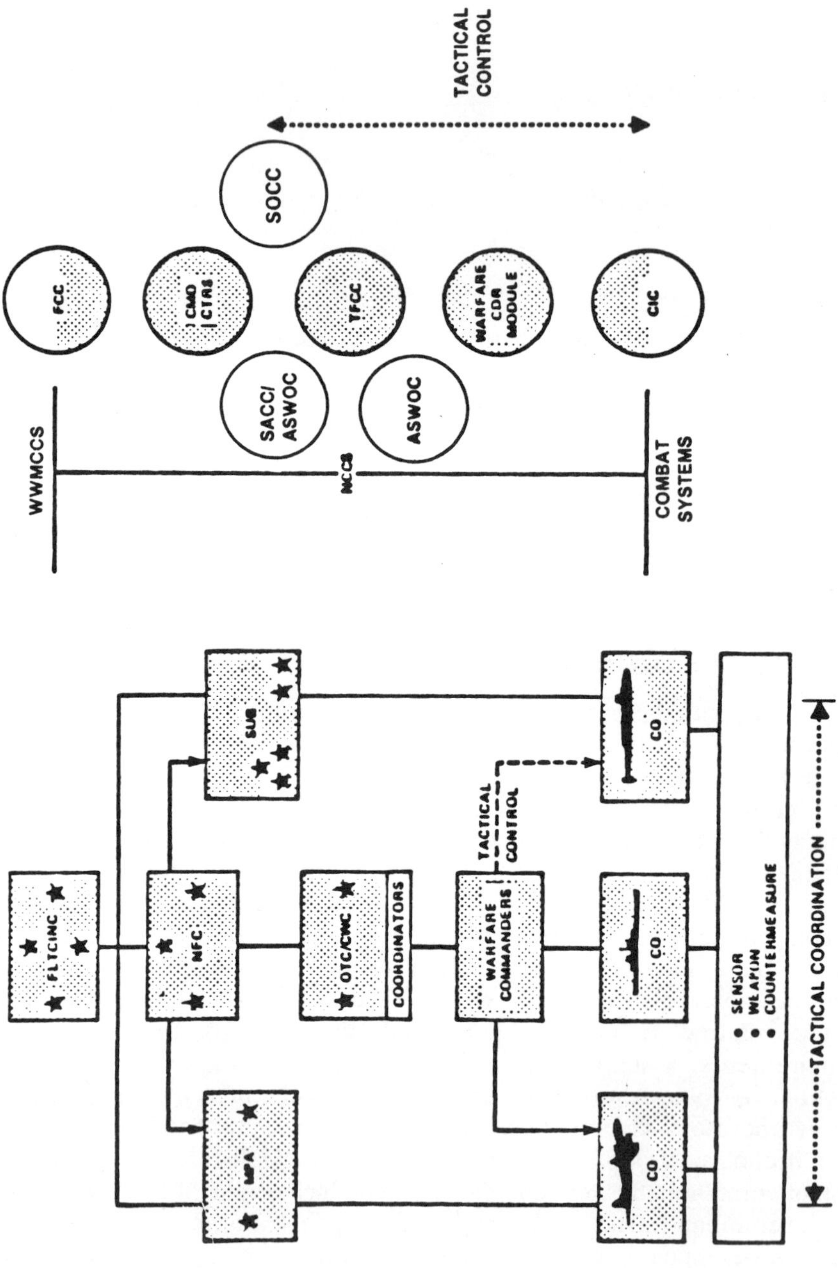

Figure 2. Command structure.

The Battle Force/BG command structure begins with the OTC/CWC, his supporting coordinators and warfare commanders. The Battle Force ships and embarked aircraft will be tailored to the assigned missions. Force composition may vary during the mission execution or mission redirection. Maritime air and submarines may be tasked in direct or associated support or they may be placed in a barrier or surveillance role supporting more than one BG. The location of warfare commanders and coordinators will vary with the capabilities afforded by the particular group of ships comprising the BG and with the evolving mission.

It is not evident in Figure 2 that the ship level block actually represents almost all of the BG tactical assets. Dozens of ships and hundreds of embarked aircraft may be involved here. Multi-mission ships and aircraft may be tasked by, and must respond to, more than one warfare commander and they must be responsive to guidance from several coordinators. This structure and supporting doctrine are intended to:

- promote the ability to conduct multiple concurrent warfare operations;
- have the capability to meet surges;
- facilitate centralized/decentralized control; and
- be enduring and flexible.

The information network requirements to support these are impressive.

Within the command structure (Figure 2), the sensor, weapon and countermeasure box represents a whole array of tactical unit interactions ranging from passive sensor cross fixing to third party targeting. Here, varying combinations of tactical units are carrying out a wide and dynamic range of activities requiring information to be exchanged, commands to be given and coordination to be accomplished. This is an often underestimated information dimension.

The Navy C^2 Plan describes the information that must flow to support the command structure as:

Command—operation orders, directives, force status reports, tasking requests;
Coordination—allocating, controlling, tasking, monitoring assigned forces and;
Information—sources and sensors, contact reports, individual tracks, environmental support information, specific intelligence reports, parametric data, status reports, encyclopedia data.

We define the terms of tactical control and tactical coordination in Figure 2. What is described above from the Navy C^2 Plan is the inter-command structure information flow that is primarily tactical control oriented. The Joint Chiefs of Staff (JCS) Pub. 1 defines tactical control as:

> "The detailed and, usually, local direction and control of movements and maneuvers necessary to accomplish missions or tasks assigned."

However, this describes only one portion of what is necessary to support Battle Force C^2 and it describes little of what is required for support of combat systems. Distinct from tactical control is the requirement for inter-tactical unit tactical coordination information exchange. Here, we refer to the usually highly interactive transactions that take place between a wide range of operators to execute the details of mission accomplishment. The air intercept controller and the aircraft under his positive or advisory control is an example. The information that should flow between passive sensor operators on cooperating tactical units to facilitate sensor cueing, cross fixing, track development and weapon delivery represent another dimension. This collection of operational requirements, well-known to operators and underpinning task execution, is not defined in JCS Pub. 1. It was defined in connection with the requirements definition phase of the tactically oriented Coordination in Direct Support (CIDS) program and was called tactical coordination. It is defined as ". . . detection, classification and localization of the threat and delivery of attack. . . ."

This latter definition is descriptive of what is required to support combat systems and cooperating sensors. This concept is important because it introduces an entirely different level of operational user within the command structure, but with information exchange requirements different from those in the command chain.

Battle Force Geometry

The requirement for defense in depth, the demands of counter targeting, response to multiple threats and base lines for better passive sensor utilization have led to major tactical disposition revisions. As can be seen in Figure 3, the dispersal is much more radical than what is done in response to the nuclear threat. Distances are not shown and the disposition is only conceptual and not necessarily representative of the current fleet doctrine. However, it can be seen that the four surface ship task units are not within visual or UHF radio range of each other. Units in the individual task units are within UHF range, and they are usually always within visual range of each other. Fighter and surveillance aircraft UHF radio ranges are mission profile dependent and therefore they are dynamic. The SSN is in a moving sanctuary area outside of UHF range. Doctrine will embark the OTC/CWC in the carrier in task unit (TU) 1; the AAW commander may be in the AEGIS cruiser in TU2; and the anti-surface warfare commander may be in the TOMAHAWK equipped cruiser in TU3. While the antisubmarine warfare (ASW) commander probably will be embarked in the carrier, he also could be embarked in a destroyer in TU4.

Figure 3 shows that while the demands of defense in depth counter targeting and passive sensor base line are met by this disposition, there are some potential penalties. Among these is the need for continuing, timely and accurate data on friendly unit locations and their operational activity. In addition, there are major tactical connectivity challenges due to the operationally dynamic mix of line of sight (LOS) and extended line of sight (ELOS)

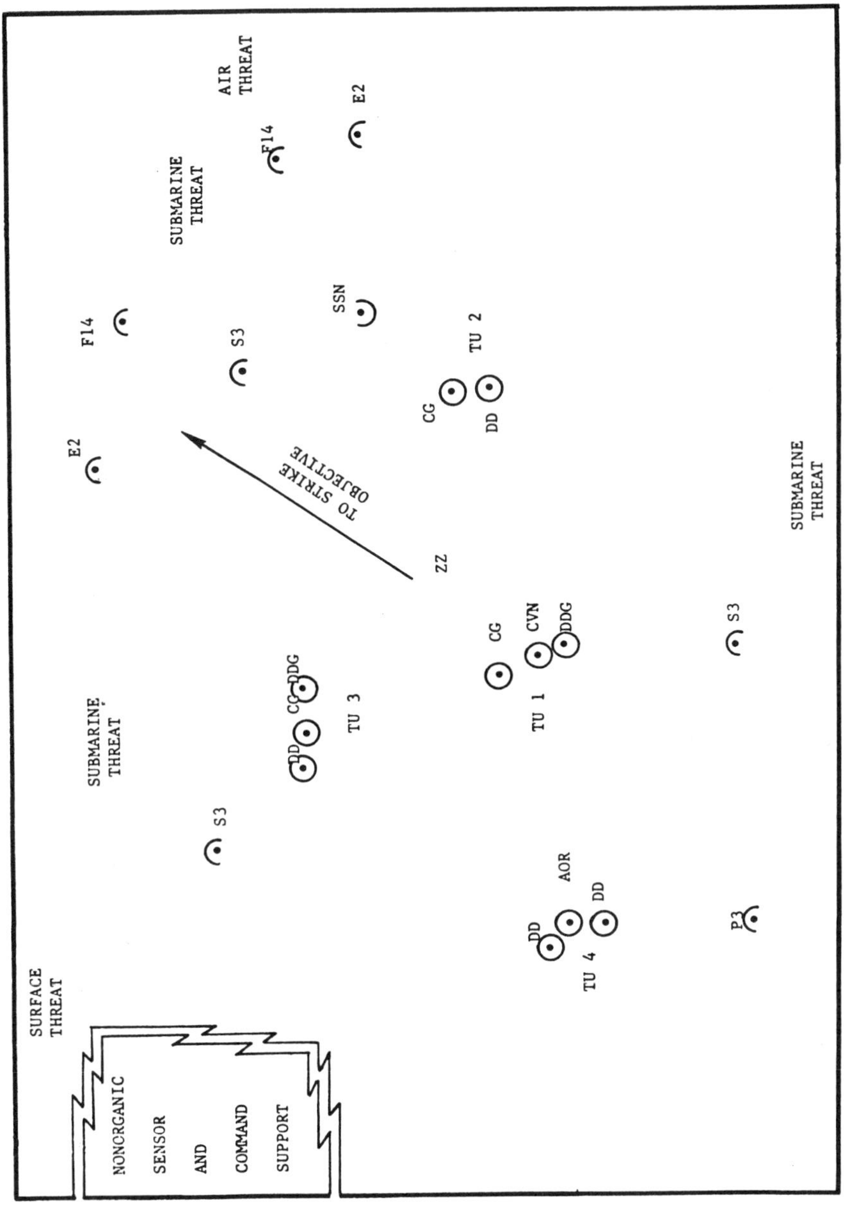

Figure 3. Single Battle Group disposition.

ranges between the various tactical units. ELOS is defined as being out to 300 nautical miles.

Timeliness Criteria

Figure 1 compared the operational phases within operational time lines as viewed from the C² or the combat systems perspective. Maximizing decision time and the ability to take effective action against the enemy before this action preempts yours is a given operational goal. A good Battle Force example is the destruction of an enemy SSN in an ASW engagement before his missile launch transforms the threat into an AAW action.

The stochastic and dynamic nature of multi-warfare operational events is what drives the command structure requirements for decentralized control, the capability to meet surges and flexibility. The timeliness criteria for countering a MACH 2 missile launching aircraft and the criteria for countering a missile launching SSN obviously are quite different before missile launch, but they are strikingly similar after the launch as the missile becomes the threat.

Timeliness requirements for the individual functions that make up a Battle Force action are difficult to relate as a coherent whole. Figure 4 outlines them in a composite battle management information warfare time line. The 10 minute boundary between first and second domain is not necessarily fixed, but the depiction is more helpful than the term "near real time." The concept does provide a basis for a better appreciation of the timeliness requirement relationship of the functions shown. Targeting, counter targeting, point defense and engagement are primarily combat system functions and they involve intensive tactical coordination. Force redirection and the second domain functions are fundamentally C² oriented and primarily involve tactical control. The point is that in any battle, concurrent and stochastic events are taking place in both time domains and the command structure together with the various involved tactical unit sensor and weapon system operators must function in both. This two domain reality should influence functional allocation between combat and C² systems and it demands an integrated tactical information system capable of supporting both.

Required Information Flow

The sensor to user or shooter data network and data flow is shown in Figure 5. We will focus on the afloat user and consider information characteristic drivers, user services and available transmission systems.

Mission information requirements are theater unique and peculiar to the warfare area. The information characteristics are scenario, battle phrase and threat type/target dependent. These factors govern data structure, perishability and the correlation challenge. The connectivity requirements are the-

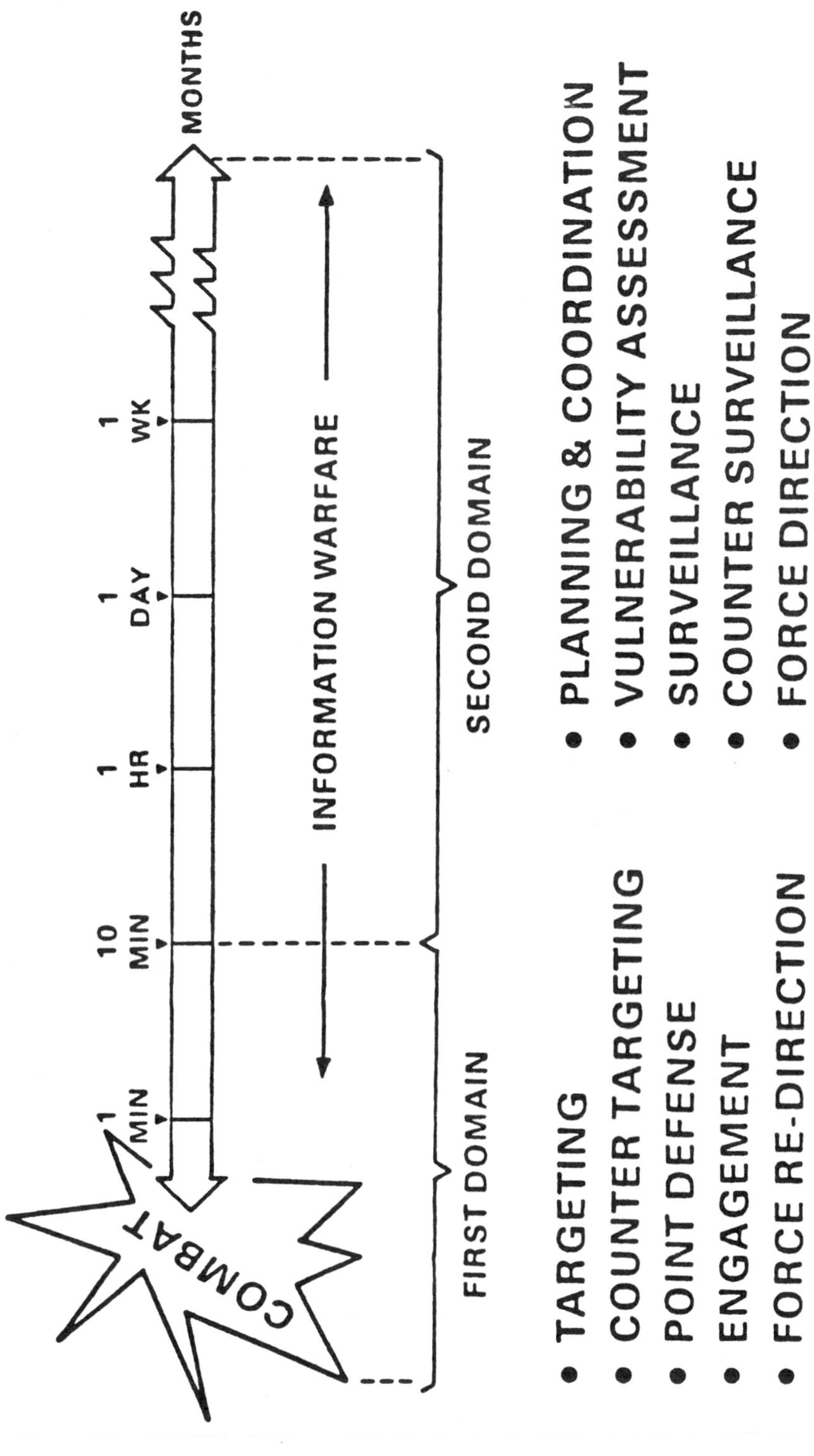

Figure 4. **Battle management time line.**

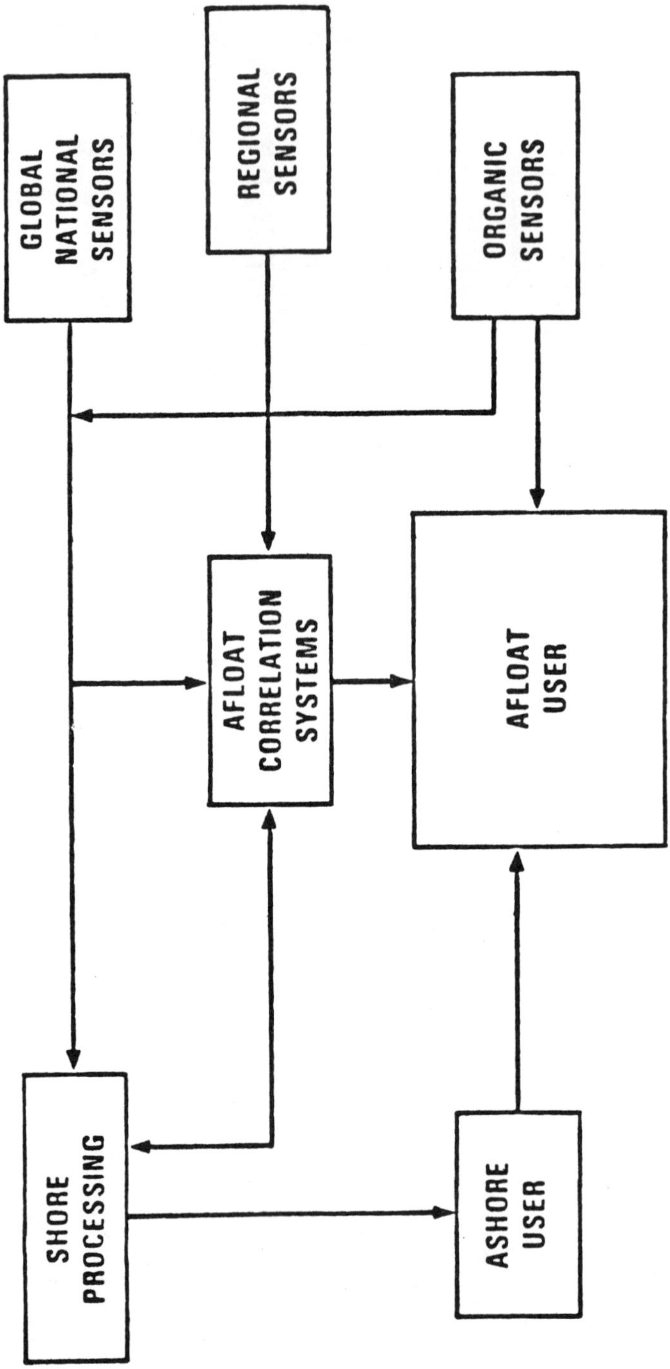

Figure 5. Sensor to user or shooter information flow.

ater and mission/task and BG geometry dependent. The tactical message, data and voice mix are dependent upon the user interface and they are doctrine and procedure sensitive. If we relate these requirements in the BG shown in Figure 3 and to the command structure shown in Figure 2, we start to define a very complex and dynamic information network that must accommodate a wide range of operational tasks and one that must function in both time domains.

Global and regional sensors and associated intelligence are essential to providing the indications and warning (I&W), early warning (EW) or early detection if the afloat user is to be given acceptable reaction time. The afloat user would prefer direct access to the appropriate nonorganic global and regional sensors and minimum reliance on shore processing. Theater uniqueness, the support of other forces and processing realities make reliance on some nonorganic processing a requirement. The afloat user also must have some control over the quality and quantity of the relevant tactical information received. The Ocean Surveillance Product (OSP) is intended to package and provide the required support.

Going back to the MACH 2 attack aircraft and the SSN both carrying anti-surface missiles and to the time line comparison in Figure 1, we can trace and compare parallel actions. I&W on the activity leading to an attack aircraft threat launch is based on intelligence plus global and regional sensors. It is second time domain tactical control information that probably is best processed ashore and provided to the BG commander in narrative message or computer readable format. Flash precedence and standard operating procedures will cause this information to be delivered to both afloat and ashore users. As the action develops and the EW phase is reached, the transition to organic sensors must begin. BG organic surveillance and force direction, together with some direct inputs from regional and possibly global sensors, are moving toward the first domain timeliness side of the information warfare time line. Events still are described as tactical control actions. The commander will become more reliant on tactical data driven displays and he will use voice for the transmission of orders and for coordination. To support the threat evaluation phase, increasing numbers of operators from Air Intercept Controllers (AIC) to active and passive sensor operators become involved. We are now at the boundary of the two time domains and currently only voice and tactical data systems can support the correlation and subsequent targeting functions. These actions are tactical coordination dominant. It is apparent that we need to provide regional sensor track information on MACH 2 threats directly to the afloat users' correlation system to facilitate organic sensor acquisition.

The parallel requirements of this network for countering an SSN threat are different than the requirements for the MACH 2 air threat. First, submarine tracks are not a well-defined series of positions. Rather, they are an assembly of clues of varying degrees of confidence. Processed regional sensor information and the information from cooperating maritime air or direct support SSN's are critical to the optimum deployment of BG organic sensor assets.

I&W, EW and threat evaluation all may require shore processing support in support of afloat correlation systems and for transition to organic sensors for targeting. Both tactical control and tactical coordination come into play, but most activity, short of targeting, takes place in the second time domain. Tactical messages and data will be the primary user interfaces, with tactical voice required for some interactions, particularly targeting.

The requirements of the hostile operating environment demand that this information network and its segments be secure, resistant to enemy intercept and exploitation and resistant to jamming. In addition, if threat driven emission control (EMCON) is imposed, it will deny the use of certain frequency bands under certain operational conditions. The electromagnetic characteristics of communications systems make high capacity and long-range service mutually exclusive without some type of relay.

All of these factors complicate efficient operability and interoperability. For example, a functional network such as AAW coordination and reporting must be supported by two separate communications systems, one HF and one UHF, to provide the required response under varying EMCON conditions and jamming threats while attempting to accommodate the dispersed BG geometry shown in Figure 3. A second example is that of parallel user service. Here, tactical data link, computer to computer, must be paralleled by an operator to operator link to provide the required degree of human interaction and a query response mode. As a result, the current composite network is characterized by dedicated circuits/links, which are transmission system unique and often with low individual circuit utilization.

Based on the preceding battle force command structure, geometry, operational timeliness and information flow considerations, several observations are made:

- No single type of current voice, data or message user service meets all tactical control and tactical coordination demands required by the battle force.
- No single type of transmission system can provide the tactical connectivity required to support user connectivity demands in the tactical unit geometries, EMCON conditions and electromagnetic threat dynamics that characterize the multi-warfare environment.
- The tactical afloat user should be able to dynamically control his information system assets, including control of information quality and quantity, in a user format and system configuration most supportive of both his tactical control and tactical coordination requirements in any battle phase.
- A properly balanced mixture of voice, and computer to computer data and message service with a selectable mix of tactical transmissions and the ability to dynamically manage them is required to adequately support the full range of battle force users.

The Navy C^2 Plan provides a functional architecture and conceptual system description that is intended to bound the required information system capabilities. The combat system developers have a parallel but slightly different

ELEMENTS OF THE C² AFLOAT ARCHITECTURE

- TACTICAL FLAG COMMAND CENTER DISPLAY AND DECISION PROCESSING
- HOST SHIP TACTICAL DATA SYSTEM
- AFLOAT CORRELATION PROCESSING
- COMMUNICATIONS PROCESSING
- EW SENSOR CORRELATION
- SHIP SIGNALS EXPLOITATION SPACE
- OTC/CWC LEVEL INTELLIGENCE

COMBAT SYSTEM MAJOR FUNCTIONAL ELEMENTS

- TACTICAL DATA INTEGRATION/MULTISOURCE TRACK MANAGEMENT
- TACTICAL MISSION PLANNING SUPPORT
- WEAPON AND AIR CONTROL/COORDINATION
- ORGANIC TEST AND TRAINING CONTROL
- DATA LINK MANAGEMENT AND CONTROL
- SYSTEM CONTROL

Figure 6. C² and combat system comparison.

approach (Figure 6). In the latest edition of the Navy C² Plan, C² functions are grouped under sensor coordination, information processing and dissemination/connectivity. Warfare specific and supporting operations requirements are defined and the architecture is further developed from this requirements base. The methodology is capable of accommodating both tactical control and tactical coordination information exchange requirements concepts. The challenge is to better define these concepts, to better understand them and, most importantly, to more effectively manage the resulting information network.

Information system design criteria must focus on the battle force operational requirement at the user level in his most stressed mode. The user interface and his full spectrum of requirements (to include both tactical coordination and tactical control) must be considered. A more adequately stated concept of communication asset management is required.

Currently, the required level of tactical communications management is envisioned only at the Joint Tactical Information Distribution System (JTIDS) subsystem level. The C² Plan architecture and its concept of asset management provide an excellent basis for the expansion of essential tactical communication management concepts beyond that of the JTIDS subsystem. This asset management concept, together with the tactical control and the tactical coordination dynamic communications requirement context, can assist in achieving an operationally viable subsystem balance. This will help achieve the best affordable mix of JTIDS, high frequency antijam and satellite systems required to support the battle force. The concept of the additional user community embodied in tactical coordination and the command function of tactical communications management are key to supporting the general-purpose battle force.

During the Battle of Okinawa, the destroyer picket controlled fighter aircraft was essentially an extension of the traditional fleet gunners' envelope of capabilities. It required the development of tactical control concepts, information exchange procedures and real time tactical coordination beyond what had existed before. Extensive CIC to CIC and CIC to aircraft tactical interactions between operators paralleled and often exceeded the traditional bridge to bridge command interactions. Innovative operational concepts, taking full advantage of newly fielded technology, were able to counter the threat. The tactical user system provider iterative dialogue worked effectively.

The technology is available today to field the counter to the anti-surface missile threat. The challenge is to implement a balanced and affordable counter capability. Tactical user implementer dialogue and sound operational concept development are essential steps in the required process. An understanding of tactical coordination and tactical communications management concepts by both gunner and commander operational communities will contribute to fielding the required battle force information systems.

C^3I in Support of the Land Commander

Anthony G. Bohannan

Introduction

This paper is concerned specifically with the tactical C^3I requirements of the land commander of the European battlefield in the last decade of this century and how today's advanced electronic technology can be used to support him in his task of winning the land battle. The paper proposes that the land commander's requirements in this respect differ from those of his air or maritime counterparts in a number of significant ways, sufficient to justify them being a major factor in the design of integrated C^3I systems.

This is not to argue that the sea, air and land battles are divisible. Together, they form one overall campaign. The outcome of one battle always will have a direct impact on the outcome of the other, and aircraft in particular always will take a direct part in all three. The three battles must be planned, controlled and fought as one and are therefore indivisible; and so therefore must the overall C^3I system that supports them be equally indivisible. But a common system to a common design concept which is required to serve three separate commands (air, maritime and land) must be a compromise, and it is the danger of a poor compromise that this paper addresses from the land commander's point of view. That there also are separate naval commander's and air commander's requirements that need to be met individually and that there may be points of conflict in system design are not in question.

**This article was originally a speech presented at the Fifth AFCEA European Symposium and Exposition held in Brussels, Belgium on October 24–26, 1984.*

Furthermore, in what follows, it is in no way intended to imply that the air or maritime battles are less difficult to fight or less hostile than the land battle. However, there are some fundamental differences in the way the three tactical battles are fought:

1. The command base is different. In the naval battle, command is exercised from a weapon platform (ship) with few extra overheads or from a distant fixed installation. The air battle is commanded from a fixed distant base. The land battle, on the other hand, is commanded from a highly mobile base well within the battle area under direct enemy air and ground fire and close to the enemy's Electronic Countermeasures (ECM).
2. The speed of events on land is generally slower than in the air or at sea.
3. The geographical coverage of the land battle is relatively small but much more concentrated with many more separate fire units.
4. Once battle is joined, the air and naval commanders are more divorced from the individual combat units than the land commander.

It has been said that in the land battle we equip the man, whereas in the air and naval battles we man the equipment. All this must have implications for the design of C^3I systems.

Aspects of the Battlefield

Uncertainties surrounding three aspects of the modern European battlefield will have (or should have) a major impact on C^3I systems being designed today:

1. No one can predict with any certainty the degree of mobility that will be possible. Will it be best to sit tight or move? Nevertheless, small size, agility and the avoidance of unique signatures undoubtedly will be of paramount importance to survival. It is inconceivable that the land headquarters of a size still fielded today can have any hope of surviving more than a short time beyond H hour.
2. ECM is another uncertain factor ahead of the event. But again it is inconceivable that communications, unprotected by advanced Electronic Counter Countermeasures (ECCM) techniques, will survive long enough to be of value to the commander.
3. The battle itself often is depicted with a clear-cut Forward Edge of the Battle Area (FEBA) between the opposing sides. Will it actually be that way, and, if not and if the opposing forces are confused and mixed together in some depth, what effect will this have on the land C^3I system?

Control of the Land Battle

There is, I believe, a growing popular image of a star wars concept where computers receive directly the intelligence from sensors, control the local deployment of resources and decide when and what weapons to launch against the enemy. In star wars, human intervention (the commander) is confined to the strategic deployment of resources and, when time allows, the ability to override the decisions made by the computer.

This possibly may be the future in those situations where the speed of events is beyond human management; and where the units involved on each side are relatively few and relatively easily distinguishable one from the other. In these circumstances, the battles may well be largely a matter of the computation of various speeds and directions, leading to a decision as to where and when to move tactical platforms and to launch missiles and the like.

But the situation on the land battlefield is unlikely to be like this, certainly not for the foreseeable future:

- First, because the equation of what, where and when and the factors that lead to a decision to launch or withhold forces is far more complex. The number of units and their minute by minute status; the terrain; and the many imponderables are so numerous and indefinable as not to be open to straightforward computation. The weight of information and intelligence, especially in the local battle area, is too great and its collection and analysis too difficult.
- Second, and above all, the human factor still has a decisive impact on the outcome of the battle. Human judgments, intuition, instincts, gut feelings and the like all have such a significant part in the land battle that they infect the whole business of command and control. These instincts may on occasions prove wrong but, nevertheless, the outcome of the land battle is still very much influenced by human initiative, both individual and collective. It is still very much a matter of the commander on one side putting his wits against those of the commander on the other. Military history tells us time and again that it is the commander who has the courage to defy conventional logic and precedence and who reacts to his instincts contrary to the available information and consensus, who more often suprises his opponent and wins.

It may be that expert systems with a built-in intuitive capability will one day take over, but not within the timescale of this paper.

The Land Commander's Requirements of C³I

No matter what function it serves, the general requirements of a C³I system such as survivability, flexibility, reliability and the like are well recognized. These apart, the land commander requires of today's C³I systems the following major attributes:

1. The ability to stay close to the battle and be able to feel the way it is going; to be able to judge for himself the state of his individual units

and formations, and what they are capable and not capable of at any one moment; similarly, to make judgments as to the enemy; to learn the terrain; and to be ready instinctively to grab that fleeting moment when success can be his for the taking. It is a rare commander who does not see a need to visit each of his subordinates at least once a day and make firsthand judgments. And if this is not possible because of the situation on the modern battlefield, then the technology of C^3I must serve the commander in such a way that this immobility is not a constraint. He must be able to move about the battlefield electronically and make face-to-face contact with those he commands.

2. That said, and again it is impossible to predict with certainty the degree of mobility that will be feasible on any future European-type battlefield, the C^3I systems must give the commander the ability to be able to move about the battlefield undetected, or at least unidentified as a high value target by the enemy. The commander and his staff cannot afford to be distracted from their C^2 task by being overly concerned about being identified and attacked. They must feel free to sit tight or move at will.
3. It follows that the C^3I system must allow the commander to exercise command from anywhere on the battlefield without being dependent on extensive command facilities and without being constrained by the limits of his communications.
4. The C^3I system must allow the commander to get the best out of his inevitably limited *combat* resources without inhibiting what he can do with them or putting at risk his control of them. In the past, the size of headquarters and inadequate communications have been major constraints in battle, particularly in terms of risk taking and responsiveness. Today's technology must be exploited to remove, for all practical purposes, these constraints completely.
5. The C^3I system must assist the commander's staff by analyzing large quantities of often conflicting intelligence and by exchanging processed data between command levels as quickly as possible with a minimum load on the communications themselves. However, the ability to handle and exchange large quantities of information is not an end in itself. Indeed, we need to reverse the tendency to accept that just because the information is there, the commander must have it. The need is first to reduce the strain on commanders and staff (not least the strain of knowing whether or not the information is reliable and up to date, and that the right deductions have been made from it); and second, to delay the making of difficult decisions to the last possible moment, thereby increasing the likelihood of their being the right decisions at the right time. This means giving greater emphasis to the reliability and immediacy of information than to the need to handle and display increasingly large and overwhelming quantities of data. Quality before quantity. The last possible moment is not intended to imply a last minute decision. Time must be left, postdecision, to implement that decision, but the more immediate the information, the quicker and easier that decision.

6. The C^3I system must allow the commander direct access to raw data, for he is as much interested in this as in the processed product, in that it is often raw data which triggers his instinctive judgments. It also must allow for filtering data both within a headquarters and between command levels.
7. The C^3I system must have the ability to provide all commanders at every level with common data and an agreed intelligence analysis.
8. The C^3I system must allow the commander to exercise command in a way that suits his character, his personality and his general modus operandi. The command structure and command procedures are a tool to help him do his job. Like all craftsmen, the commander needs to feel comfortable with that tool. Awkward tools produce only inferior results, and a C^3I system that requires a commander to command in a way that does not suit him personally will, in that commander's view, be inferior. This is not to advocate taking this to what would be absurd limits in allowing every commander to organize and equip his headquarters and develop unique procedures to meet his own idiosyncrasies. There must, of course, be a large degree of commonality and total compatibility throughout the command structure. But C^2 also has to be flexible enough for each commander at each level to feel comfortable with and have confidence in the effectiveness of his command tools. The C^3I system must support the commander, not dictate to him.
9. Finally, and most important, the C^3I system must allow the commander to project his personality down through the chain of command so that he can directly influence events in every corner of the battlefield. Most commanders will say that the best way to do this is with his voice and through his presence and personality, in face-to-face contact. He also will say that the best way to absorb information and assess a situation is with his own eyes and brain.

In sum, to serve the land commander effectively, the designers of modern C^3I systems must recognize the need for those systems:

1. To enable the commander to put across his personality; to influence the battle;
2. To adjust to his personal command techniques (and of course, to those of commanders who come and go in war and to those of different commanders at each level);
3. To allow the commander to move freely, to command from anywhere in his area of responsibility and to stay close to the battle electronically if not physically (The system must permit the commander to make personal judgments and not divorce him from assessments, events and subordinates through a chain of data banks and displays.);
4. To use advanced data handling techniques to improve the reliability and immediacy of information, if necessary at the expense of quantity, to safely postpone decision making to the last possible moment;

5. To provide to all commanders at every level, common data and an agreed intelligence analysis—in particular, agreement as to what intelligence to ignore; and
6. To enhance command and control and not inhibit it by having unduly large and immobile headquarters, vulnerable communications and ponderous staff procedures. Command by exception itself reduces significantly the need for C^3 facilities.

C^3I System Requirements

All this leads to a number of system requirements of particular significance to the land commander:

1. *Robustness:* First, in terms of individual equipment and its ability to continue to perform satisfactorily despite everything the battle, the environment and the soldier himself can do to and with it. "For want of a nail, a shoe is lost. . . ." Second, robustness in terms of the system overall; diversity, redundancy and the like.
2. *Responsiveness:* The ability of the system to respond to a constantly changing situation, and thereby changing requirements. In air and maritime battles, the units are like pieces on a chessboard. In the land battle, the chessboard itself changes; the battle itself alters the terrain and the environment.
3. *Dispersion.* The C^3I system must allow the various elements of a headquarters (itself far smaller than the headquarters we see fielded today) to be dispersed in permanently separated packets. Today, they just move in packets and come together at each new location because this is the only way the staff can operate together effectively. In the future, they must be able to operate permanently dispersed. One bonus of this will be that a step-up facility is no longer required. Another reduction will be in overheads.
4. *Mobility:* Mobility, of course, is a requirement of any tactical system. In the context of this paper, mobility is the ability of the C^2 elements (the commander and his various command/staff cells at each level) to carry out their functions from anywhere within the formation's allotted area. This requires very small, lightweight and highly agile components. It also requires an integrated communications and data system able to provide the necessary command terminal facilities at these small, dispersed command elements, and the interlinking communications. Everything must be almost entirely automated if it is to be manageable and if overheads are to be reduced, not increased.
5. *Modularity:* Modularity increases robustness and reduces the risk of abrupt system failure.
6. *Invisibility:* This is the ability of the C^2 components to get lost in the crowd; to be indistinguishable from any other component of the for-

mation; to have no uniquely distinguishable signature—physical or electronic. Each C^2 requirement (operations, artillery, intelligence and the like) must have a common solution. Their C^2 communications must not stand out from the bulk of other traffic such as infantry manpacks. One distinct benefit of modern digital Large-Scale Integration (LSI) technology is the affordability of universal, built in, 100 percent crypto protection.

7. *Invulnerability:* This will come in part from the robustness, responsiveness, dispersion, mobility and invisibility of the C^2 elements—their chameleonlike qualities, which should ensure them an adequate degree of invulnerability to physical attack, and to a lesser degree, electronic attack.

   a. The latter can be achieved by advanced ECCM. First and foremost, the various links must be indistinguishable one from the other; operations from logistics; one command level from another.
   b. Equally important, the communication links must be used to the minimum extent practical.
   c. It goes without saying that the communication links must be totally, cryptographically secure, jamming resistant and strictly emission controlled.

All this points to two objectives of system design:

- Reduce the load on the system.
- Reduce the size of the components of the system and its overheads.

At the battle of Waterloo, the Allied soldier literally in the front line was able to look over his shoulder and see his Army Commander (Wellington) and virtually the complete C^3I system. Our aim today should be to return to this situation, not literally but in the electronic sense.

We have to reduce drastically the commander's entourage of men and vehicles, e.g., a divisional headquarters of 700 men and 200 vehicles! This can be achieved by reducing considerably the command functions (greater automation, delegation and command by exception); by reducing the size of the physical components of the system; by dispersing on a permanent basis the elements of a headquarters over a far greater area than currently is the practice; or by a combination of all three. Today's technology allows us to do all this. Nodal points and large headquarters must virtually disappear.

First and foremost, we need to look for ways to reduce the amount of information that needs to be exchanged, to reduce reporting and recording times and to speed the passing of orders and directives. The users must be disciplined in their use of the system. The commander in battle is constrained by the finite quality of men, weapons, ammunition, fuel, rations and the like that he is given to fight with. However, he and his staff often behave as if communications and data processing capacity are infinite. The Parkinson's Law of C^3I, whereby usage increases as the capacity of the system increases, must be reversed. This in itself will:

1. Reduce the size of the command/staff cells and their overheads, make them indistinguishable and able to operate from anywhere on the battlefield;
2. Reduce the need for high bit-rate channels leaving more room for error correction, ease frequency management and reduce the impact of jamming; and
3. Hamper enemy ECM.

Command and Control

In the land battle, manual staff procedures are still very much with us. Information is received from many sources at each level of command. It is collected, sorted and recorded on the battle map and in the statistical records manually. It is then distributed around the different staff cells in the headquarters, where more often than not, the process is repeated again. It is then passed up one command level where the process is again repeated manually. The load on the staff system and the scope for errors and delay are very considerable.

It also has much to do with the unacceptable size of the present-day headquarters and the reluctance to disperse them, the excessive demands on the communications, and the unresponsiveness of the staff procedures. If we can reduce the scale of the job of reporting and recording, then we can reduce the staff and thereby the size of the headquarters, the communications and all the other overheads.

In all this, there is a place for voice and a place for data. Voice must remain the primary means of command for a number of reasons:

1. First, because of the need, mentioned earlier, to protect the personality of the commander. Only with the voice is it possible to express effectively immediacy and concern. Written information tends to lose any real sense of urgency and expression.
2. Speech is quick and simple; it is the language in which we think. It leaves our hands free and involves no interfacing keyboard or the like.
3. Speech carries much more information than the equivalent written words.

On the other hand, the eyes are by far the most effective and quickest way of absorbing information, of appreciating and assessing situations and the like. Furthermore, visual data can be absorbed and reviewed at will.

This suggests that, at least for the command function, the most effective and least resource demanding means of intercommunication is "out by the voice and in by the eyes." On the other hand, for the control function, data is by far the most effective way of passing large quantities of information and of reducing the load on the communications system, even if the data input is directly by voice.

REQUIREMENTS	TECHNOLOGIES
Preserving the commander's personality and his ability to influence events.	High quality Voice Systems. Visual Transmission and Displays ECCM. Size and Weight of Equipment
Reducing the load on commanders and staff, and thereby the size of headquarters and the other command overheads	Voice Processing and Recognition Data Displays Processor Control Digital Technology ECCM Cryptography and its Management Expert Systems
Making headquarters indistinguishable from the rest	Digital Techniques Automation Size, Weight and Power Consumption Less Environmental Control Burst Transmissions Emission Control ECCM
Improving the reliability and immediacy of information	ECCM Improved Sensors Information Processing Voice Processing Signal Processing Error Reduction Automation Cryptography and its Management

Figure 1. Requirements and Relevant Technologies

The Technologies to Achieve This

In the first column of Figure 1 below are listed the above requirements and in the second column are the relevant technologies. One technology can meet more than one requirement. The list is not intended to be exhaustive.

Voice Processing

To the battlefield commander, speed and quality of information are of overriding importance, including voice authenticity and intelligibility. The current 16Kb Continuous Variable-Slope Delta (CVSD) standard meets these requirements and actually improves intelligibility at limited range in modern equipment optimized for this mode. For rearward communications, where satellite or HF communications may be used, immediacy and preserving voice personality are less important than security and technical performance, which may be achieved, for example, by the use of vocoders. While packet switching data systems look attractive, the use of these for conveying real time digital voice does not seem practicable, at least within the timescale of this paper. Overcoming the delay problems involves great cost with little apparent benefit. A more economic means of utilizing voice channels better or reducing the load on the communications, and with ECM protection, is likely to be frequency hopping.

The difference between wideband (CODEC) and narrowband (VOCODER) secure voice quality is very noticeable. But the high quality is only won by using more bandwidth. Therefore, to conserve spectral occupancy without reducing quality, we must look for new voice coding techniques. When ECM and ECCM are considered, it is clear a different measure must be employed; for if we reduce bandwidth occupancy by removing redundancy, we may have made ourselves more vulnerable to ECM, then requiring error correction coding or other bandwidth widening techniques to restore it. On a statistical basis, it is possible to have more frequency hopping systems hopping over 50MHz, than fixed frequency radios having bandwidths of 25KHz.

A disadvantage of direct voice communications is that the recipient has to be there at the time to receive it without interruption. An advantage is that the spoken word carries far more information than its written equivalent and is far quicker. This points to the advantages of a voice store made possible by digital technology and its ability to time-shift data. The commanders and staff can have the facility of canned verbal reports that, while still in progress, can be reviewed from the beginning. A speech store controlled by a voice recognizer offers a very attractive C^2 facility.

Data and map displays are becoming more commonplace in advanced C^3I systems, but the need to punch a keyboard to input or recall information is wasteful of time and resources and calls for skills (albeit relatively simple) that not all are prepared to acquire. Keyboards usefully can be replaced by voice commands. Reliable and cheap voice recognition devices are now within the state-of-the-art and will find an increasing and valuable use in forward area C^3I systems.

Better still, if commanders and staff are prepared, in certain circumstances, to discipline themselves to a vocabulary of say 200 words, then the scope of voice activated C^2 systems is greatly increased. Such discipline, and indeed a determination to reduce the load on the system instead of increasing the system to carry the load, would have an immense, advantageous impact on system design overall.

Data Displays

A visual display is by far the quickest and most effective way of presenting information. A symbol can tell more in a moment than many lines of writing.

The land commander's primary work tool is the large-scale survey map. Nothing else can replace it. The near ideal, therefore, is a digital symbol display superimposed on a survey map; the map itself being able to be moved left and right, forward and back, and changed in scale at will. If this is coupled with an ability to input data by voice to appear instantaneously on all the operations cell maps throughout the formation, then what a saving in resources this would give. It is well within the state-of-the-art technology today.

Video

Another means of projecting the commander's personality is video despite the penalty of a wide bandwidth signal which could be otherwise employed carrying many speech and teleprinter channels. Perhaps to the land commander, one video channel is worth more than all the other channels it replaces. In any event, the technology is available now to fully encrypt such a signal. One solution might be the use of high capacity, totally secure fiber optic links. But like cable and over the distances envisaged in a widely and permanently dispersed C^2 system, they may be too vulnerable and too inflexible. Small SATCOM terminals could well be the solution.

ECCM

ECCM contributes especially to the reliability and immediacy of information and the invisibility of headquarters. However, it defeats the object if specific ECCM techniques are used only on a selective basis. Many technologies are involved, but two are paramount, i.e. spread spectrum, notably frequency hopping and advanced cryptography.

Frequency Hopping

Frequency hopping is appropriate for all types of communication links. The amount of protection is governed chiefly by the bandwidth available for spreading. Where it is acceptable, reducing the capacity requirements ensures communications even in the face of the most sophisticated and heavy jamming.

With frequency hopping, the transmitter and receiver both select simultaneously and pseudo randomly, the same frequency from a set of available frequency channels and pass data for a few microseconds before changing frequency again. This is no different from fixed frequency communications where every net has a unique frequency but for hours or days, not microseconds. The radio receiver design is such that it can tolerate a large percentage of individually jammed communication channels, each for a very short period. Frequency hopping also reduced the probability of the enemy detecting traffic flow or even that a radio transmission is actually taking place, thereby enhancing the land commander's ability to remain hidden and indistinguishable. Indeed, when all radios change frequency hundreds of times per second, the enemy is presented with a totally confusing electromagnetic picture.

There are conflicting views about whether frequency hopping improves or aggravates the frequency congestion problem. That is why true orthogonal hopping has been built into equipment like the Marconi SCIMITAR (an ECM jammer) range of radios. Nonorthogonal hopping radios also may have their uses, particularly on nets which have a low duty factor, allowing more of them to share a given number of channels than with fixed frequency radios. But undoubtedly the jamming resistance is less than with orthogonal hopping.

However, the arguments about congestion really reflect the concern to conserve communications, and it is in this context that the combining in one channel by statistical multiplexing, separate low volume C^2 functions, like logistics and fire control, can help. More optimally encoded data at source is needed with all redundancy identified and stripped off before addressing the communications problem. Redundancy still needs to be added in the communication process, but this can best be done much more intelligently with error correction coding (for example minimum for line of sight systems: maximum for HF) than with the accidental redundancy, as with full bandwidth voice that the source produces.

The optimum example of this is the use of speech recognizers, which offer the prospect of reducing voice to rates in the order of hundreds of bauds. Considerable redundancy and ECCM protection could then be obtained by error correction coding the voice into 2,400 bit/sec EUROCOM voice channel transmissions. The resulting redundancy could, for example, be used to transmit the data many times over on different frequencies. Even if there were only a 10 percent chance of getting through on each frequency, the message would get through almost every time.

Advanced Cryptography

Advanced digital cryptography, particularly if it is coupled with fast full-band frequency hopping as in SCIMITAR, not only provides total, long-term security but also contributes significantly to making the headquarters indistinguishable (as long as all or the majority of users are similarly encrypted). Developments in LSI technology offer encryption modules sufficiently small in size to be integrated into the transmission or terminating equipments.

Undoubtedly, LSI development will lead to even more secure systems as defined, for example, by length of nonrepeated key stream. However, since existing systems already provide key streams of lengths which represent between 100 and several thousands of years' security, dependent upon data rate, little improvement is required. Most effort must be concentrated on reducing the size of the equipments.

The other major factor in achieving crypto security is the number of independent key streams available. The larger the number, the less vulnerable is the system, but the problems of key management increase proportionately and could become unmanageable especially when universal encryption is employed. Hence, only marginal increases in key setting options will be cost effective, but to achieve a balance between the key management task and security overkill, more automated mechanisms must be employed for key distribution. The most secure means will remain physical distribution by fill gun, but the sheer numbers of crypto units involved will require the introduction of remote key setting. If a link is properly secured, it can be used to convey a key setting update.

High voice quality presents no problem cryptographically, since no quality impairment is caused by encryption per se. However, in narrowband applications such as HF, voice quality remains a problem. Secure HF coupled with vocoders and/or 2.4 Kbit/sec error corrected data and rugged modems offer only some limited improvement in HF voice quality and recognition.

Processor Control and Digital Technology

The size of a headquarters and hence its mobility and vulnerability are largely determined by the overheads of staffing, communications, power and administration. The number of people actually involved directly in C^2 is small. Since the design of such systems as PTARMIGAN and RITA over 10 years ago, advances in the technologies of solid-state Complementary Metal Oxide Silicon (CMOS) circuitry, processor control, frequency synthesis and digitization have led to very remarkable reductions in the size, weight and power consumption of telecommunications and related equipments. It is now practical to fit comfortably in one Land-rover-sized vehicle a high capacity trunk switch, message switch and their associated encryptors and multiplexers, as well as two radio-relay terminals; a complete COMMHEAD in one ¾-ton vehicle, compared with, say, three 4-ton trunks in the first generation system.

Modern equipment is not only far smaller and less demanding in terms of environmental control and power, but it is considerably more rugged, reliable and simple to use and maintain. This process will continue and the savings compounded until it will be possible to conceive a return, figuratively, to Wellington's C^3 requirements at Waterloo—himself and five or so staff officers and no overheads.

Expert Systems

I have argued against, in the foreseeable future, a star wars concept applicable to the land battle. However, that is not to say that the concept of expert systems cannot be used to assist the land commander in the process of decision making. This emerging computer science, whereby machines are capable of manipulating information as opposed to just straightforward numerical processing, has an application on the battlefield. The expert system can provide a number of decisions with differing degrees of confidence. A novel feature of the expert system enables the user to ask for justification of decisions. The machine can trace its own progress explaining how a particular decision was arrived at.

This said, the land battle presents a particular problem. The decision making and self-learning capability of an expert system could represent a quantum leap in the land battle C^2. But to gain this intelligence, the expert system needs to be fed initially with detailed analysis of past decisions and events, and there is a danger here of investing the computer with the same deficiency that generals in the past have so often been accused of—fighting the next war in the way they ended the last. Undoubtedly, computers with an artificial intelligence and self-developing intuitiveness could be a significant aid to C^2 on the battlefield, but this is a long way off. In the meantime, existing C^3I systems need to be designed to enhance the commander's own human intuitive instincts.

Conclusions

I believe that in C^3I design for the land commander, we must reverse a trend which appears to see the need to handle, manipulate and display ever increasing amounts of data as the primary aim. This inevitably means a more complex infrastructure, more overheads and less mobile C^2.

It is possible to visualize instead a situation whereby greater importance is given to voice and where through the exploitation of technology and through a determination to reduce the load on the system (instead of enlarging the system to carry an increasing load), it will be possible for the commander with a small staff entourage to roam the battlefield undetected—not least because he is indistinguishable from the rest and not dependent upon a large supporting "tail"—and make his own assessments of the situation and bring his influence to bear whenever and wherever it is required.

This command cell will be small enough to live-off an adjacent unit wherever it may be. The equipment to achieve this can be so small, lightweight and undemanding in power and so reliable and so automated as to reduce drastically the overheads needed to support it in the field. The field headquarters and the communications nodal networks that support them, as we see them today, can electronically and physically, literally disappear.

This concept need not mean that the commander is cut off from events and out of touch, or that his access to the C^2 system be significantly reduced. State-of-the-art technology makes possible a concept of large quantities of indistinguishable, secure, low cost distributed capabilities (data handling, transmission and switching) spread across the whole army area and easily accessed.

What is required is:

- Exploitation of state-of-the-art technology—digital switching; processor control; voice processing and recognition; ECCM—frequency hopping; advanced cryptography and the like. So many of the systems being fielded today are of a technology at least a generation old.
- Discipline in seeking ways to reduce the load on the system through a better man-machine interface (voice in/eyes out) and in the selection and filtering of the data to be processed and displayed.

In the end we must strive to optimize human performance and make the C^3I system match the man.

Intelligent Aids for Tactical Planning

Stephen J. Andriole, Harlan H. Black, Gerald W. Hopple and John R. Thompson

Introduction

Tactical planning is a perennial defense process. Commanders at all echelons must plan optimally if they are to survive. The processes by which commanders plan, however, have not yet been influenced by the revolutions in analytical methodology or microcomputing. This article looks at two aids designed to support corps commanders in the generation and evaluation of alternative plans or *concepts of operation.*

Tactical corps planning is characterized by a great deal of uncertainty about adversary intentions, likely "Red" courses of action and possible responses. The challenge we faced several years ago involved the design of aids that would amplify expert judgment in a way that would reduce battlefield uncertainty. TACPLAN and INTACVAL, the two planning aids discussed here, evolved over time in response to a steadily growing body of literature that we collected and data that we extracted from subject matter experts. This data enabled us to validate requirements iteratively, and also permitted us to develop strawmen "storyboards" (screen displays of the aids before they were programmed) that our planners could work with directly.

This article describes the steps that were taken to develop TACPLAN and INTACVAL as well as the aids themselves.

Systems Design Methodology

We approached the planning support problem with the systems design methodology that appears in Figure 1. We assumed from the outset that,

**This article was originally published in IEEE Transactions on Systems, Man and Cybernetics,* Vol. SMC-16, No. 6, November/December 1986.

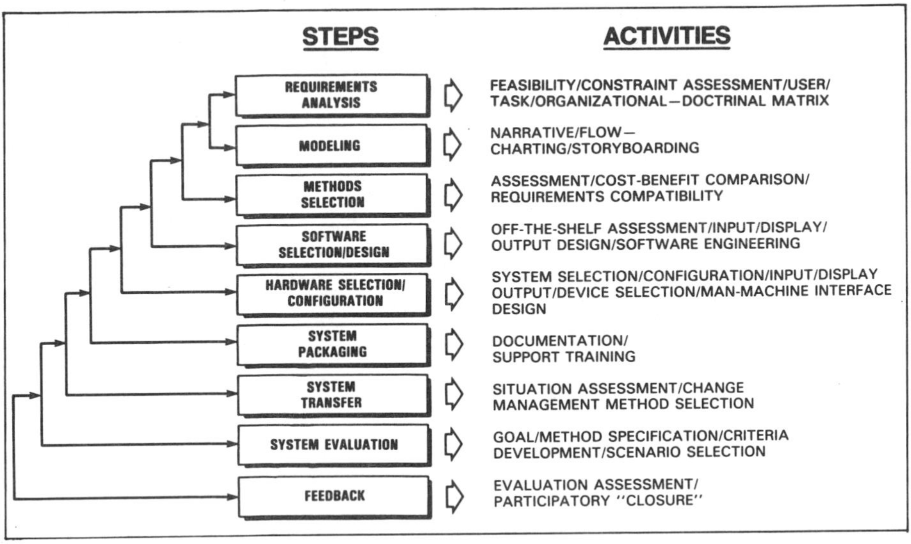

Figure 1. Structured systems design methodology.

since the domain was primarily cognitive, iteration would be necessary; hence, the rapid prototyping methodology depicted in Figure 1.

The methodology called for a detailed and systematic requirements analysis, requirements validation (in this case, via storyboards) and requirements/analytical methods matching—all before programming began.

The methodology permitted us to take small steps toward the development of the aids. In fact, the methodology itself led us to INTACVAL, the aid that evolved directly from TACPLAN (TACPLAN was conceived in 1983; INTACVAL in 1985).

Planning Requirements Analysis

Figure 2 presents the tactical planning process. Note the distinctions among staff and commander actions. Note also the planning *process* depicted in the figure. We began with an assessment of the elements and steps in the planning process and, for analytical and bureaucratic reasons, focused on the corps planning process. Corps planning is more abstract than lower level planning though just as goal-directed and hierarchical. The systems design methodology applied to our corps planning problem could be applied just as well to any echelon.

Several experiments were conducted at the Army War College in Carlisle, Pennsylvania, to identify the processes by which corps plans are generated and evaluated. We used several War College scenarios to identify requirements, which we gathered via videotaped exercises of actual planners solving a planning problem. As the planners formulated and evaluated alternative plans, we requested that they "think aloud" the protocols used to assess terrain, capabilities and courses of action. We studied the tapes and, in

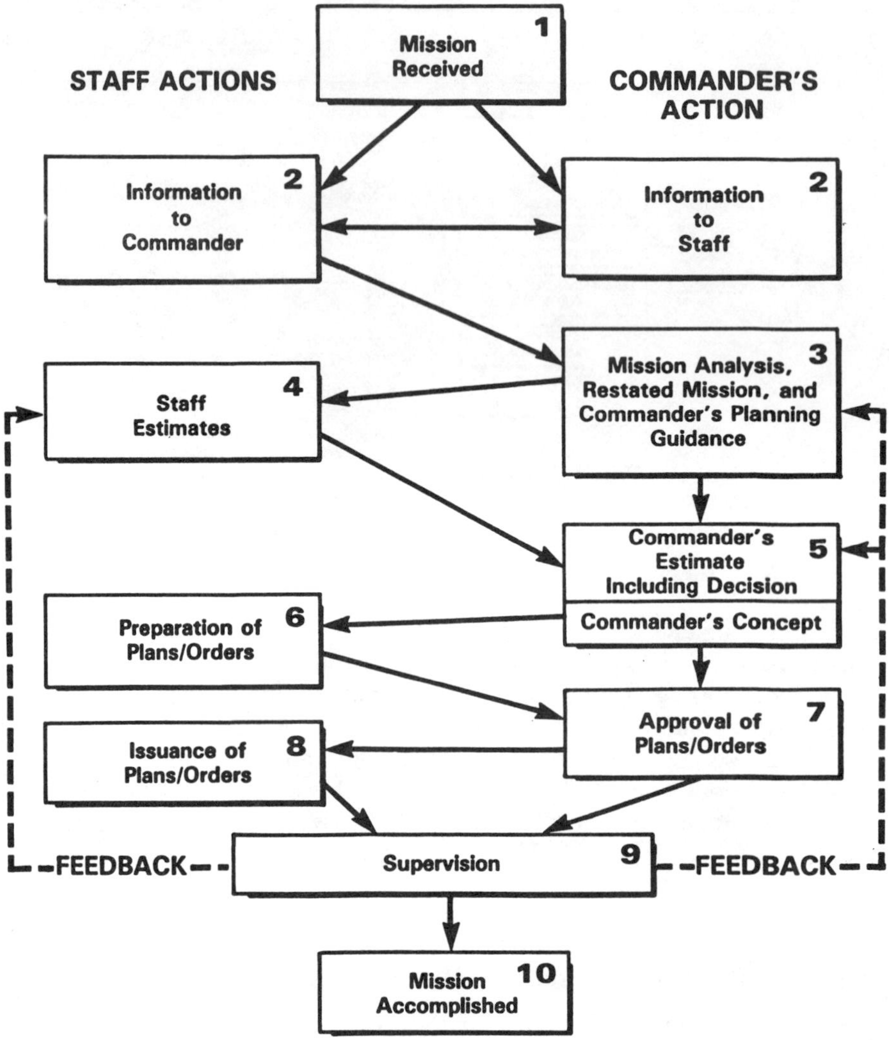

Figure 2. Military planning process.

conjunction with codified planning doctrine, developed a functional concept for TACPLAN and then INTACVAL.

Figures 3 through 6 describe the planning experiments and present the results. Two groups of planners participated. One group—students from the War College—comprised three Lieutenant Colonels (05s), while the other, three Colonels (06s who were also planning instructors at the College). They were given a scenario and asked to formulate a tactical defensive plan given a likelihood of a Warsaw Pact invasion of Western Europe (as the map in Figure 4 suggests). The scenario, of course, was fictitious in content, though it permitted us to observe the planning process in great detail.

The results of the experiments appear in Figures 5 and 6. We noted the differences between the groups as well as their similarities. We discovered,

- **The Purpose:**
  - *To gain insight into the MILITARY planning process; to develop a model of military planning*
- **The Participants:**
  - *3 (06) expert tactical planners from Army War College's (AWC) Land Warfare group; 3 (05) planners from the AWC student body*
- **The Scenario:**
  - *Imminent red (Soviet) invasion of western Europe*
- **The Mission:**
  - *Develop a tactical plan for the defense of western Europe; deploy blue (Allied) forces optimally*
- **The Method:**
  - *Filming/recording of planners during the planning process; analysis of data; conversion of data into planning protocols; conversion of protocols into functional model of the military planning process*

Figure 3. Carlisle planning experiments.

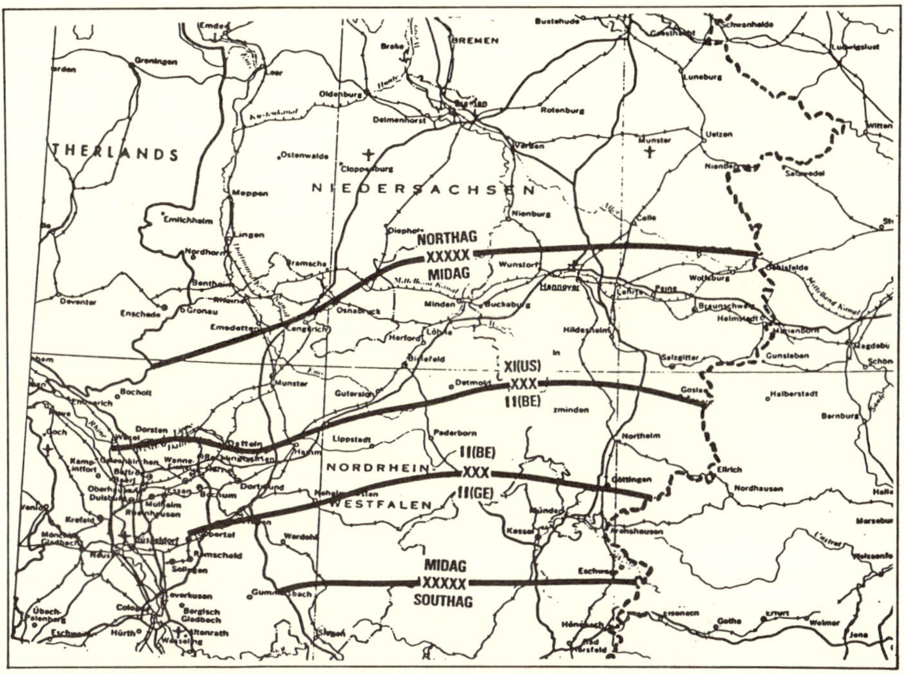

Figure 4. Scenario map. (MIDAG denotes Corps Command).

• THE MILITARY PLANNING PROCESS	• MILITARY VS. NON-MILITARY PLANNING (MP/NMP)
• **Highly structured** • *Sequential* • *Procedural* • *Doctrinal* • *Integrated . . .* • **Highly repetitive** • *Planning, re-planning, & contingency planning* • **Mission-oriented** • **Verbal, graphic, non-numeric process**	• **Planning guidance originates from above in MP, but from within in NMP** • **MP is usually adversarial; NMP is usually not** • **MP is accompanied by high accountability, while NMP is usually not** • **MP is explicitly goal-directed, while NMP is frequently opinion-driven** • **MP planning receives near-immediate, highly structured feedback, while NM — especially personal — planning is often stimulated from directly within the planner** • **MP is by nature & definition distributed, while NMP is frequently localized with one or two planners**

Figure 5. Results of the Carlisle experiments: military planning process.

GROUP #1 VS. GROUP #2

- **Group #1 (06 expert planners) considerably more risk averse than Group #2 (05 student planners); Group #2 expressed greater certainty re adversary intentions**
- **Group #1 more deliberate (i.e., more options, more dimensions of value) than Group #2**
- **Group #1 expressed considerably LESS confidence in their solution (plan) than Group #2**
- **Group #1 institutionalized "Devil's Advocate" role, while Group #2 avoided structured challenges to solutions to various parts of the planning problem**
- **Group #1 relatively more devoted to military (Army) doctrine than Group #2, perhaps as much a reflection of familiarity with doctrine as devotion to it**

Figure 6. Results of the Carlisle experiments: Group 1 versus Group 2.

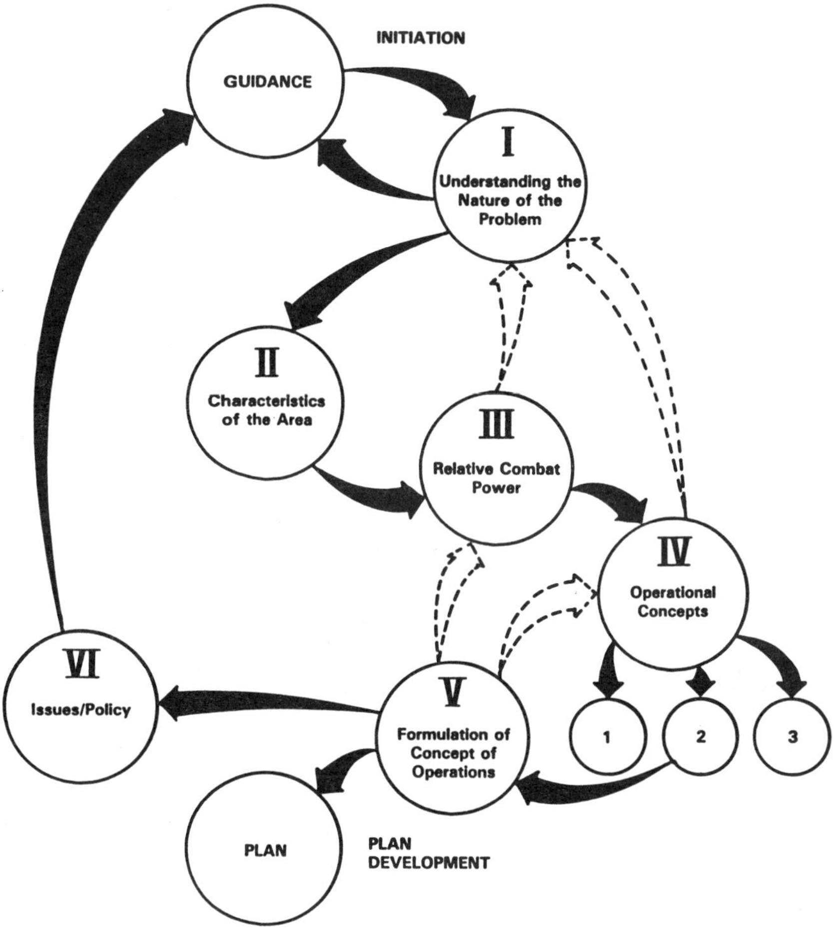

Figure 7. Iterative planning process.

for example, that regardless of the planner or the group, planning was inherently graphic and nonnumeric; that without their "props"—maps, acetate overlays and grease pencils—they simply could not plan. This finding had great implications for the subsequent design of the aids (See below).

Also, the experiments validated the planning process that we had discovered via other codified means, such as the one that appears in Figure 7 [3]. TACPLAN and INTACVAL both adhere to the process represented in Figure 7.

Functional Modeling via Storyboarding

Following the requirements analysis, we modeled the planning process (procedures, functions and tasks) in a storyboard depicting how the aids might function in operation. Storyboards are nothing more than screen displays of the functions and tasks that the aid might perform when activated by a user. However, they are extremely powerful vehicles to requirements validation, system "sizing," and man–machine interface development. We developed a number of storyboards for TACPLAN and INTACVAL [2] and presented them to our prospective users for comment and criticism. TACPLAN went through two reviews while INTACVAL went through six. (TACPLAN was a much simpler aid than INTACVAL, in concept and operation.)

The storyboarding exercise enabled us to validate requirements, identify some totally new ones, experiment with alternative man–machine interface (MMI) techniques, and, most importantly, select the analytical methods most likely to help drive the aid. After an analysis of the range of methods (computer science, decision analytic, operations research, and artificial intelligence), we opted initially for a hybrid decision analytic/artificial intelligence methodology for TACPLAN. As the research progressed, however, we moved further toward the artificial intelligence side of the hybrid [2].

Functional and Systems Architecture: TACPLAN

TACPLAN was conceived after the first round of experiments at the Army War College and after an initial study of planning.

We relied as heavily upon planning doctrine (at least as far as we had gone with it at the time) as upon expert planner judgments (as captured in the experiments at the Army War College). The Carlisle experiments, refined analytical framework, the TACPLAN functional description, and the TACPLAN concept of operation all determined the TACPLAN *delivery system*. There were also a number of overarching technical and applications guidelines that determined the configuration. Compatibility across these two data bases yielded the delivery configuration described below.

Personal Planning Aid: A common technology need these days that calls for the design and development of low cost, portable and distributed personal decision aids. All of these terms are, of course, relative. Low cost generally means less expensive than a minicomputer. Portable usually means *transportable,* and distributed usually means *accessible* and sometimes *networked*. However, *personal* has some interesting connotations beyond the literal derivatives. Personal should mean controllable and manageable. It should mean nonthreatening and supportive; and it should mean flexible and adaptive. The planning process is certainly compatible with these definitions.

Our observation of the planning process suggests the appropriateness of personal aiding. Also, there are no computational or display requirements that cannot be satisfied with a personal aid. Low cost, portable, distributed and personal suggested to us the use of a microcomputer with interactive graphic display capabilities. We selected the IBM Personal Computer (PC) with ample storage and memory capabilities (IBM PC/XT). The TACPLAN prototype is easily implemented on this microcomputer (though INTACVAL

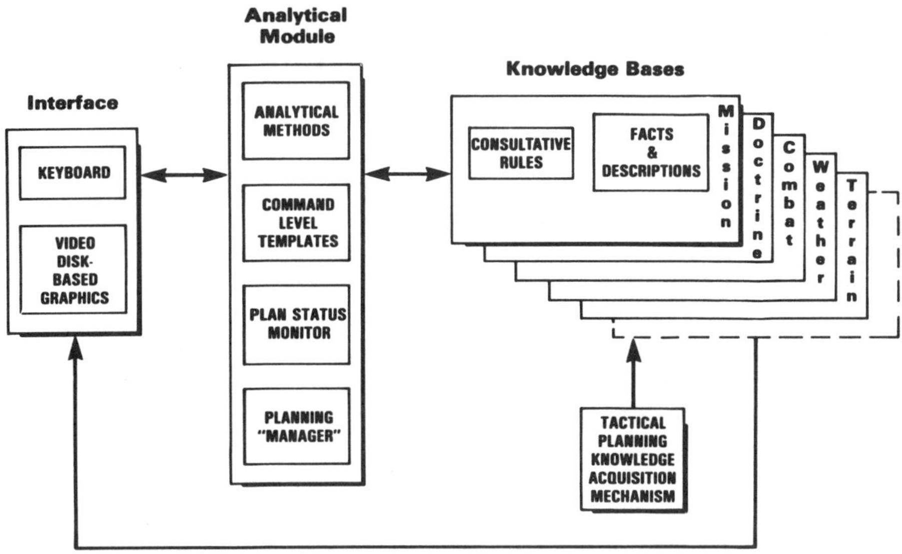

Figure 8. TACPLAN.

did indeed require the capabilities of an IBM PC/AT, which has larger memory and storage capabilities). This system gives us the option of color or black and white (monochrome) displays as well as several input options. The prototype permits keyboard and joystick/mouse input; it also has been configured to interact naturally and synergistically with the interactive graphic interface.

The Graphics Interface: As Figure 8 suggests, TACPLAN has a graphics interface comprised of an interactive video disk, video disk player, and video disk/mapping control panel. Video disk based maps are made by first filming the area that is to become the disk based map. The team already has filmed the areas necessary to implement the TACPLAN prototype. The NATO/Pact areas of Western and Eastern Europe are already on disk at multiple scales.

The video disk displays actual maps of the Corps area of responsibility at near perfect resolution. In fact, the video disk-based map is clearer and much more flexible than any conventional paper map. It permits planners to annotate what appears on the display (which is located right alongside the IBM PC/XT display) with nodes and symbols for later reference. It also enables planners to create their own personal symbols for annotation purposes. It supports decluttering, a simple operation that removes (stores or erases) annotations.

The video disk based mapping system also permits planners to "fly" around a map, which, if laid out, might cover a warehouse floor. Under joystick control, planners can fly across great spans of maps, switch off to other kinds of data (photographs, films, weapons descriptions) and even zoom in on an image or location of particular interest.

The most important capability of the video disk based mapping system, which has been linked to the (IBM PC resident) TACPLAN analytical system, is its ability to communicate symbolically with both the planner and the analytical side of TACPLAN. When a planner illustrates an enemy course of action by drawing it on the video display, TACPLAN immediately "knows" something about this course of action. For example, it may know about the terrain that might constrain or propel the course of action; it may know about the relationship between the course of action and force structure; and it may know about the relationships among doctrine, orders of battle and trafficability. Coordinates on the video disk based map are linked to analytical routines and knowledge bases stored in TACPLAN. When planners illustrate courses of action, annotate on the disk image, or erase an idea, TACPLAN knows what has happened and reacts accordingly. The nature, timeliness and depth of the reaction is determined by the analytical routines and knowledge bases resident in TACPLAN not by capabilities of the interactive video disk.

In time, it should be possible for a planner to draw a complete route across a map, or to deploy several divisions along a border, and expect TACPLAN to react immediately, consistently and substantively about the implications of the route or plan. (See the sections below on the capabilities of INTACVAL.) Once the technology that links locations on the video disk to analytical routines and knowledge bases is fully developed, then the possibilities are unlimited. Procedural and substantive rules can be invoked spatially. Alternative plans can be generated once a planner has entered his first candidate, and narrative criticism and advice can be offered—all from what the planner does via TACPLAN's graphic interface.

TACPLAN Knowledge Bases: TACPLAN's knowledge bases are primarily oriented to the relationship between terrain type, maneuverability and unit type. If a certain kind of unit tries to cross a specific kind of terrain TACPLAN "knows" whether or not the route makes sense, that is, whether a planner has driven an armored unit into a forest. When a relationship between unit and terrain type is identified, TACPLAN either remains quiet or warns the planner about a rule violation.

The rules should be used as constraints, not prescriptive planning tools. Plan formulation remains in the hands of the expert planner, not in the system's software. The concept behind the TACPLAN aid was to marry elements in decision analysis with powerful AI representation techniques. In operation, TACPLAN permits planners to make wholesale judgments about area characteristics, mission and doctrine and then subjects these judgments to what is contained in the knowledge bases.

The knowledge bases themselves were developed using a scaling technique for maneuverability and terrain. If certain classes of units are expected to move along certain terrain lines, then certain maneuverability constraints will restrict such movement. Units may be lightly, moderately or highly constrained by terrain types. A window appears on the screen of TACPLAN when a rule is violated and the outcome calculated. In operation, then, the planner would see a list of constrained units as well as their score (light/

medium/heavy). Then, he would be permitted to change his judgments or assumptions about the planning situation.

The important aspect of this capability is the direct link between the graphic interface and the knowledge bases, which can be triggered by drawing a line (with the joystick/cursor) directly onto the video disk image as a designated hypothetical course of action. TACPLAN then calculates the location of the units designated as part of the course of action and determines how constrained they are by the terrain. All of this is done instantly without the planner's intervention. In other words, all the planner has to do to implement the knowledge bases is interact with the video disk based display.

Another important aspect of the knowledge base is the way its output is displayed to the planner. Instead of complicated formal rule structures (such as IF–THEN), TACPLAN displays a simple explanation of the problem, such as "You have just moved armored units into a forest and they are highly constrained. Your mission is in jeopardy." This subtle change in the way rules and rule violations are displayed to users (from the way conventional "expert systems" display rules) is critical to the success of systems like TACPLAN. Note that our intended users are not experienced systems users nor are they methodologically sophisticated. Therefore, they need (and have a right to expect) systems to support the way they actually do business. A simple change in the way rules and rule violations are displayed to users will improve performance immeasurably.

TACPLAN Analytical Modules: TACPLAN permits the evaluation of alternative plans (concepts of operation) via several embedded multi-attribute utility routines that are linked (like the knowledge bases) directly to the graphic display system. Planners are simply asked to make judgments about the elements of tactical planning: the mission, area characteristics, blue and red capabilities, and red and blue courses of action. These judgments are not quantitative-empirical, that is, the planner does not have to come up with scores for each element of tactical planning. Rather gross judgments are required to a) feed the embedded MAU routines and b) fire the constraint knowledge bases. The aid is decidedly interactive and works with the planner to fine tune judgments and hypotheses. The aid very much assumes the value of an intelligent assistant and behaves accordingly.

System Configuration: Figure 8 suggests how TACPLAN is configured and also suggests the components of its architecture. One of the ways we were able to satisfy the low cost portable constraint was first to develop the knowledge bases symbolically and then to convert them to numeric form so that we could implement them on the PC/XT. Then, we developed conversion files for rule and rule violation explanation purposes. This approach not only supported our goals for smooth man–machine interaction, but it also permitted us to use conventional programming techniques instead of alternative recursive/search techniques. Our primary motivation for this approach was not to avoid symbolic processing, but to make certain that the system would run on the least expensive configuration possible.

We were pleased with the mileage we were able to get from off-the-shelf software and software utilities. The Lattice C windowing system, for example, served our purposes very well. Also, we were able to exploit existing Perceptronics video disk control software. The TACPLAN project demonstrated that it is indeed possible to utilize off-the-shelf software productively and thereby to cut design, development, and especially programming costs dramatically.

A TACPLAN Session: TACPLAN walks through the planning process by asking the planner a series of questions about his planning problem. First, it asks him to describe his mission. Then, it asks him how much time he has to solve the planning problem as well as the problem's geographic location.

Then, it asks the planner to describe the area characteristics in terms of advantages and disadvantages to the planner. He is required to decide if friendly or adversary forces have advantages in aspects of terrain by simply checking blocks that ask: "favor NATO?/favor PACT?" The embedded MAU routines test these judgments against what is stored in the knowledge bases and if no violation occurs, TACPLAN remains silent. However, if a violation does occur, then a window appears to the planner that describes the violation.

The planner has to assess relative positions for area characteristics, combat capabilities and courses of action. As the planner makes these judgments, the knowledge bases are checked for violations. More importantly, the course of action (COA) assessment is done completely on the video disk display, where the planner actually draws hypothetical courses of action and then waits for TACPLAN to respond. These COAs can be labeled and stored for later comparison. They can be overlayed for highlighting and comparative purposes. The ability to write directly (via a digital overlay onto the video disk based image) on the graphic display permits the planner to recreate many of the same capabilities with which he or she is so familiar.

After the course of action is selected, TACPLAN provides a summary of what the COA and overall concept of operation assume about the situation. As is true throughout the TACPLAN aided planning process, the planner is free to disagree with the system's judgments.

TACPLAN is a simple assistant in the planning process. Via some rules about the relationships among unit type, terrain and maneuverability, it advises and constrains planners about what they can and perhaps should not do. It places a great deal of input burden on the planner, requiring him to make many (nonnumeric) assessments about opportunities and constraints.

At the same time, TACPLAN successfully demonstrates the utility of "graphic equivalence" and the benefit to be gained from a flexible graphic representation of the planning environment. It also demonstrates the utility of hybrid analytical methods. INTACVAL, on the other hand, expects less from the planner and tries to make the leap from "assistant" to "associate."

Functional and Systems Architecture: INTACVAL

INTACVAL was conceived after TACPLAN was tested in prototype form with actual planners. We discovered that the "aiding paradigm"—or the role

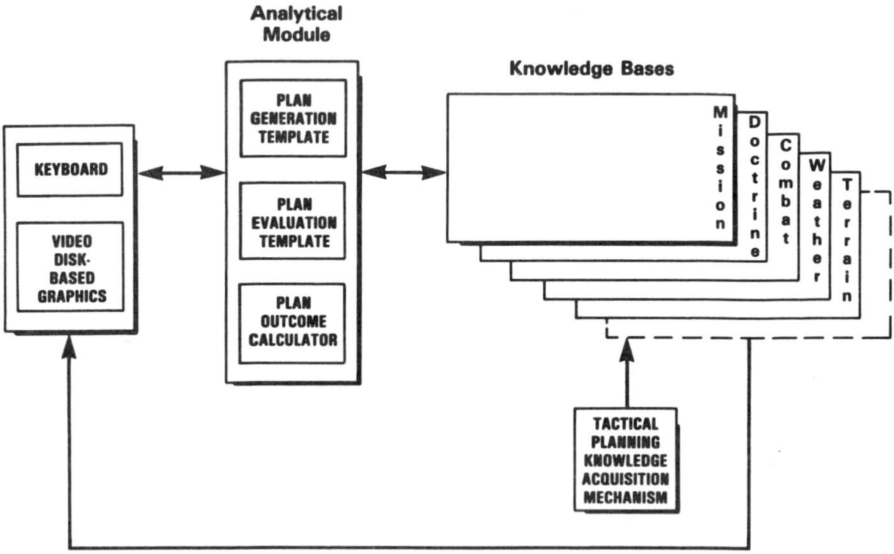

Figure 9. INTACVAL.

that the aid should play—had not been specified well enough; hence INTACVAL. The aid itself was built upon TACPLAN, but is fundamentally different in operation and function.

Functional Components: INTACVAL, as Figure 9 suggests, comprises a planning interface, knowledge bases, plan formulation and evaluation templates, and a plan outcome, or battle, calculator. Note that INTACVAL is an interactive aid, not an automated expert system. The planner is intended to be an integral part of the aiding system.

Knowledge Bases: INTACVAL's knowledge bases are not comprised essentially of rules (like TACPLAN's). In fact, the content and depth of its knowledge bases are determined by the depth identified within the overall object–attribute–value structure that we have adopted [4].

Objects–attributes–values, or O-A-Vs, enable a great deal of information about planning to be encapsulated in a flexible yet formal, structure. As Figures 10 and 11 suggest, O-A-Vs represent a technique for knowledge representation that is a derivative of larger frame and network based approaches.

In the planning domain, O-A-Vs comprise planning objects that correspond to the familiar elements of tactical planning. These objects have attributes which, in turn, have values. (See figures.) Thus, it is possible to identify a set of objects, attributes and values that pertain to all of the elements of tactical planning and then to look for patterns among the values to determine which, in effect, "go" together. Various combinations of values relate to specific concepts of operation that, in turn, can be used to guide the planning process, check planner judgments and even generate strawman concepts of operation.

O-A-Vs lie at the heart of the INTACVAL knowledge base. They are used to write rules about value combinations, to display planning logic to planners,

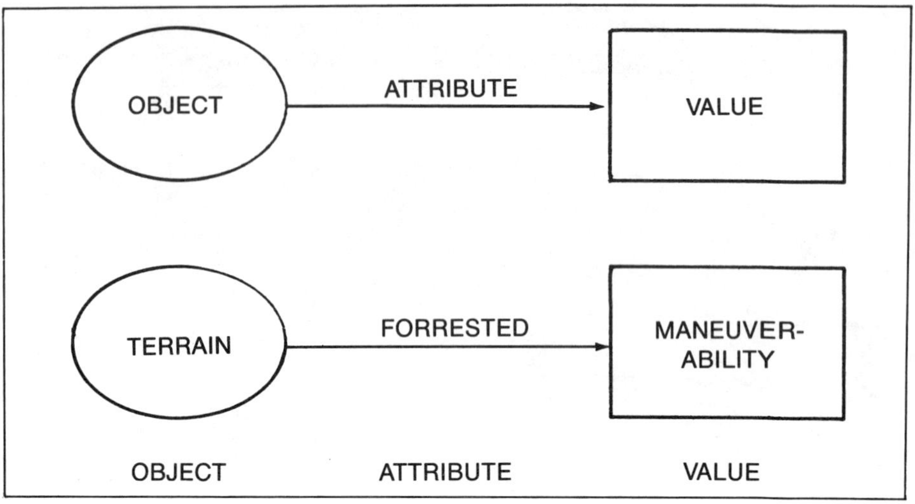

Figure 10. Basic object–attribute—value triplet.

and to update what is learned about tactical planning over time. A large O-A-V triplet structure has been built by International Information Systems, Inc. (with help from expert planners and doctrine). A list of the objects and attributes in the structure appear in Figure 12. The values, along with definitions, all appear in [4].

- *OBJECTS* ARE PHYSICAL ENTITIES OR CONCEPTUAL ENTITIES (e.g., "UNIT" OR "PLAN")
- *ATTRIBUTES* ARE GENERAL CHARACTERISTICS OR PROPERTIES OF OBJECTS (e.g., "ARMORED" OR "READY")
- *VALUES* ARE NATURES OF ATTRIBUTES (e.g., "MANEUVERABILITY")
- CAN BE USED TO DIRECT INFERENCES IN RULE/NETWORK STRUCTURES:

	OBJECT	ATTRIBUTE	VALUE
IF:			
	Terrain	Forrested	Low-Maneuver
	______	______	______
	______	______	______
	______	______	______
	UNIT	ARMORED	CONSTRAINED

Figure 11. O-A-Vs explained.

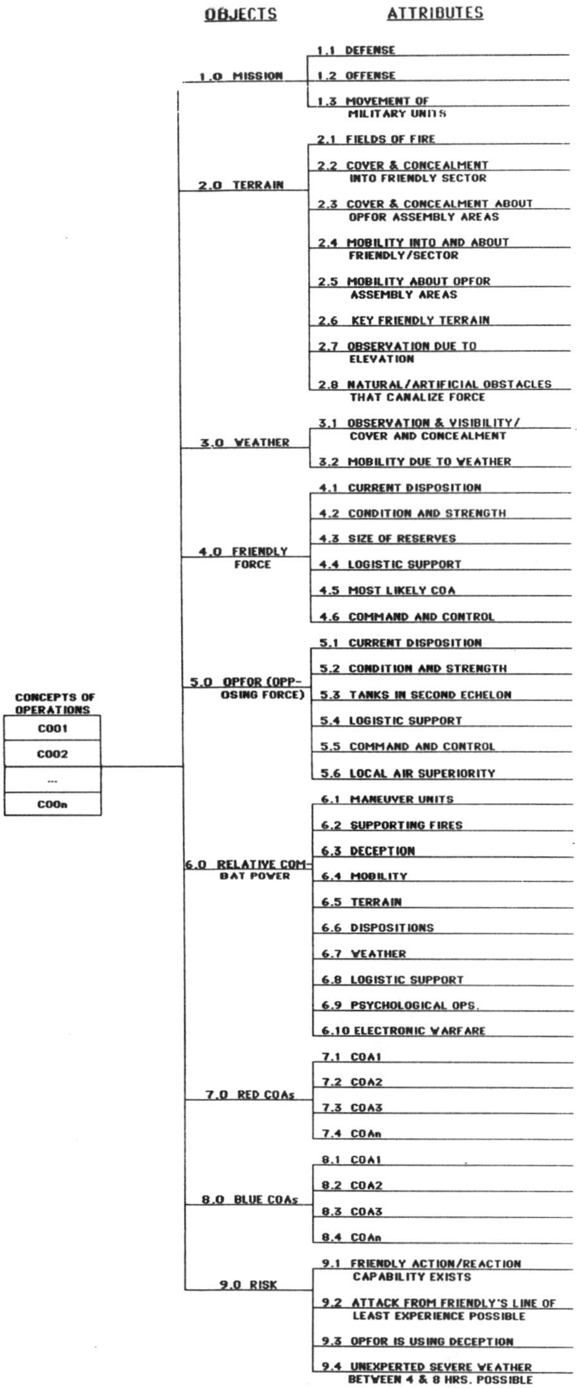

Figure 12. Objects and attributes from INTACVAL knowledge base.

The Tactical Battle Calculator: INTACVAL's knowledge bases will support the development of alternative plans and their evaluation. The aid permits a planner to formulate a concept of operations and then to activate a tactical battle calculator that will play out the plan vis-à-vis the terrain, the force structure, the mission, logistics and, of course, the adversary. This battle calculator is model *and* knowledge based. It is model based to the extent that there is a set of battle patterns resident in the battle model; it is knowledge based to the extent that the model accesses INTACVAL's knowledge bases in order to determine the plan's most likely outcome.

INTACVAL displays the processes by which alternative plans are executed both graphically and alphanumerically. The graphic display consists of digital overlays onto a video disk based map image, while the alphanumeric display presents INTACVAL's reasoning. Thus, the planner is able to "see" how the plan does and then receive guidance about why it is likely to succeed or fail. (See the section below on how INTACVAL actually behaves in operation).

The battle calculator does not "decide" which plan is best and which plan is worst; instead, it "checks" the plans to see if they violate any of the constraints and prescriptions in the knowledge bases. It does not numerically rank order the alternative plans, though it will suggest how the plans compare with one another along consistent logic/reasoning paths. For example, when a plan is "run" through the battle calculator it will be assessed according to what INTACVAL knows about tactical planning. When several plans are run through the calculator INTACVAL will permit the comparison of the plans according to how well or badly they each did in all of the appropriate categories, for example, terrain, unit types, relative, and absolute combat capabilities, and doctrine. Thus, it is possible for the planner to make meaningful comparisons across the alternatives. Some techniques for facilitating comparison include listings of "strong" and "weak" plan components, side-by-side comparisons of plan features, and the capability to respond to queries about individual or several plans.

Man–Machine Interface: The primary interface device is the video disk based map display. Through this display, the planner is able to build tactical plans (by actually moving unit symbols across a map), label them, overlay them upon each other, make notations about their features and "see" the plans executed in real time animation.

Use of the video disk based map display may be justified by several experiments conducted over the years regarding the use of spatial/graphic versus alphanumeric information processing and our experience with the TACPLAN prototype [1]. Some tasks are inherently spatial, graphic and nonnumeric, while others are more naturally solved through mathematical or quantitative means. Our research suggests that tactical planning is inherently graphic and non-numeric. When planners plan, they move icons, refer to illustrations, draw courses of action and argue via references to pictures, graphs and lines. They do not perform mathematical calculations or convert their judgments into numeric form; hence, the use of an interface device capable of supporting nearly all aspects of the non-numeric problem solving process.

The issue of using video disk based versus computer generated imagery (CGI) displays is also worth noting. In many applications, CGI displays are appropriate, especially when one-to-one realism is unnecessary. However, in those applications where it is important to present near-perfectly real information, then the video disk alternative makes sense. What about tactical planning? Without question, tactical planners prefer the actual map image to a computer generated one. Real planners are trained on real maps, understand real maps and solve problems efficiently with real maps. Perhaps more importantly, INTACVAL's video disk based interface actually improves upon conventional (paper) map displays by permitting the planner to scroll around the map, zoom in or out for different scale views of the area and separate out the features of the map (terrain, rivers, cities) for "decluttered" viewing and problem solving. Since INTACVAL's interface permits digital annotation of images, lines and text directly onto the map image, planners are able to annotate alternative courses of action and store them for comparison.

There is also the issue of cost–benefit. Unless the application specifically calls for—or can tolerate—CGI based maps, there is every reason to opt for the video disk based map simply because of the enormous cost of CGI maps relative to their video disk counterpart. CGI systems are also expensive to maintain and field.

The arguments about flexibility are also worth noting. In the not too distant past, video disks were criticized for their lack of flexibility, but since the establishment of production centers throughout the United States, the criticism is no longer valid. Video disks now can be produced in a matter of days instead of the months that were previously required.

Finally, the age of read/write video disks is all but upon us. Within a few years, it will be possible to add or delete images from the same disks. When this technology matures, the last functional shortcoming of video disk based technology will disappear.

INTACVAL, like TACPLAN, has dual screens. One display presents the map(s), while the other presents the alphanumeric aspect of the planning process (for example, displays of reasoning); however, actual development of alternative courses of action will all be done using the interactive video disk.

Actual manipulation of the map and related images is accomplished through a trackball integrated into INTACVAL's keyboard. Nonvideo disk input is by way of a conventional keyboard with graphic and nongraphic menus organized around the planning process.

System Configuration: INTACVAL is configured much like TACPLAN, except it resides on an IBM PC/AT with enhanced graphic display capabilities. It also has a different menu structure from TACPLAN's, relying on a pull-down/windowing type.

INTACVAL was programmed in C (like TACPLAN) and makes heavy use of interactive graphics to display the components of the knowledge base, avenues of approach and the like. INTACVAL also has a modified animation capability.

An INTACVAL Session: INTACVAL was designed to support corps commanders and their G-3 (Operations) staff. G-2 (Intelligence) input to the process is simulated in INTACVAL. The commander and his staff thus receive information about area/terrain, adversary disposition and strength, and the planner is expected to process it into a concept of operation, or plan.

When all of the planning "data" has been received, the aid generates a series of strawmen options for the planner to consider. Note that this represents a radical departure from the aiding paradigm in TACPLAN. The planner is then free to query INTACVAL about why it recommended the strawmen that it did. The O-A-V triplets become very important in this iterative process. Recall that the triplets are used to determine the relationships among specific courses of action and values in the structure. After receiving all of the planning data, INTACVAL searches for pattern matches among the values until it finds one. It then recommends the course of action that its rules (firing on the O-A-Vs) tell it is valid. The planner then can ask INTACVAL to display the values that led to the plan's generation and acceptance. If he does not like the values in the pattern, he can change them immediately to what new rule might be triggered by the change. This on-line knowledge base interaction capability links the planner directly with the essence of the aid, and supports knowledge based iteration on strawman candidate plans.

As the candidates are displayed, the planner also can ask INTACVAL to play out its implementation quickly. This modified war gaming capability gives the planner another look at the plan; INTACVAL also calculates a kind of balance sheet for how well (or badly) the plan did vis-à-vis the prescribed mission.

The planner's primary function in the INTACVAL aiding process is to iterate on strawmen generaged by INTACVAL, to query INTACVAL about how and why it recommended what it did, and to play out candidates in real time for further inspection. In spite of the system-generated solutions, INTACVAL is not an expert system. It is intended to amplify planning expertise, not to replace it. It is a constraint checker and an "associate"; thus it performs like TACPLAN but is far more intelligent. Final judgments about the acceptability of plans lie with the planner.

Conclusion

INTACVAL and TACPLAN represent attempts to aid a complicated analytical process. They are both prototypes, far from operational application. At the same time, they successfully demonstrate the integration of advanced methodology and microcomputing for support purposes. They also suggest that the systems design methodology used—especially with its emphasis on requirements definition and functional modeling through storyboarding—can be expected to yield continued dividends over time.

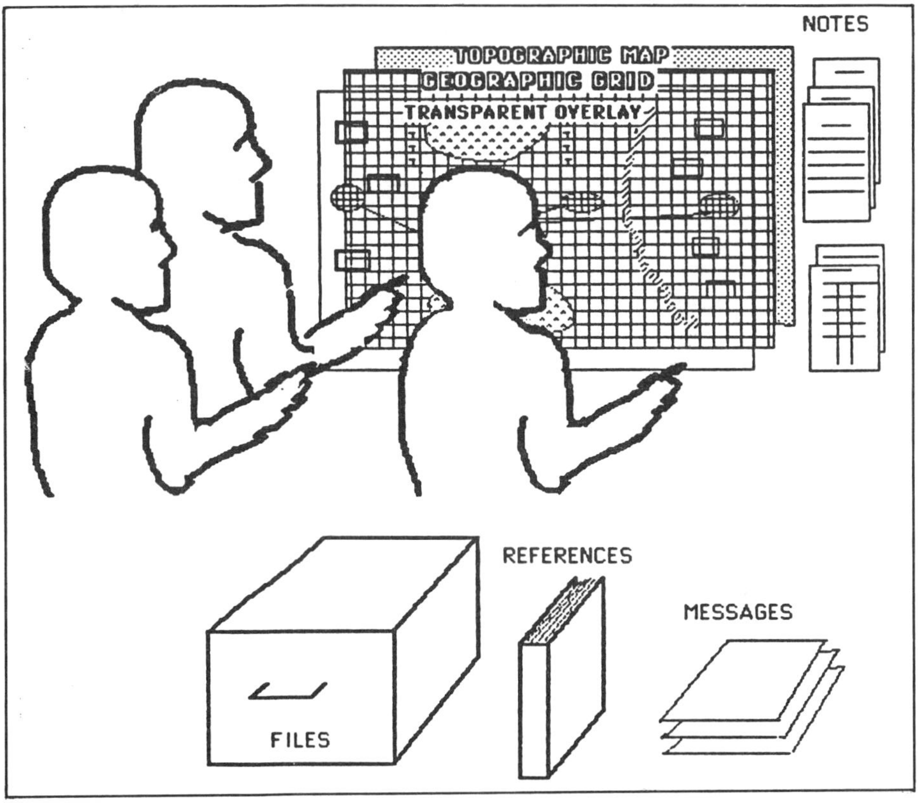

Figure 13. Manual planning process.

The objective of our work can be seen graphically in the final two figures. We clearly are trying to move from "manual" planning to computer aided planning—without disturbing or distorting the process itself. While we may have taken some small steps in that direction via TACPLAN and INTACVAL, the real work lies ahead.

Manuscript received May 20, 1986; July, 1, 1986. This work was supported in part by the Army Research Institute and the U.S. Army's Communications-Electronics Command, and under subcontract (to International Information Systems, Inc.) from Perceptronics, Inc.

References

1. S. J. Andriole, *The Design and Development of an Intelligent Planning Aid: The TACPLAN Prototype*. Marshall, VA: International Information Systems, Inc. 1986.
2. ———, *INTACVAL Storyboard (Version IV)*. Marshall, VA: International Information Systems, Inc., 1986.
3. Army War College, *Sound Military Decision-Making*. Carlisle, PA: Army War College, 1984.
4. G. W. Hopple, *Alternative Defensive Plan Generation and Evaluation: Object-Attribute-Value Knowledge Base*. Marshall, VA: International Information Systems, Inc., 1986.

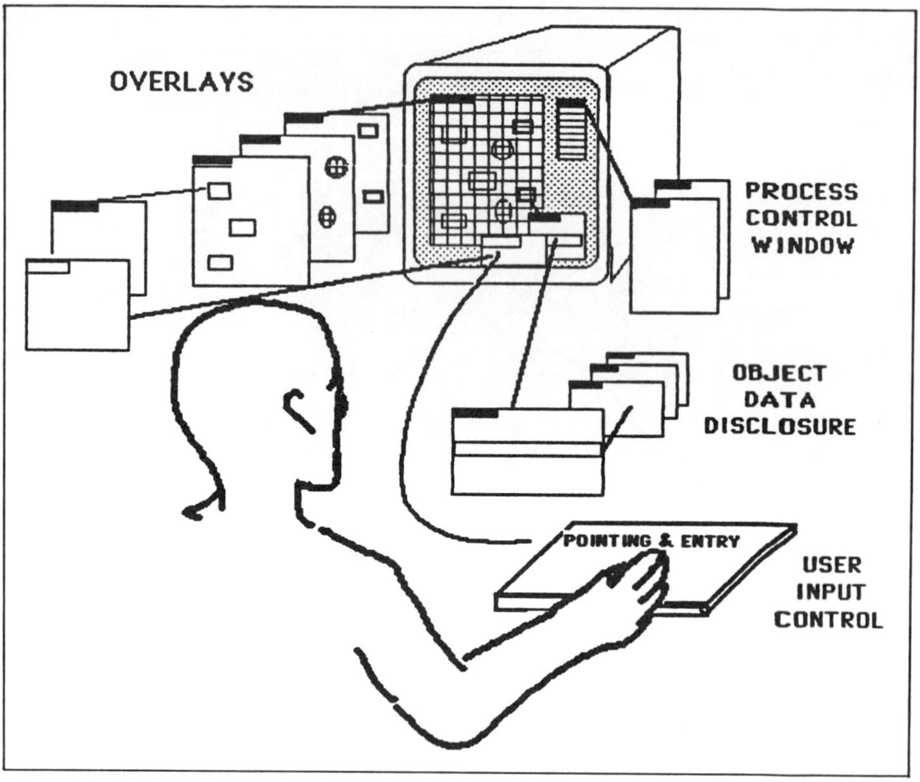

Figure 14. TACPLAN and INTACVAL aided planning process.

Data Fusion and Decision Support for Command and Control

Edward L. Waltz and Dennis M. Buede

Introduction

An important development in the automation of command and control (C^2) functions is the implementation of systems which collect and merge information from diverse sources into an accurate representation of the tactical situations. They also must provide the commander with recommended alternative military actions and expected outcomes. These systems, often called battle management systems or combat decision aids, require the interaction of the following two primary functions.

1) *Data fusion* collects information from a variety of sensors and sources to develop the best possible perception of the military situation. This situation includes the orders of battle of friendly and hostile forces; locations of material, equipment, weapons, and their movements; events of tactical interest; and intelligence data as it relates to past, present and predicted future behavior of the enemy. The fusion process includes the collection, association, aggregation and merging of data to create and display current (and past) situations.

2) *Decision support* performs the creation and quantitative evaluation of alternative estimates of the real military situation and the responses available to the commander. This evaluation process includes the consideration of

**This article was originally published in IEEE* Transactions on Systems, Man and Cybernetics, *Vol. SMC-16, No. 6, November/December 1986.*

Parameter	ASW	TACAIR	Land Battle
C^2 command level	Battle Group Commander	Air Component Commander	Division Commander
Surveillance volume sensor systems	2000 × 2000 km 4 surveillance aircraft 12 ASW ships 2 ASW subs	800 km² × 20 km 6 airborne warning system aircraft 50 ground radars 100 fighter A/C	500 × 200 km 4 surveillance aircraft 20 outposts 6 ground radars 1 RPV battalion
Targets in track (max)	100–200	500–1000	100–250
Reports/min	1000–5000	50, 000–100, 000	100–500
C^2 decisions/min	1–5	25–50	10–25

Table 1. Typical information requirements in three tactical C^2 areas.

target selection options, resource allocations and employment timing for both autonomous and coordinated force actions. Candidate response options and predicted results are provided to the human commander requiring modeling and evaluation of the effects of optional actions that can be taken.

As tactical and strategic warfare has increased in speed, complexity and scope (Reference 1), the requirements imposed on data fusion and decision support techniques have exceeded the capabilities of traditional manual techniques (plot boards, cross-check lists, display overlays, etc.). The increase in information flow from multiple sensors and reduced reaction time requirements have dictated the automation of the collection and merging of information as well as the generation of candidate actions and modeling of their expected outcomes.

Table 1 depicts the magnitude of data and decisions required to be processed by the C^2 function in three tactical warfare areas. In each case, the volume of information to be collected, stored and acted upon imposes a formidable task for the area commander and staff. Automation of the fusion of data, and quantitative evaluation of alternative actions are necessary tools for the decision makers, the human commanders.

A Model of the C^2 Fusion/Decision Support Functions

The paradigm for C^2 decision making proposed by Wohl (Reference 2) includes four elements as depicted in Figure 1.

1) Stimulus: The initiation of the decision making process is the data fusion function which provides information on the current situation with associated uncertainties. These uncertainties result from lack of sensor coverage, ambiguities in report, conflicts in reports or inaccuracies in measured data.

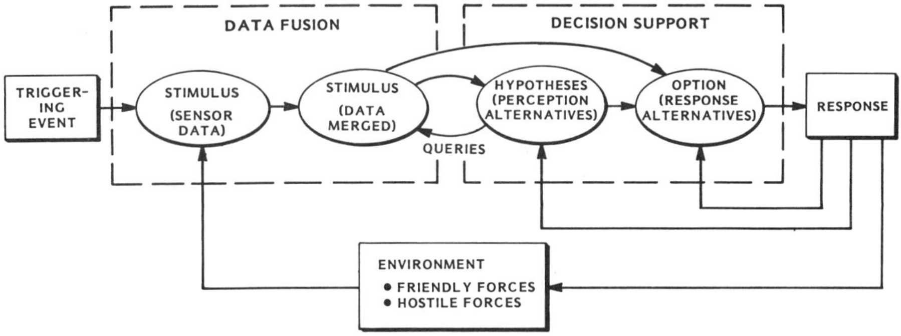

Figure 1. Dynamic relationships in tactical decision process.

2) Hypothesis: Because of uncertainties in the correlated sensor data, the data fusion and decision support processes must jointly create hypotheses (perception alternatives) that are the best candidate real world military situations that explain the fused data.

3) Option: For each hypothesis, options (response alternatives) are created and expected military results are modeled. These options and results are provided to the human decision maker for selection and action.

4) Response: Finally, the selected military response (action) is taken by the forces under command. This, in turn, creates new stimuli to the overall system.

The feedback and feedforward loops in the Wohl model emphasize the complexity of real world, dynamic C^2 systems. Figure 1 adds another feedback loop to the Wohl model; the ability of the hypothesis process to query the data fusion process to search for data that may support candidate hypotheses under consideration. It is this interface between data fusion and decision support that must be carefully considered in the design of both functions to ensure that all available information is available to support timely and accurate decisions. This interface also must be able to couple the data fusion and decision support processes as current sensor systems, characterized by hard decision signal processors and visual displays, giving way to soft decision data fusion techniques. The human interfaces will remain at the outputs of both functions: 1) combined sensor data will be displayed to the commander as the perception of the military situation, and 2) response options will also be enumerated to the commander.

Benefits of Data Fusion/Decision Aids in C^2 Systems

In addition to the obvious improvement in the rate of information processing accrued by automation, several other benefits are provided by integrated data fusion/decision support systems. Table 2 summarizes the inherent ben-

Improvement Category	Specific Characteristics	Tactical Benefits
Improved spatial/ temporal coverage	extended spatial coverage provided by multiple, overlapping sensor fields-of-regard	increased probability of threat detection
	increased detection probability due to synergy of multiple sensors	enhanced situation awareness and mutual support
Improved measurement performance	reduced ambiguity in data with multiple measurements	target acquisition, track and ID at longer ranges
	improved detection, tracking, identification when data is integrated from multiple looks at the target	accurate discrimination of target type and military capability (threat)
	reduced uncertainty in data when multiple independent measurements are integrated	
Enhanced operational robustness	redundancy of multiple sensors may increase mission reliability	reduced vulnerability to denial of a single sensor's data
	degraded (sensor subset) modes available	availability of alternate sensor modes to reduce observability
	robust performance provided by multi-spectral sensors	

Table 2. Benefits provided by multiple sensor systems over single sensor systems.

efits of multiple sensor systems over traditional single-sensor approaches. The fusion of multiple sensors can include multiple different sensors (such as, fire control radar, IR search and track, electro-optical imager, and passive electronic support measures) or multiple, common sensors (such as, surveillance network of identical radars). In both cases, the use of multiple sensors may enhance overall C^2 performance in terms of spatial/temporal coverage, measurement performance and operational robustness. These performance benefits, of course, must be weighed against the additional cost, complexity and interface requirements introduced in any given application.

The benefits of automated decision support extend beyond the obvious benefits of rapid evaluation of multiple complex alternatives. Objective evaluation and consistent performance that is independent of battle stress is provided to the commander. As addressed by Wohl (Reference 2) these objective outputs may require output tailoring to suit the decision making style of individual commanders.

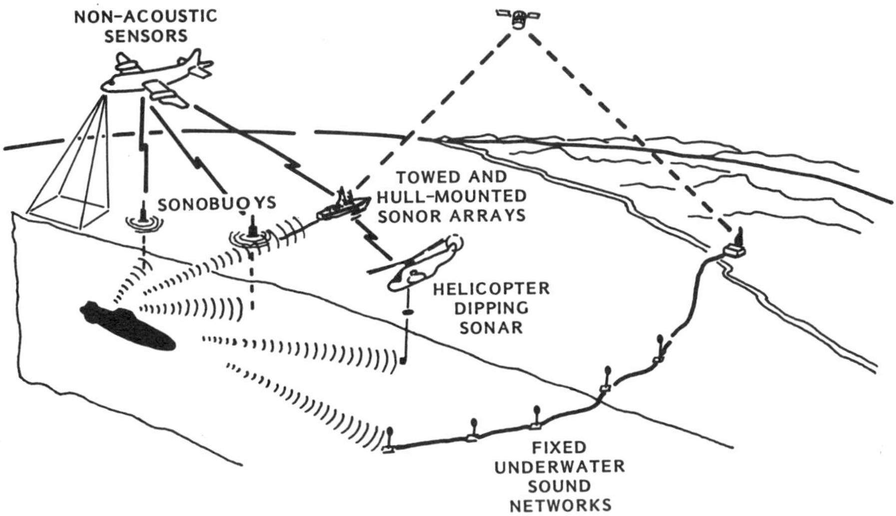

Figure 2. Antisubmarine warfare (ASW) C² mission.

Data Fusion/Decision Aids in C² Systems Applications

Virtually every C² application employs a variety of sensors (e.g., radar, sonar, IR/EO detectors, seismometers) and sources (e.g., humint, photoint, data-linked reports) to collect information necessary to develop a perception of the military situation. In each C² application, the fusion requirements may vary widely because of the unique characteristics of sensors, collection system and target or event behavior. The primary mission-unique parameters that characterize each application include the following:

1) Target quantities (number of targets to individually detect/identify/track) and categories (level of distinction between targets) vary for each application. These factors influence the size of the target data base and the processing requirements to update the data base.

2) Detection and decision rate requirements are a function of the relative temporal behavior of targets (speed, rate of fire, rate of activities, etc.) and reaction time requirements of the C² system. These requirements dictate the necessary update rates for sensors to acquire and track target behavior.

3) Quantity and variety of sensors greatly influence the architecture and processing requirements on the data fusion algorithm. Sensors may be collocated on a single weapon platform or they may be distributed. Differences in sensor characteristics (such as resolution, FOV, range, target revisit rates) influence the data association and sensor management functions.

Figures 2 through 4 characterize three different C² applications that employ automated data fusion techniques to combine information from multiple sensors and sources. The diverse requirements and characteristics of each mission area are highlighted in the figures.

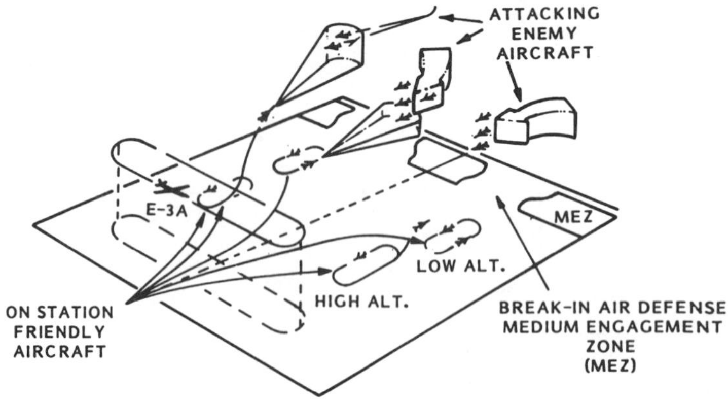

Figure 3. Tactical air warfare (TACAIR) C^2 mission.

Antisubmarine Warfare

The antisubmarine warfare (ASW) mission is crucial to the survival of the Navy task force. The coordination of numerous sensor systems is necessary to ensure that hostile attack submarines do not enter the outer defense zone undetected or unidentified. Aircraft (carrier based and land based) provide the over-the-horizon ASW sensing capability of the fleet, using air-dropped sonobuoys and nonacoustic sensors (References 3 through 5) to detect and track submarines. (See Figure 3.) Towed array and hull mounted sonars of ASW destroyers, deployed at the perimeter of the task force, form the outer surveillance barrier.

The task force commander also must rely on probable target tracks relayed by other naval forces and ground based centers to predict threats approaching the outer perimeter. These reports, combined from fixed underwater sonar sensors (deployed at necessary passage points), visual sightings, periscope detections, emitter detections, national intelligence sources, and other task

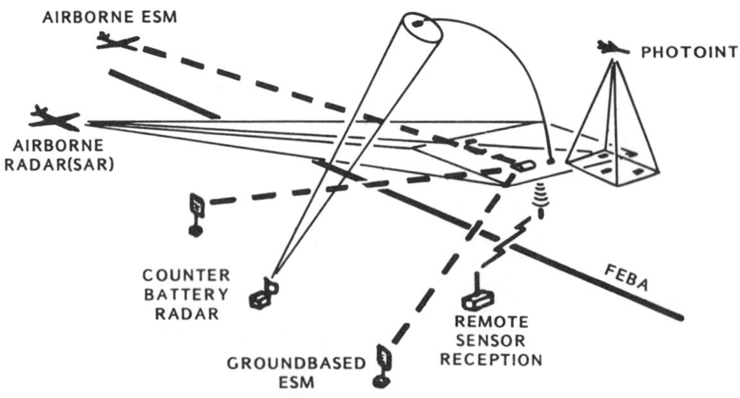

Figure 4. Landbattle battlefield C^2 mission.

force detections are used to cue the local task force's sensors. The fusion of data from these external sources, from organic sensors detecting possible coordinated hostile force movements, and from the task force's own ASW sensors is a difficult task. It requires coordination of a wide variety of platforms to detect, track and identify elusive targets hidden in such a vast surveillance volume.

Tactical Air Warfare

The defensive counterair mission depicted in Figure 3 shows fighter aircraft flying "lanecap" patrol with long-range surveillance support from airborne warning and control system (AWACS) aircraft (Reference 6). Two sources of data are available for data fusion on each fighter aircraft: 1) ownership sensors may include fire control radar, infrared search and track, radar/IR/EO warning sensors, identification friend or foe (IFF), and EO (electro-optic) sensors and 2) tactical data links provide the ability to pass target tracks and identities from AWACS to fighters or fighter-to-fighter. The fighter requires a moderate volume (10–30 tracks) data fusion/decision support capability to rapidly acquire, track, identify and hand off targets as well as to recommend attack options to the pilot. Such air battle management systems require the ability to operate independently or in a cooperative mode, in which cooperating fighters share sensor data and coordinate a multiple target attack. In this case, the data fusion/decision support function is effectively distributed among the air battle management algorithms, located in separate fighters, interconnected by the tactical data link.

Landbattle Battlefield Warfare

Only the land battle portion of the total airland battlefield scenario is depicted in Figure 4, emphasizing the sensors and fusion necessary to detect and locate ground based targets. These targets include C^2 modes, weapon systems (stationary and moving), troops, transportation systems, material and geographical features (e.g., bridges, roads, railways and terrain features influencing troop movement). Detection and identification of these targets are achieved by a variety of phenomena of which the most significant are movement, RF/EO/IR emissions, and weapon engagement related events (battlefield targets are characterized as "movers," "emitters" and "shooters," respectively). The wide variety of sensors employed (Reference 7) include the following:

1) Airborne radars detect and locate stationary and moving targets (MTI radar) at slant ranges extending well beyond the Forward Line of Own Troops (FLOT).
2) Airborne passive emitter detection/location systems provide the location of RF emitters (e.g., air defense and surveillance radars, communication nodes) beyong the FLOT.
3) Ground based counter battery radars track incoming rounds and compute the source weapon location.

4) Airborne remotely piloted vehicles (RPVs) as well as manually piloted reconnaissance aircraft provide targeting beyond the FLOT using TV/IR imaging (photoint).
5) Remotely deployed ground sensors (magnetics, seismic, acoustic) report the detection of ground events, troop movements, formation and assembly activities, beyond the FLOT.

In addition, verbal reports from field commanders, outposts, patrols, and intelligence reports from the interrogation of prisoners or exploitation of captured communication, are used to develop a comprehensive picture of the battlefield situation.

Table 3 summarizes the differing characteristics of these three areas of warfare and highlights the similarity in the basic requirements imposed on the C^2 data fusion/decision support functions.

Use of Sensed Information

Data fusion uses a combination of sensors and sources to collect information on the tactical situation. This information includes: reports of events or target detections, target tracks (location plus derivatives to protect future behavior) or factual statements. This data is used to detect, locate and discriminate between tactical targets and events.

Discrimination between targets of a similar type or discrimination between targets of different types (i.e., identification) can be achieved by any one or a comparison of sensed variables. The discrimination process can infer identity by measured target attributes, target behavior or contextual clues provided by multiple sensors.

Directly measured features include attributes of the target (spectral signatures, spatial characteristics) or of phenomena that can be associated with the target (effects on the environment, secondary effects or events linked to the target). These attributes are measured by the sensor directly or result from preprocessing operations (filtering, integration, clustering), which refine and combine the raw measurements into a single attribute.

Behavior of the target includes temporal behavior (velocity, acceleration, maneuvering, direction of travel, etc.) and tactical activities (emitter status/mode, and hostile or friendly acts such as jamming, deception or engagement actions). In most cases, it can be used to discriminate an identity only when performance of a target exhibits unique characteristics. Doctrinal procedures, however, can be established to permit behavior to be independent of target performance and to allow unique target discrimination. Air corridors and restricted zones are examples of procedures used in tactical air warfare to provide discrimination of foes by restricting the behavior of friends. These procedures must be maintained secure, of course, to achieve friend-foe discrimination.

Context includes the total spatial, spectral and temporal situation in which the target (or event) is found. Spatial context includes location and relation-

Characteristics	Antisubmarine Warfare	Offensive/Defensive Tactical Air Warfare	Battlefield
Targets	submarines submersibles mines	aircraft missiles	vehicles (movers) C^2 centers (emitter) air defense units missile, artillery units (shooters)
Events	surface activity torpedo launch missile surface broach	aircraft maneuvers missile engagements ECM (jamming)	weapon firing nuclear bursts engagements troop assembly, movement
Sensors	sonar signature —land based —towed array —hull mount —sonobuoy —helo dipping passive ESM satellite surveillance nonacoustic sensors —Magnetometers —IR/EO visual sightings	radar —fighter fire control —airborne surveillance IR search/track IFF passive ESM EO imager visual sightings	airborne surveillance —SAR/SLAR —passive ESM emitter location —photorecon recon teams counter-battery radar ground-based ESM (DF) remote ground sensors —acoustic —seismic RPV —EO/FLIR —ESM
Sources	intelligence national sources	pre-flight intel tactical data link —passed tracks —friend locations	intelligence national sources
Command/ control decisions	submarine —which tactical approach —when to initiate tactical approach —when to engage	DCA —air order of battle aggregation —allocation of A/C —assignment of A/C to raids —weapons choice OCA target prioritization OCA route planning C^3CM planning	central battle —which avenue of approach second echelon attack —where, when —which targets —with what weapon

Table 3. Characteristics of three C^2 mission areas.

Discriminant Categories	Tactical C^2 Mission Areas ASW	TACAIR
Directly measured features	acoustic signatures magnetic signatures diesel fume spectra measurable phenomenologies —surface wave effects —biological effects —atmospheric effects —nuclear effects	radar signatures IR signature electronic emissions (PRF, pulse characteristics) IFF replies acoustic signature imagery shape/size signature (IR/EO)
Behavioral characteristics	speed sustained speed depth maneuverability hostile act (cruise missile launch, torpedo launch) communication activities with hostile forces	velocity/altitude envelope maneuverability operation with an air corridor (friend) or a restricted volume (foe) hostile act (ECM, missile engagement against a friend, etc.) flight path tactics executed
Contextual information	origin of mission (homeport) or ports of call location relative to friendly surface vessels	origin of flight from hostile or friendly base flight path ECM activities relative formation with other knowns

Table 4. Examples of measurable discriminants for antisubmarine and tactical air warfare missions.

ship to other targets or participants in the scenario. The spectral context may include sensed attributes in relation to other activities in the same spectrum (e.g., communication exchanges, countermeasure activities, levels of noise). Temporal contextual information includes the relative timing of sensed events or target activities and their implications of coordinated group behavior.

Table 4 provides typical examples of measured features, behavioral characteristics and contextual information that can be used to discriminate submarine and aircraft targets in the ASW and TACAIR missions, respectively.

Data Fusion/Decision Support Architecture

The functional elements of the data fusion and decision support subsystems (DFS and DSS, respectively) of a generic C^2 system (Reference 8) are depicted in Figure 5 and closely follow the Wohl model. A variety of sensors collect data from the combat environment and pass them on to the DFS. The sensors may provide different kinds of data: 1) detections (reports) may include time-tagged events or targets with available location or identity information; 2) tracks include temporal models of target behavior (position plus time history or derivatives of position) and may include identity information; and 3) statements of fact are typically provided by intelligence sources and usually are related to current or predicted (probable) behavior of hostile forces.

The raw sensor data must be associated, that is, correlated in space and time, to form common files for each sensed event or target. As these reports are accumulated in time, temporal tracks are maintained to model the dynamic behavior of each target. These tracks of targets, histories of events/detections, and statements of fact form the situation data base that is used to develop a perception of the combat environment.

Numerical data combination algorithms and symbolic reasoning processes refine the data in the data base to a model of the real world conditions. This model includes:

- locations of all targets and events (and the uncertainty associated with each),
- identification of targets (if known),
- predicted behavior of targets (tracks),
- situations of military importance resulting from individual target locations, behavior (e.g, threat weapon systems in attack positions) or aggregate behavior of multiple targets.

The DFS passes this model to the DSS for further analysis. The modeled situation includes, of course, uncertainty arising from limitations in the sensor measurements and coverage. As a result, this uncertainty allows for multiple hypotheses concerning many of the elements of the situation model (e.g., target locations, identities, behaviors). The situation assessment function refines these hypotheses using doctrinal templates and probabilistic inference to select a finite set of important alternatives for analysis. The set of possible situations, plus uncertainties of the state variables are passed to the alternatives analysis function to create, analyze, and rank alternative courses of military action.

The human commander (pilot, division commander, task force commander and tactical air control commander) views the situation display (range/azimuth display, map display, etc.) with symbology overlaying target identities, velocity vectors, weapons envelopes, and other information relevant to each target. In addition, the commanders may view the alternative courses of action and expected results. The commander's action commits the weapon systems under command and completes the feedback loop by affecting the combat environment. Planned actions and procedures are also entered into the situation data base to aid the combination and reasoning process.

The three major functional elements of the DFS are described below, summarizing the major implementation alternatives available.

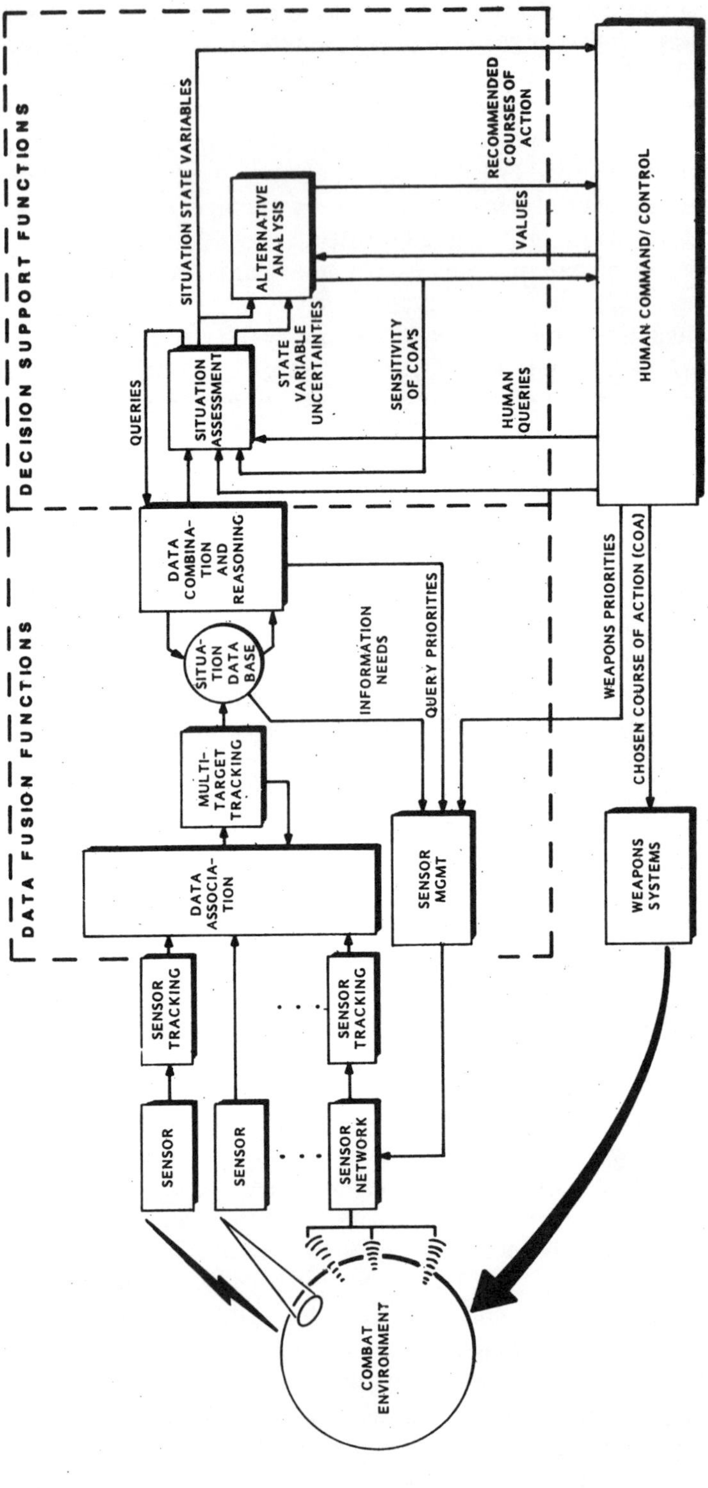

Figure 5. Data fusion and decision support architecture.

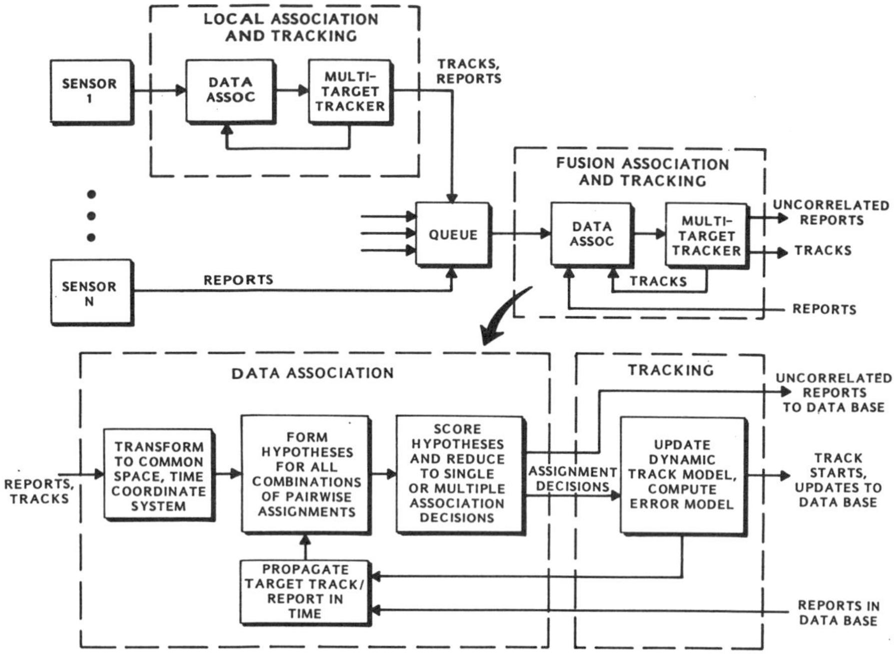

Figure 6. Data association and tracking functions.

Data Association and Target Tracking

Detections (reports) and tracks (from sensors which maintain local track files of targets) must be associated in time and space to assign sensed data to individual target files. Because the reports and tracks arrive from sensors with differing geometries (relative to the targets), resolutions, fields of view, measurement spaces and measurement times, it is necessary to transform all measurements to a common spatial reference (i.e., latitude/longitude, range/azimuth/elevation, X–Y–Z, etc.) and temporal reference. Once the measurements are transformed to a common measurement space, association hypotheses can be developed and scored to determine probable report-report or report-track assignments.

Figure 6 depicts the typical association/tracking functions for a DFS which includes both tracking and nontracking (report-only) sensors. Each new report on track is first transformed to a common spatial coordinate system before hypothetical assignments (pairings) are formed to existing reports or tracks in the situation data base. Each hypothesis is scored on the basis of a metric which can include the following: 1) spatial distance measures between the sensed report and hypothetical assignment and 2) feature data measures that characterize the sensed report and hypothetical assignment. These hypotheses are then evaluated to select a single pairing or to minimize the set of acceptable pairs if multiple hypotheses are to be saved for further iterations when more data is available.

When unique assignments of new report data are made to existing tracks, the dynamic tracking filter is updated to improve the estimate of the targets' behavior. Uncorrelated reports are passed to the situation data base for possible association in the future. Because of the uncertainty in sensor measurements and density of targets, probabilistic methods are often required to score and evaluate the many hypotheses that can result in multiple sensor systems (References 9 through 11). Hypothesis management over time and dynamic tracking of maneuvering targets are the most difficult challenges to the association/tracking function.

Data Combination and Inferencing

The core function of the DFS is the process of combining the collected sensor data on a single target to infer its identity and even higher order attributes (e.g., intent, future behavior, threat capability). Conventional, structured algorithms and expert systems are both candidates to mechanize these combination/inference functions.

Expert system implementations offer the advantages of partitioning the discriminating information into a distinct knowledge base and providing powerful reasoning capabilities using any combination of directly measured features, behavioral characteristics and contextual information.

Expert system reasoning functions employ a variety of categories of knowledge to combine multisensor data and infer higher levels of information about targets than contained in any individual element of the raw data.

1) *Structural models* are maintained to define the relationships between measured target attributes (e.g., acoustic/IR/radar signatures, RF emissions, kinematic envelopes and component parts) and targets or target classes. These relationships between attributes and entities are defined by the so-called IS-A and IS-A-PART-OF hierarchical connectives in a frame-like data base.

2) *Behavioral models* relate the temporal behavior (e.g., aircraft velocity/altitude flight envelopes, ground troop assembly/attack profiles) as well as spatial behavior (e.g., target formations, weapon ranges) of targets or events.

3) *Heuristic inference rules* including simple IF-THEN and Boolean (AND, OR, COMPLEMENT) relationships, provide a general structure to construct inference templates. These rules include an antecedent (the IF conditions that must be satisfied by the sensor data) and consequent (the THEN inference that can be concluded when the antecedent conditions are present).

4) *Numerical algorithms* combine incomplete, ambiguous sensor data by computing composite hypothesis scores which use the independence of multiple measurements to reduce the ambiguity in the measurement data.

The focus of much of the recent data fusion work has been to develop these numerical combination algorithms. When sensors provide hard decisions (declarations), the numerical combination may be a simple algorithm to sort declarations to search for a unique target or target set which describes

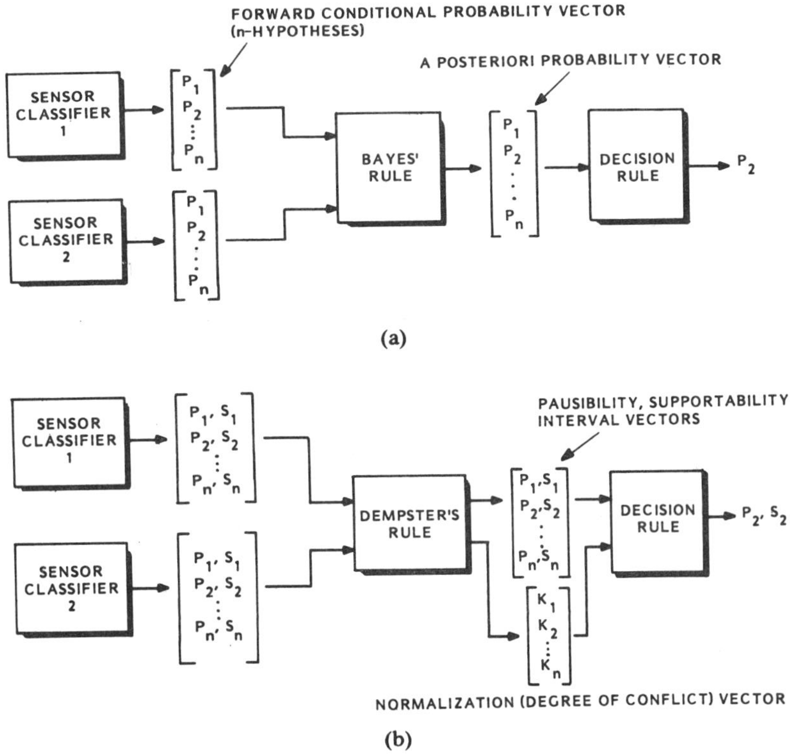

Figure 7. Comparison of two data combination approaches: (a) Bayesian combination. (b) Dempster-Shafer combination.

the set of data, or to accumulate declarations for combination by majority vote or weighted summation decision rule.

Because sensor measurements and intelligence data are often characterized by uncertainty, a number of methods may be employed to quantitatively represent the uncertainty in sensor and intelligence data. Probabilities (Reference 12), fuzzy variables (Reference 13) and probability intervals (References 14 through 17) are candidates for representation of the uncertainty associated with sensor data. Numerical combination algorithms may be used to provide a more accurate composite estimate of the measurements, reducing uncertainty in target identity or location.

Figure 7 compares the structures of two numerical combination algorithms that merge multiple hypothesis data vectors from N sensors. In each case, the sensors include statistical classifiers that classify measured features into a vector of parameters. These parameters quantify the certainty (likelihood, degree of fit, etc.) that the measurements represent each hypothesis. For the Bayesian inference case, these parameters are forward conditional probabilities. Bayes' rule is applied to compute a composite *a posteriori* probability. A decision rule, such as the maximum *a posteriori* (MAP), is then applied to select the most likely hypothesis.

Because probabilities mathematically represent events of chance rather than events measured with uncertainty, the alternative Dempster-Shafer representation uses probability intervals to describe sensor data. Each hypothesis is represented by two parameters: 1) supportability variable ($0 \leqslant S(x) \leqslant 1$) describes the degree to which measurements support the hypothesis and 2) a plausibility variable ($0 \leqslant P(x) \leqslant 1$) that represents the degree to which the evidence fails to refute the hypothesis. The difference, $P(x) - S(x)$, is a measure of ignorance about the hypothesis. When $P(x) = S(x)$, the probability interval collapses to a single probability equivalent to a forward conditional probability.

Dempster's rule of combination, analogous to Bayes' rule, provides a means of computing composite supportability/plausibility intervals (credibility intervals) for each hypothesis, reducing the uncertainty in the measured data. An appropriate decision rule is applied then on the basis of both supportability and plausibility.

The inference process can be driven by the arrival of new sensor data or by queries fed back from the DSS. The process of combining new data with the existing situation data to determine if ambiguity is reduced or consequents can be achieved, is referred to as *forward-chaining*. The arrival of new associated report data (antecedents) on a given target will cause the combination algorithm to be iterated to refine the estimate of target state and determine if new inferences (consequents) can be drawn. The situation data base is then updated with any new consequents. In contrast, the queries from the DSS cause a *backward-chaining* process that searches to determine if the antecedents in the data base support a hypothetical consequent supplied by the query. The DSS query may request, for example, the degree of support for a consequent: "What is the certainty that multiple hostile bombers with fighter support (a strike package) are enroute to attack the airfield at location *X* within ten minutes?" Such a query may use each of the types of knowledge to verify the consequent. This process could be as follows:

- Behavioral models define airfield attack profiles and airspace regions in which potential strike packages must be located.
- Situation data base is then searched to locate hostile and unknown aircraft meeting the strike criteria and any related events that may indicate the presence of a strike.
- Structural models then can be compared with target identities or attributes, flight patterns, related electronic warfare events, etc. from the data base to determine if a strike package can be located.
- Numerical combination algorithms compute the certainty for each hypothesis (each candidate strike package located).

The DFS responds to the queries with multiple hypotheses, each represented by state variables (target locations, identities, etc.) and associated certainties. These are formatted as vectors, with the degree of belief for each hypothesis being an element of the vector.

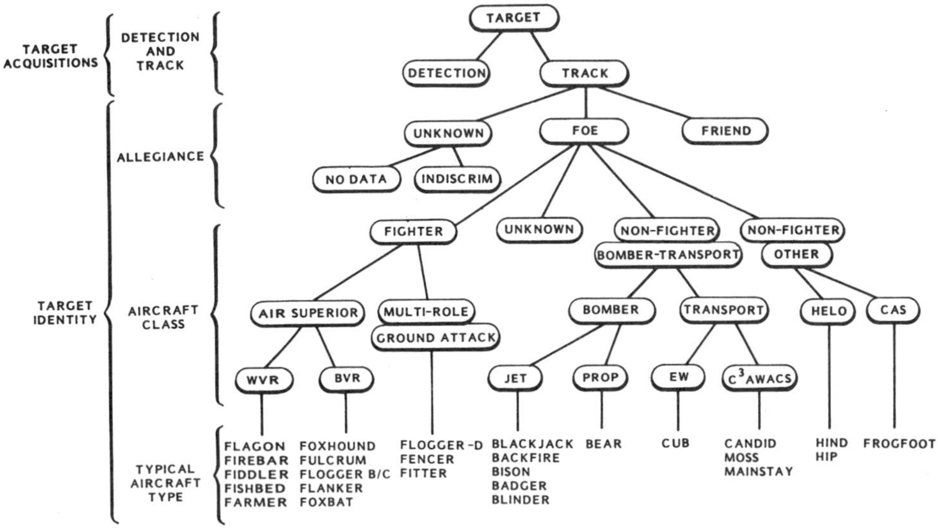

Figure 8. Hierarchy of data fusion information on sensed aircraft target.

Sensor Management

The allocation of sensors to targets refers to the ability to control the pointing and dwell time of sensors on detected targets or regions of the surveillance volume. Management of the sensors is based upon the following driving functions.

- Information needs include detected or tracked targets that require additional data to refine the track, resolve ambiguity in target location or discriminate target identity. These needs may be ranked by degree of threat to friendly forces.
- Query priorities are data required for situation assessment. For example, this may include sensor searches to reinforce developing hypotheses.
- Weapons priorities affect those sensors which directly support weapon systems, such as fire control sensors. These priorities result from C^2 decision to engage targets.

The sensor management function merges these (possibly conflicting) requirements to allocate sensors to targets. The functions also include sensor cueing, handoff and human override.

Characteristics of Fused Information

The information provided by the data fusion function is characterized by multiple levels of information content and a representation of the degree of certainty (confidence) in the information provided. Figure 8 depicts an example of the levels of identification information needed by a TACAIR decision support system. The hierarchy of aircraft identification includes four general levels of information content.

1) Detection and tracking of a target is the first level of information provided. Detection information can include three-dimensional spatial coordinates (e.g., range, azimuth, elevation) or single dimensional data only (e.g., passive detection of an emitter with azimuth information only). When repeated detections are associated and a time series representation of the target's dynamic behavior can be determined, a track can be established.

2) Aircraft allegiance is the traditional friend-foe-neutral identity which provides the first, and most important level of situation awareness. If foe, the aircraft is a potential threat and a candidate to engage.

3) Aircraft class provides a general identity of the aircraft mission, weapons (degree of threat) and performance capability. The discrimination of fighters from bombers is a critical capability for friendly fighters to allow them to avoid threats (hostile fighters) and to engage strike targets (bombers) selectively.

4) The highest level of information is the discrimination of aircraft type. Soviet aircraft selected from (Reference 18) are depicted in Figure 8 to illustrate the variety of aircraft types that are desired to be discriminated. For an identified foe, this information will aid the DSS to assess the degree of threat imposed by a fighter or to select the highest value bomber/fighter bomber to attack.

Additional levels of information that may be provided on targets by the DFS include: weapons status (stores remaining), estimated fuel status (inferred from type and point-in-mission) and predicted mission intent (e.g., intended targets, next-action).

The various levels of information supplied to the DSS can be reported in one of two methods:

1) Hard decisions (declarations) are statements of fact where the combined evidence exceeds a predefined threshold of certainty (i.e., greater than 0.99 probability of occurrence).
2) Soft decision data includes a representation of uncertainty in the information using probabilities, possibilities/supportabilities, or fuzzy variables for each hypothesis.

Hard decisions are reported as single statements or ambiguity lists (lists of possible alternatives, each with an equal likelihood of occurrence) and soft decisions are provided as multiple hypotheses, each with a representation of the uncertainty associated with the hypothesis.

The information supplied to the DSS results from queries initiated by the DSS and by new sensor reports which result in changes in the knowledge of the tactical situation. For example, such changes can include:

- new targets detected;
- refined/revised identities, unknown targets identified;
- changes in target behavior or location;
- loss of targets from track; and
- detection of events or aggregate target behavior that infers opponents' tactics or future behavior.

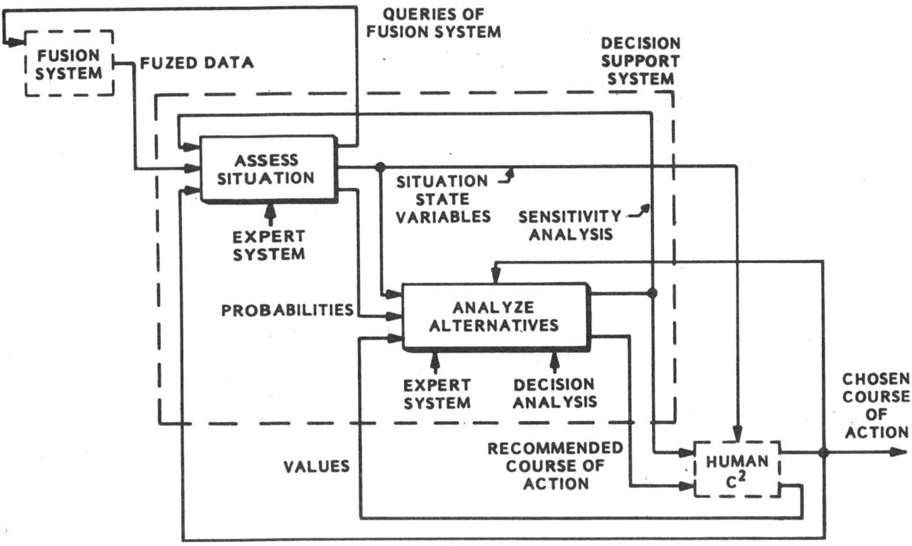

Figure 9. IDEF diagram of decision support functions.

C^2 Decision Support and Fusion

As described earlier, the two major functions of decision support are situation assessment and alternative analysis. The relationships among these two decision support functions, the fusion system and the human based command and control are depicted in the IDEF diagram of Figure 9. (IDEF terminology is as follows: processes are encapsulated in boxes, inputs to the process enter from the left side of the box and outputs exit from the right, guidance for performing the process enters from the top, and a mechanism that is needed to support the process enters the box from below.)

Fused data from the fusion system is the primary input for situation assessment. Also utilized are the results of sensitivities derived from the analysis of alternatives and the chosen course of action determined by the commander. Outputs include the situation state variables (the descriptors necessary to define the situation for decision purposes, for example, the identification of all aircraft); certainty vectors for the possible values of the state vectors (e.g., the measure of certainty or probability that a given aircraft is friend, foe, or neutral); and additional queries of the fusion system that help determine the priorities for management. These queries may be needed to determine the degree of belief of specific hypotheses or to distinguish between two competing hypotheses of the entire situation. The major mechanism of this process is a goal driven (backward-chaining) expert system that probes the fusion data base to define the situation, its state variables and their probabilities.

The analysis of alternatives operates on three inputs: the state variables and their certainties from the situation assessment and the values expressed

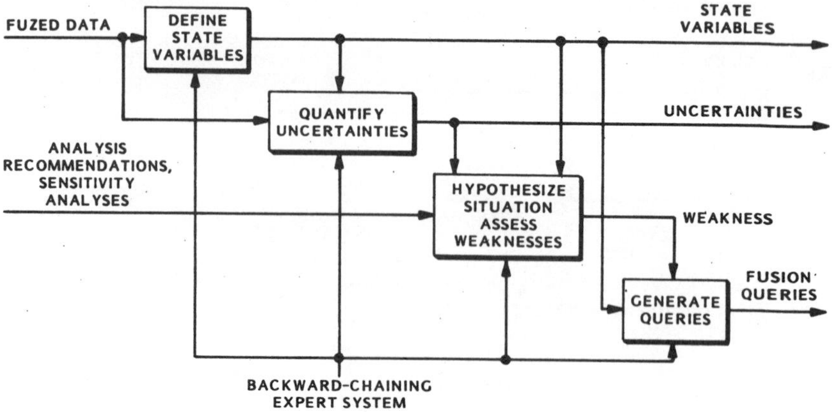

Figure 10. IDEF diagram of situation assessment.

by the commander and his subordinates for specific situations. The generic course of action that has been chosen by the commander provides guidance for this analysis process. The analysis of alternatives will generate and analyze alternatives for microdecisions within the generic course of action, but will also review the generic course of action and recommend when it should be revised. These inputs, coupled with a decision analysis mechanism, produce recommended alternatives and associated sensitivity analyses for the decision makers to review and accept or reject.

The commander and his subordinates receive the recommendations and sensitivity analyses of the analysis function, use the state variables of the situation assessment function as guidance, and have the ultimate say in alternatives chosen. There will be provisions for the analysis process to make some automatic decisions which are either routine in low periods of stress or time critical in high periods of stress.

To observe the situation assessment and alternative analysis processes more objectively, it is necessary to look at a lower level of detail. Figure 10 depicts the four processes that comprise situation assessment: 1) define state variables, 2) quantify probabilities of state variables, 3) hypothesize situation assessment weaknesses, and 4) generate queries of the fusion system. The backward-chaining expert system serves as a mechanism to all of these processes. Backward-chaining is a general strategy used in expert systems when it is more efficient to choose a possible answer or proposition ("the enemy's course of action is. . ."), and then to select those rules whose conclusions deal with this situation.

By accessing DFS, antecedents of these conclusions can be examined to determine which enemy course of action is most likely from this conclusion. Data from the fusion system are used to produce the state variables of the situation; and also, with the state variables as guidance to estimate probabilities for the state variables. However, there will be many uncertainties for which there is little and weak data for the estimates of the hypotheses. These

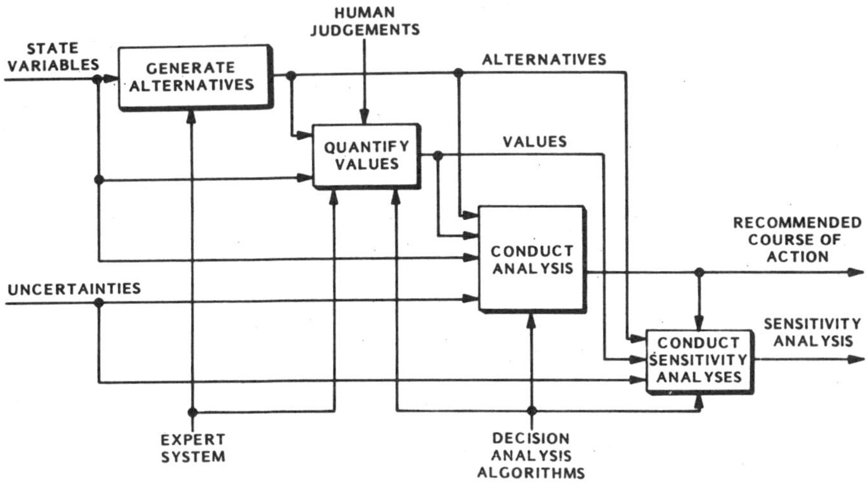

Figure 11. IDEF diagram of analysis of alternatives.

weaknesses would be established by the third process within situation assessment. This process of hypothesis generation and evaluation would also be implemented by a goal driven expert system. The inputs for this process would arrive from the alternative analysis outputs or recommendation and sensitivity analysis. These analysis outputs guarantee that the potential weaknesses of situation assessment that are probed via backward-chaining are important in the overall decision process and not trivial items of interest that have no impact. Finally, the query generation process uses the state variables as input and the weaknesses as guidance to specify sensor collection requests that are output to the fusion system.

The alternative analysis process also contains four processes: generate alternatives, quantify situation values, analyze alternatives, and conduct sensitivity analyses (see Figure 11). The alternative generation process uses an expert system with inputs from the state variables to generate alternatives within the context of the developing situation. This expert system can apply forward-chaining, with potential state variables as the antecedents followed by possible alternatives as conclusions. The quantification of values can be accomplished either via decision analytic assessment questions directed at the commander or his subordinates for critical decisions or through the execution of rules embedded in an expert system. The state variables and alternatives would be necessary inputs to this quantification in either case. The actual analysis of alternatives would use the alternatives, state variables, probabilities and values within the decision analysis paradigm of maximizing expected value to produce a recommended alternative. The sensitivity analysis would vary probabilities and values to find those key factors in the decision context that would change the decision maker's choice. Algorithms have been developed to vary up to three parameters at a time in conducting policy region analyses of the available alternatives.

	ARMY ATTACK OF 2ND ECHELON	NAVY SUBMARINE VS. SUBMARINE	AIR FORCE DEFENSIVE COUNTER AIR
BEGIN ENGAGEMENT TACTICS OR GATHER MORE INFORMATION	ISSUE ORDERS TO ARTY/ AIR ASSETS INFO: LOCATION SPEED	BEGIN APPROACH TACTIC INFO: LOCATION DIRECTION	ASSIGN F-15 TO SPECIFIC TARGET INFO: LOCATION FRIEND/FOE
CHOICE OF WEAPON TACTIC FOR ENGAGEMENT	SMART ANTI-TANK VS. AREA WEAPONS INFO: SIZE COMPOSITION	LEAD/LAG VS. DIRECT APPROACH INFO: COURSE CAPABILITY	LONG VS. SHORT RANGE MISSILE INFO: FIGHTER/BOMBER
FIRE OR CONTINUE TO IMPROVE POSITION	FUTURE VS. CURRENT FORCE SURVIVABILITY P_k (TARGET)	OWN-SHIP SURVIVAL P_k (TARGET)	OWN-SHIP SURVIVAL P_k (TARGET)

Figure 12. Decision sequence examples.

It is necessary to examine specific situations and decisions in which this combination of fusion and decision support could operate. Three time phased sequences for the decision dimension are used: initiation of engagement tactics, choice of weapon or tactic for engagement and the timing of the attack. Overlaid on this decision sequence are diverse examples from Army, Navy and Air Force situations: the corps' attack of the second echelon, submarine versus submarine, and the fighter aircraft's defensive counter air mission. Figure 12 shows this matching of decision sequence with specific applications (References 19 through 25).

The first decision in this sequence involves the balancing of a need for more information combined with the potential of tipping one's hand versus waiting too long and jeopardizing the survival of key elements of one's force and/or losing key opportunities. In looking across the three applications of this decision in Figure 12, information regarding location of the threat is present in every situation. Other information for specific applications ranges from speed to direction and intent to friend/foe identification.

Information regarding the capability of the enemy force is more relevant for the second decision in the sequence—choice of weapon or engagement tactic. In choosing between smart antitank and area weapons for attacking the second echelon, the size and composition of the enemy force at key locations as well as the choice between delay versus destroy missions are key components of the decision process. The choice of the approach tactic by the friendly submarine depends on the course and capability of the enemy submarine. The fighter pilot will have a choice between long and short range missiles and would like to know if the foe aircraft is a high performance fighter or bomber in making this decision. Information about the speed of the foe aircraft can be conclusive about this distinction.

Finally, the timing decision of when to fire balances survival versus expectations of kill probability. For individual platforms (such as the submarine and fighter) opportunities for reattack are included in this decision. For larger force considerations such as the attack of the second echelon, the opportunity cost of diverting the attack resources from the first echelon to the second echelon must be balanced in terms of immediate impact versus future impact.

Conclusion

The integration of automated data fusion and decision support systems holds the promise of more effectively refining and analyzing the military meaning of the vast amounts of information collected by modern C^2 systems. The benefits of such automation, however will not eliminate the central element of command and control—the human commander. The real role of data fusion and decision support is that of converting data into meaningful and timely information that will enhance the process of human judgment. The current concerns that such automated systems will be vulnerable to deception (Reference 26) are based, in fact, on the premise that countermeasures and tactics often rely on unpredictable behavior. The human commander is best suited to cope with these higher order reasoning processes and is the appropriate authority to commit lethal weapons to engage the adversary.

References

1. C. A. Fowler and R. F. Nesbit, "Tactical C^3, Counter C^3 and C^5," *J. Electron. Defense*, Nov. 1980.
2. J. G. Wohl, "Force Management Decision Requirements for Air Force Tactical Command and Control," *IEEE Trans. Syst., Man, Cybern.*, vol. SMC-11, pp. 618–639, Sept. 1981.
3. N. L. Stone, "ASW, the Soviet View," *Military Electronics/Countermeasures*, pp. 34–73, June 1979.
4. R. J. Starkey, Jr., "Anti-submarine Warfare Oceanography," *Military Electronics/Countermeasures*, pp. 67–72, July 1981.
5. R. Wohl, "Ocean Transparency: Impossible or Inevitable?," *Defense Sci.*, pp. 20–28, Feb. 1984.
6. D. Anderegg and B. Payne, "Tactical Relevance—A Fighter Pilot's View," *Defense Syst. Rev.*, pp. 36–40, Sept. 1983.
7. F. K. Mahaffey, "Intelligence, Surveillance and Target Acquisition in the 80's—The Army Concept," *SIGNAL*, pp. 39–42, Oct. 1979.
8. E. L. Waltz, "Data Fusion for C^3I Systems," in *Int. C^3I Handbook*. Palo Alto, California: EW Communications, 1986.
9. T. Fortman, Y. Bar-Shalom, and M. Scheffe, "Multitarget Tracking Using Probabilistic Data Association," in *Proc. 19th IEEE Conf. Decision Contr.*, vol. 2, pp. 807–812, Dec. 1980.
10. M. Kosaka *et al.*, "A Track Correlation Algorithm for Multisensor Integration," in *Proc. Digital Avionics Syst. Conf.* (DASC), 1983, pp. 10.3.1–10.3.8.
11. I. R. Goodman, "An Approach to the Data Association Problem through Possibility Theory," in *Proc. 5th MIT/ONR Conf.*, Aug. 1982.
12. P. J. Nahin and J. L. Pokoski, "NCTR Plus Sensor Fusion Equals IFFN," *IEEE Trans. Aerospace Elect. Syst.*, vol. AES-16, pp. 320–337, May 1980.

13. E. H. Ruspini, "Possibility Theory Approaches for Advanced Information Systems," *Computer,* pp. 83–91, Sept. 1982.
14. G. Shafer, *A Mathematical Theory of Evidence*. Princeton, New Jersey: Princeton Univ. Press, 1976.
15. R. P. Bonasso *et al.*, "ANALYST II: A Knowledge-based Intelligence Support System," The MITRE Corp., Tech. Rep. MTR-84W00220, Apr. 1985.
16. H. E. Rauch, "Probability Concepts for an Expert System Used for Data Fusion," *The AI Magazine,* pp. 55–60, Fall 1984.
17. P. Cohen *et al.*, "Representativeness and Uncertainty in Classification Systems," *The AI Magazine,* pp. 136–149, Fall 1985.
18. J. W. R. Taylor, Ed., *Jane's All the World's Aircraft 1983–84*. London: Janes Pub. Co. Ltd., 1984.
19. R. C. Bromage, R. V. Brown, J. O. Chinnis, M. S. Cohen, and J. W. Ulvila, "Decision Aids for Submarine C^2, Phase III Concept Implementation," Decision Science Consortium, Inc., Falls Church, Virginia, Tech. Rep., Mar. 1983.
20. J. W. Ulvila and R. V. Brown, "Attack Planning Decision Aid for Submarine C^2," Decision Science Consortium, Inc., Falls Church, Virginia, Tech. Rep. 83–84, Oct. 1983.
21. W. H. Gossard and G. R. Linsenmayer, "C^3I Systems Requirements to Support C^3CM," *SIGNAL,* pp. 18–28, Apr. 1985.
22. T. Garvey and T. M. Strat, "AI for Command Decision-making," in *Proc. ADPA Artificial Intelligence Robotics Symp.,* 1985.
23. E. Wanner *et al.*, "The Application of Decision Aid Technology to Command and Control," *SIGNAL,* pp. 43–46, Mar. 1984.
24. J. R. Slagle *et al.*, "BATTLE: An Expert Decision Aid for Fire Support Command and Control," in *Proc. 49th Military Operations Res. Symp.,* 1982.
25. "Decision Aids for Target Aggregation," Technology Corp., Decisions and Designs Inc., Systems Control Technology, Tech. Rep. TR-84-108 PAR, May 1984.
26. D. Greenwood, "Strengthening Conventional Deterrence," *NATO Rev.,* vol. 32, no. 4, pp. 8–12, Aug. 1984.

Some Strategic Implications of Breakthroughs in C^3I

George H. Quester

The most important result of continuing breakthroughs in the application of electronic and space technology to military purposes is the tremendous increase in information available to the commanders on both sides, including even a fair amount of real-time information. This article will try to discuss some of the impact of this revolution on wars within the earth's atmosphere as we have known them, as well as sorting out types of future combat which may be fought in space. The possibility will be explored that such technology may make it difficult in the future to separate "arms races" from "wars."

Students being introduced to the philosophical nature of war are often asked to think about comparisons and analogies with the game of chess. Chess, given the shapes and names of its pieces and the shape of its board, would seem to be a very natural war game. Yet, the more reflective student quickly spots the weakness of the analogy, for in chess each player can see the location of all the opponent's pieces.

Lunchtime at the Rand Corporation has long featured a modification of chess quite aptly named "Kriegspiel," a version of chess played with two separate boards and a monitor/umpire, as each player knows only the location of his own pieces, and has very limited information about those of his opponent. As in real warfare, therefore, uncertainties dominate the game, and pieces are at times moved merely in an attempt to gain information.

Kriegspiel is thus much better than chess as an approximation of war as we have known it. Yet, the major point of what we will be discussing here is

**This article originally appeared in the March 1982 issue of SIGNAL.*

that real war, under the umbrella of space-based systems, will be becoming more and more like chess and less like Kriegspiel. Chess finally will become a more apt analogy and model for war, as war becomes a contest with much higher levels of information available to both sides.

Information

Perfect information will, of course, never be assured to both the sides. As before, each side will have to assign some effort to collecting information, and devote other efforts to keeping the enemy from acquiring information. Intelligence about the enemy's dispositions will continue to be a battlefield asset as before.

Yet, the important difference will come in the ranges of uncertainty over which this battle of information can be fought. Where previously the range on each side might have wandered between 50 and 90 percent reliability, the ranges now might be between 90 and 100 percent. Each side may be assured of a much clearer picture of the battlefield than before, no matter what the adversary does to try to bring back the fog of war. Each side will have far more limited options for fogging up the enemy's version of how the war is going.[1]

The classical categories of military intelligence activity and counterintelligence will continue to apply. Indeed, much of what we will be analyzing here as future possibilities of war in space will come precisely in the categories of obtaining information for yourself, and denying information to the enemy.[2] Yet the form of such activity and counteractivity will be far more technological and automated than it was in the past, and the degrees of accuracy obtained in the information gathering process will be considerably higher.

The extended range of functions for the application of information to war will thus be as before. Each party to the conflict will want to enhance its information about the deployments and movements of the forces on the other side. Each combatant will wish to withhold such information about its own deployments and movements. As part of protecting its own sources of information, each armed force will try to withhold information from the enemy about the technology and nature of its information-gathering system; each armed force will try to acquire information about that system, about its technology and nature, thus applying methods of blinding it. We will be discussing two parallel questions through the rest of this article.

First, given our premise that the levels of information about the battlefield will remain high despite the best efforts of each side to blind the other's intelligence systems, what will this do to the likelihood and nature of war as we have known it? We will have to address conventional wars here for land, sea and air, and also the likelihoods and nature of various forms of nuclear war.

Second, given the continuing effort of each side to try to maintain its own intelligence while degrading that of the adversary, what kinds of new high-technology warfare might this produce? These contests of capability might develop a life of their own quite apart from the wars just cited. Besides contests of intelligence gathering, what else should we anticipate in wars in space?

Impact on War Close to Earth

With regard to warfare as we have known it in the past, a first important question about the impact of space-based surveillance systems is whether these tend to favor the offense or the defense. Given the confusion sometimes surrounding such terms, some definitions should first be offered.[3]

What will be treated as weapons systems favoring the offense here are those which reward the side which is moving as it engages an adversary who is sitting still: those weapons systems which reward the taking of the initiative and punish whomever waits for the enemy to take the initiative. What will be regarded as a support for the defense here would rather be developments with the opposite impact, imposing additional costs (in terms of casualties or prospects for victory) on the side which is moving or which has initiated the battle, rewarding instead the side which elected to sit still in a crisis, letting the battle come to it.

What will be the impact, in terms of these considerations, of space-based systems which render a battlefield situation less foggy and opaque? In many ways, such additions of information reduce the temptation toward the offensive. Surprise, in terms of the timing and the location of an attack, in the past often played a role in facilitating an offensive breakthrough. It is widely stated that the attacker typically needs a three-to-one advantage to break through a line. But this has not meant that aggressors always needed to outnumber their adversaries by this ratio all along the front.

In the absence of reliable information about force deployments, the attacker could group his forces along a particular sector of the front, to generate a local superiority well in excess of three-to-one, and then could break through. The introduction of satellite surveillance, with real-time scanning of the force deployments on both sides of the line, might work to make this impossible in the future and might thus reduce the general prospects of the offensive.

A few examples can be offered working the other way, but they are probably not sufficient to outweigh the general tendency just outlined. Forces sometimes shield themselves against an enemy's offensive attack mainly by concealing their position. This has classically been the protection of guerrilla bands, and of submarines carrying ballistic missiles. A more reliable scanning capability based in space could thus strip the cover from some forces and tempt an enemy to attack them. Surface naval forces in particular may become very vulnerable to attack by ballistic or cruise missiles once their exact location is known on a real-time basis by means of reconnaissance satellite.

These latter cases apply most, however, to situations where no defensive line is being maintained, e.g., as by the guerilla forces in Afghanistan, El

Salvador, Angola or Thailand. The typical situation of concern for the West may be more like the NATO central front, where the real estate on both sides is so valuable that forces have had to be arrayed to man the line.

The U.S. Position

Which way do American interests work here? Being basically a defensively minded status-quo power, the United States would have to welcome those changes in military technology which strengthen the defense. More generally, war becomes less likely when both sides see less opportunity for taking the offensive and are less in fear of such initiatives from the other side.

World War I is a very clear example of a crisis leading to armed conflict because each side could not be sure of how the other was moving its troops and could not wait to see what the other would do. International inspection is often proposed as a means for policing disarmament agreements, (i.e., mutual commitments not to procure arms) and some of such inspection in the 1960s and 1970s have been accomplished by the national technical means of space-based satellite reconnaissance.[4] Such inspection by national technical means, by space-based intelligence systems, may similarly now serve to support mutual commitments not to *use* weapons.

With regard to warfare as we have known it in the past, a second question on the impact of space-based surveillance systems would be whether these work to the advantage of the larger or smaller forces.

Where the difference between forces is enormous, the smaller force typically has had to fall back on guerrilla tactics in the past, and the addition of space-based surveillance would thus help the larger force. (Consider the impact of simple aircraft surveillance on the guerilla operating in the desert, as PLO guerrillas cannot hope to match the effectiveness of the forces that fought with Lawrence of Arabia.) Where the difference is of a smaller magnitude, perhaps less than three-to-one overall, the addition of surprise has in the past (as just noted) often allowed the larger force to assemble significantly larger superiorities on designated segments of the front. The addition of satellite surveillance should generally make this more difficult, and thus would help the *smaller* force.

Having classically been the side with smaller totals of ground force troops and tanks, the United States would seem to have had a clear interest in developments which favor the smaller force. Being generally more committed to a defense in line than a defense by guerilla harassment, the United States again therefore would have reason to welcome the addition of space-based intelligence sources.

Likelihood of Nuclear Weapons

A third important question about the impact of the new means of scanning the battlefield is whether this will increase or decrease the likelihood that nuclear weapons will come into use, whether it will increase or decrease the chance of limiting war at the conventional level.

The impact here is again mixed. It should be noted at the outset that all of the nuclear/conventional distinction hangs somewhat together. For a nuclear warhead to be used anywhere in anger will thereby increase the chances that nuclears will be used elsewhere. The nuclear taboo which has precluded all such usage since Nagasaki has been self-sustaining and self-enforcing; each year that goes by without any such use of nuclear warheads gives all sides a greater reassurance and greater vested interest that they not be so used.

Because war can become so much more destructive once nuclear weapons come into use, the weighting of the interest of the United States should be fairly clear. Barring some very special circumstances, the United States would prefer to win wars, or to avoid even having wars fought, while maintaining this nuclear taboo.

The bulk of the impact of space-based intelligence systems is thus again probably positive. Because surprise attack will become less likely, defenses will generally be less threatened, so that the threat of nuclear escalation will not have to be brought to bear as much as a way of reinforcing such defenses. Surprise attack in the past entailed marshalling masses of troops, which thus generated a logical target for tactical nuclear warheads; but such targets may become fewer when the option of surprise is less from the very beginning.

The Role of Nuclear War

Some countertendencies also must be identified. As we shall discuss, there will be some natural temptation on each side to interfere with the space-based systems of the other. At points, the most straightforward way to accomplish this might be by the use of nuclear warheads, under circumstances where no one is killed or hurt, where no impact of the explosions is felt on the surface of the earth. If a nuclear warhead were thus to be used in a satellite-killer role, would this then amount to the use of nuclear weapons "in anger"? What if such a nuclear round were only used to blind the satellite of the other side, without any further physical destruction of it?

The point being opened here is that the introduction of highly automated satellite-based reconnaissance systems may over time threaten to blur some of our clear distinctions between war and peace. The nuclear taboo may indeed by a valuable asset for both the United States and the Soviet Union. Yet there will be officers on each side watching for opportunities to employ nuclear warheads without upsetting the rest of the taboo. An explosion which killed no one, which indeed was very difficult to detect back on the earth's surface, might seem to be such an opportunity.

Apart from possibly inviting uses of nuclear warheads against the satellites themselves, the deployment of space-based intelligence gatherers might similarly sometimes also increase the attractiveness of nuclear escalation back in the traditional battles on the surface. If there is a real-time information of an enemy's massing of forces, and one is not able to exploit this with a countermassing of one's own forces, the information may tempt a use of nuclear weapons. As targets are more easily identified from space, some of them will look like appropriately nuclear targets to one and all.

Yet, the net impact, as conjectured above, is still probably in the desirable direction. Satellite-based intelligence systems will allow for much more detailed assessments of which bridges have been destroyed, and which remain standing, in the enemy's territory. Facilitating much more accurate strikes by conventionally armed cruise missiles and ballistic missiles, such space-based systems will allow for some effective counterforce moves without an escalation to the nuclear level.

Kinds of War in Space

But what are the possibilities now for war out in space? The sheer technological impressiveness of the new possibilities is fascinating enough, but of equal fascination for the strategist must be the complication of traditional categories of warfare, and the generation and elaboration of some new categories.

One can draw some interesting lessons about the future roles of space satellites by comparing such roles with the introduction of aircraft and of radio at the beginning of this century. Satellites inherently must combine some of the characteristics of radio and aircraft. Because satellites are necessarily smaller than aircraft, they normally will be unmanned as military vehicles. As a result, they are run by remote control, i.e., largely by radio. Because they can remain aloft for longer periods of time (in many cases indefinitely), they will be able to undertake a greater array of tasks than aircraft. Because they are platforms in space, above the earth, they will undertake more tasks than those simply of electronic warfare.[5]

To begin, satellites are of great military value for their enhancement of simple navigation, letting aircraft, ship captains and commanders know exactly where they are. Radio played a similar role.

Space platforms, like radio, also substantially enhance communications between elements of one's military force. But as with radio, such communications invite an enemy's attempt at interruption or interception. To keep an enemy from talking to his own forces is to enhance the chances of defeating him. To listen in on his communications is to enhance these chances even more.

Third, satellites, like the earliest aircraft, are very valuable for their ability to survey the surface of the earth and to give us a clearer picture of an enemy's dispositions. Just as French aerial reconnaissance at the Battle of the Marne may have blunted the German offensive crucially turning World War I into a stalemate, so the newer forms of reconnaissance from space may very much change war as we have known it.

Just as they can warn of the disposition and strength of an enemy, satellites can warn of his movements. Not the least in importance of such warnings would be detection of the launchings of an enemy's missile force. Such warning of strategic missile attack, could be used to evacuate national leaderships or to launch our own strategic forces on warning or, when coupled with some other vehicles which might also be based in space, to try to intercept and blunt the bulk of such a missile strike.

Electronic Technology

Electronics thus had already done a great deal to make the battlefield more of a contest of intelligence. Before radio communication was introduced, the battalions of the same army had much less capacity for staying in coordination. At the same time, the enemy had less to listen in on. The Russian Navy at the battle of Tsushima in 1904 lacked radio communication. The Russian Army had such communications links at the battle of Tannenberg in 1914 but apparently had not taken the precaution of encoding its messages, offering German listeners an unexpectedly clear view of Russian plans for the battle. The British accomplishments in breaking the German ciphers in World War II, in parallel with the American success at breaking the Japanese codes, would have been of no impact if the Germans and Japanese had not already become committed to the battlefield coordination accomplishable by radio communication.

One could not expect to win a battle in the twentieth century unless one harnessed electronic technology to coordinate one's forces. Yet this very coordination comes with a risk that the other side will also be better able to keep our battalions in view.

As noted, the most significant military role for space platforms may thus already have been demonstrated with the increase in ability to observe all the traditional battlefields on, or close above, the land and the sea. Just as the airplace made the battlefield more observable, the reconnaissance satellite is most profound in its impact here. Just as wireless radio communication facilitated electronic coordination (and simultaneously exposed the combatants to electronic intelligence operations [ELINT]), so the orbiting satellite allows more rapid communication within armies.

But an additional mission of a war in space then evolves speedily enough from all of what has just been mentioned. Satellites, like aircraft, can be directed against their opposite numbers in the enemy force (to force the analogy, radio transmissions can also be used to jam the other side's radio communication). Space vehicles can thus be designed to seek to destroy reconnaissance satellites, to keep an enemy from knowing so fully what we are doing with our land and naval forces and to frustrate his internal communications.

A pattern for likely hostile exchanges in space has already been previewed in the recent contests of electronic warfare within the atmosphere. As shown in the campaigns of the Vietnam War, and also in the 1973 Arab-Israeli War, the advances in electronic tracking of aircraft have offered substantial advantages to the defensive, advantages which at times threatened to impose unacceptable costs on an air force trying to carry through bombing missions, or at least forced that air force to commit substantial resources to reducing the flow of information which favors the defense.[8]

Radar stations thus have had to be attacked directly, or have been attacked by missiles designed to home in on the radar signal (the missile in the end being rated as a success even if it merely forced the radar station to shut down its transmissions, with the missile not otherwise scoring any kind of "kill"). Aircraft have had to be devoted entirely to jamming or to distributing chaff, the equivalent in the old days of assigning soldiers entirely to the role of counterintelligence, reducing one's battlefield forces somewhat, merely so that the enemy could not have too clear a view of the battlefield.

Clarity of the battlefield, in the air as on the ground, may favor the defense. Reducing that clarity may then be essential to making the offensive possible; but the extent to which this clarity can be reduced, to which the traditional fog of war can be recreated, is now definitely limited, and the resources that will have to be drawn into this area of electronic countermeasures will be substantial.

Attack from Space

The vehicle for watching combat often also gets used for participating in combat. The airplace in World War I soon enough graduated from the role of observation to the delivery of warheads against the enemy.

Like aircraft, satellites also can be used to attack targets on the surface of the earth. Compared to the destructive power now deliverable by aircraft, or by missiles based on earth, this may remain a minor and secondary role for space-based systems; yet, nuclear weapons based in space, or directed energy weapons based in space, still might be of considerable destructive importance.[7]

Space-based vehicles equipped with destructive capabilities could have either a counterforce or a countervalue role. For a counterforce role, they presumably would have to develop advantages in accuracy or time-to-target, surpassing what we have to look forward to in ballistic missiles and cruise missiles launched from the earth. This seems unlikely, with a single important exception, a space-based antimissile defense role.

Just as RAF fighter aircraft strove to shoot down German V-1 cruise missiles in 1944, satellites could be designed and programmed to block the passage of enemy ballistic missiles going past them in space or of enemy cruise missiles flying below them through the atmosphere. By the use of lasers or particle-beam weapons, the enemy missiles might be destroyed, or at least could be thrown off course.

Such space-based ABM systems pose all the worrisome threats to mutual deterrance of other surface-based ABM systems, for they wear away at what has been the basis for peace since the 1950s. If either side felt that it had a reliable protection against the other side's intercontinental missiles, might it not strike with its own missiles while the once-in-a-century opportunity lasted? If either suspected that the other was about to acquire this opportunity, might it not strike first to head off the inevitable attack?

Unhappily for the believers in ABM, and happily for those who wish to continue with mutual assured destruction (MAD), the likelihood is that such a space-based ABM system could not prevent nuclear retaliation by either side, as countermeasures will emerge to allow missiles to bypass it. A satellite-based laser system could destroy some of the missiles going past it, but not all, and there will have been a great number of warheads launched on each side. Some of such missiles may be deflected by such attacks, their accuracies degraded by the application of laser or charged-particle beams. While this loss of accuracy might be serious where the warhead was aimed at a missile silo, however, it might make little difference where a large metropolitan area was the target.

Happily or unhappily, space-based systems do not therefore really promise to offer perfect counterforce capabilities. If they offer additional countervalue capabilities, moreover, capabilities for inflicting additional retaliatory devastation back on the earth's surface, they may actually reinforce mutual assured destruction if any such reinforcements were needed.

For space-based destructive vehicles to be efficient in the countervalue role relates to a different set of issues. Space-basing presumably would satisfy a search for additional locations by which to escape an enemy counterforce attack. If land-based missiles and bombers and submarine-based missiles came into question at some future time, a reserve of warheads in orbital trajectories could substitute for missing legs of what had been the triad.

In summation, the issues of offense and defense are, as always, considerably more complicated for the global nuclear balance than for conventional war.[8] The total impact of the addition of space-based weaponry is thus somewhat more difficult to sort out, but the net of the impact may again be the same, to stabilize rather than to destabilize, to make a nonnuclear World War III less likely rather than more likely.

Attacks in Space

As the range of missions that can be performed from platforms in space increases, moving beyond observation to active participation in various combat roles, the motive for attacking such platforms themselves increases again. Missiles could be configured to seek out and destroy any counterforce vehicles deployed in space; and they could be aimed at any countervalue vehicles deployed in space. Such antisatellite vehicles could be destroyed on the ground, of course, or could themselves by deployed in space; our ultimate responsibility may be two space-based counterforce vehicles, each equipped for attacking the other.

Perhaps this space basing will be necessary because the laser is the most effective antisatellite weapon, but only when the earth's atmosphere does not attenuate its power.[9] For whatever reason, just as aircraft have sometimes surpassed surface antiaircraft artillery as the best vehicle for destroying enemy aircraft, the prospect looms of satellite vs. satellite dogfights in space.

In such a face-off, as in all the military face-offs of history (e.g., warship vs. warship, ICBM vs. ICBM), one important question would be whether the exchange rates were destabilizing or stabilizing, i.e., whether they rewarded taking the offensive and firing the first shot or discouraged it.

Forecasting the exchange rates of a new military technology is always hazardous. Would more than one antisatellite vehicle have to be expended in the process of destroying one satellite on the other side? Everything is imaginable. To destroy one platform in space could involve a sacrifice of something like zero (perhaps just a short pulse of a laser beam—whoever shoots first wins the round) to the total expenditure of more than one satellite vehicle (roughly analogous to the face-off of land-based missiles before the advent of high accuracy and multiple warheads—a quite stable situation).

Very tentatively, it could be predicted that a space-based antisatellite system is likely to be more destabilizing than one based on earth, because whatever vulnerability of the satellite is being exploited should pertain to the attacking platform as well. A situation of "whoever shoots first, wins" is always more encouraging to attack, and it is more likely to occur where the satellite killer is itself attached to a satellite platform. (On the other hand, while an earth-based antisatellite weapon is more likely to be stable, it brings with it the prospect of greater destruction on earth itself, as this weapon system also again becomes the target of attack.)

As with everything we have discussed here, we can only try to make net predictions, incorporating the sum of some contrary tendencies. Directing attack at platforms in space will upset a kind of sanctuary which has been observed ever since 1960. A move to conduct such attacks themselves from platforms in space will be all the more of a revolution in warfare. It could very well be the kind of revolution which makes space warfare more likely than less, which is destabilizing in the net by rewarding the offense.

But, it should be borne in mind that this would not be conventional warfare on the ground and would not be thermonuclear war within the atmosphere but a very esoteric new form of warfare, fought out entirely beyond the sight or earshot of mankind.

Painless War

As a final rounding out of the complications we have noted, the introduction of space-based systems may substantially decrease the casualties in some violent exchanges, so much so that we will have trouble deciding whether we would wish to call them war.

We have already noted how information acquisition has become a central part of the military role in space. High-grade intelligence of course, will allow each side, if it wants to, to direct its fire more at military targets and less at nonmilitary. Such care about what is being hit could apply because of moral feelings, or as part of a mutual deterrence game whereby each side was sparing the value targets of the other.

Yet, to continue some of the analogies we have been drawing, we may see not just the application of intelligence but a very much intensified contest of intelligence gathering and intelligence denial, a contest which in the past has often not been regarded as out-and-out warfare.

Espionage has classically been a part of peace as well as of war. Spies often work, and are captured, out of uniform. If they are caught and executed they are not typically listed as "casualties." Do they thus become "the first to fall" in a war, or are they instead somehow outside the range of what we define as war. Similarly, we have efforts to collect information and to prevent its collection well below the threshold of what anyone would label as violence. If we adopt new codes, because we fear that you have broken our old codes, and you then put your code breakers to work in an effort to reacquire your insight into what we are planning, is this war? No one is getting killed or wounded, except for a few nervous breakdowns among the code breakers.

The contests of intelligence gathering versus intelligence denial of the future may take very much the same form as some kinds of space-based instruments which are improved to preclude their being blinded by the enemy, while others are blinded. Some of this blinding will take the form of physical interference with the way a satellite functions, while at other times it will simply entail changing the signals emitted from what that satellite was monitoring, to frustrate the monitoring function in a more passive way. And a fair number of the blinding moves taken may be impossible to sort by any categories of active versus passive, or violent versus nonviolent.

Further Effects

Satellite technology also can make war less inhumane, and thus harder to define, because of its capital-intensive nature, for most of the vehicles involved will be totally unmanned. Such automation, the substitution of machinery for men, makes it all that much easier to fight with discretion, distinguishing carefully between value and force targets. Every human life is valuable, so that any airman or soldier killed is a value loss. A nation takes pride also in the equipment it lofts into orbit, but the pain, if such a vehicle is destroyed or rendered inoperable by enemy action, is far less.

The Soviet-American competition in space may thus in the end fall into a zone which is neither quite war nor quite arms race. An arms race consists of military efforts on each side to undo the accomplishments of the other. So does war. War is distinguished from arms races, some would say, in that weapons are employed, rather than merely procured, but this distinction is not very clear. Wars are perhaps really more distinguishable from arms races in that violence is involved, with people being killed and bombs going off.

But what do we have then, when new countermeasures to electronic or photographic intelligence continually are being deployed and tested in each side's efforts to degrade the other's intelligence, while enhancing its own? No one is being killed, and the explosions or burnings out of sensors are happening far out in space. The war that results is a totally noncounter-value war, with our robots slamming away at theirs, and vice versa.

Human beings may generally unite in condemning arms races as well as wars.[10] Yet few of us would deny that wars are worse than arms races, that we would rather tolerate such races than actual deadly combat. War in space is again something no one looks forward to. The point here is that it may nonetheless happen, in part, because it will rank with arms races, rather than with war, in terms of how much suffering it imposes.

References

1. For a balanced report on the space-based reconnaissance accomplishments to date, see: Jones, Jr., Cecil B., "Photographic Satellite Reconnaissance," *U.S. Naval Institute Proceedings,* Vol. 106, No. 928 (June 1980) pp. 41–51.
2. For analysis of the categories of military uses of space, see: "The New Military Race in Space," *Business Week* (June 4, 1979), pp. 136–139, 142, 145, 149; Hartman, Richard, "War in Space," *Defense Electronics,* Vol. 10, No. 10 (October 1978), pp. 34–49; and Andelman, David A., "Space Wars," *Foreign Policy,* No. 44 (Fall, 1981) pp. 94–106. Soviet military space efforts are outlined in: Charles S. Sheldon, II, "The Soviet Space Program in 1979," *Air Force Magazine,* Vol. 63, No. 3 (March 1980), pp. 88–93.
3. This author's definitions of offense and defense are outlined in Quester, George H., *Offense and Defense in the International System,* New York: John Wiley, 1976.
4. The evolution of space-based surveillance capabilities is recounted in: Klass, Phillip, *Secret Sentries in Space,* New York: Random House, 1971, and Greenwood, Ted, "Reconnaissance and Arms Control," *Scientific American,* Vol. 228, No. 2, February 1973, pp. 14–25.
5. See: Karas, Thomas H., "Implications of Space Technology for Strategic Nuclear Competition," *Occasional Paper* No. 25, Muscatine, Iowa: The Stanley Foundation, 1981.
6. Lake, Julian S., and Hartman, Richard V., "Air Electronic Warfare," *U.S. Naval Institute Proceedings,* Vol. 102, No. 884, October 1976, pp. 43–49. See also Dixon, Paul, *The Electronic Battlefield,* Bloomington: Indiana University Press, 1976.
7. Karas, Thomas B., "Military Race in Space," *The Defense Monitor,* Vol. 9, No. 9, pp. 1–8.
8. For a useful analysis of the impact of antisatellite weapons on the strategic balance, see: Melby, Svein, "Antisatellite Weapons and the Strategic Balance," *NUPI NOTAT,* No. 194, October 1980.
9. See: Snow, Donald, "Over the Strategic Horizon: Directed Energy Transfer Weapons and Arms Control," *Arms Control Today,* Vol. 921, November 1979, pp. 28–35.
10. For useful analyses of the relationship between formal arms control negotiations and space-basing options, see: Scoville, Jr., Herbert, and Tsipis, Kosta, "Can Space Remain a Peaceful Environment," *Occasional Paper* No. 18, Muscatine, Iowa: Stanley Foundation, 1978; Garthoff, Raymond L., "Banning the Bomb in Outer Space," *International Security,* Vol. 5, No. 3, Winter 1980–81, pp. 41–60; and Stares, Paul, "Outer Space: Arms or Arms Control," *Arms Control Today,* Vol. 11, No. 6, July/August 1981, pp. 1–3, 6–8.

C^3 Planning in Crisis Response

John Paul Hyde, Joe B. Warren, Jr., and Charles E. Kesson II

Crisis is defined as "an incident or situation involving a threat to the United States, its territories and possessions that rapidly develops and creates a condition of such diplomatic, economic, political or military importance to the U.S. government that commitment of U.S. military forces and resources is contemplated to achieve U.S. national objectives."[1]

By the measure of this definition, the occurrence of crises clearly is increasing. From the mid-1950s to the mid-1970s, crises occurred on an average of one per year. In the mid-1970s, the average jumped to three per year. So far, over the 1980s, the average is almost six per year.[2] Moreover, most of the events (more than 80 percent) have transpired in places other than where U.S. military forces already were stationed, so a military response usually had to consider taking in whatever unique command, control and communications (C^3) system would be needed.

Clearly, the C^3 demand per event has been increasing, too. Users have wanted more control, more coverage and more connectivity. Providers seek the safety of redundancy. Technology has created more sophisticated capabilities long sought but not possible before. Unfortunately, inventories have not always been able to keep up with demand.

While past crises can be lumped into some broad categories, such as evacuation of U.S. citizens, most essentially were unique in their major features (location, kinds and numbers of forces, political setting, timing), requiring a tailored, if not a totally unique, solution. As if that were not enough, they also have tended to unfold rapidly and often under the strictest secrecy. All of these factors and trends complicate C^3 planning for crisis response.

**This article originally appeared in the November 1986 issue of SIGNAL.*

Deliberate Versus Crisis Planning

The deliberate planning process is used to develop operations plans (OPLANs). The process is described in the Joint Operation Planning System (JOPS), Volumes I through III. It is driven by U.S. Defense Guidance and culminates in an OPLAN, prepared by the commander in chief (CINC) of a unified or specified command. This OPLAN identifies what forces he will use and how he is going to employ them to fight the war. Annex K of these OPLANs deals with C^3 systems, and JOPS tells the C^3 planner how to develop the annex and what to put in it.

The process that produces OPLANs ensures that they are executable. Major combat forces available for the CINC's OPLAN are identified to him, and he builds his plan around those forces. The C^3 systems available for OPLAN development also are identified for planning. In addition to the tactical communications capability available, the CINC and his components also may plan to use in-place fixed systems, including available host nation systems, as well as government owned and commercial (leased) systems.

This deliberate planning process is detailed and methodical, taking months and possibly years. The CINC's entire staff and the components' staffs are involved. Many checks and balances are incorporated in the process, including several levels of review. The final product is a polished and thorough document that provides a good starting point for a large scale military operation.

Crisis action planning is described in JOPS Volume IV. It is methodical and iterative too, but time is compressed to days, or even hours. The issues are the same: "What forces will the CINC use, and how will he employ them?" For the C^3 planner, the format for Annex K is still a useful guide, but there might not be as much time to polish it as there is in deliberate planning. The major factors that differentiate crisis action planning from deliberate planning are the number of people involved, the time within which to plan and execute the action and the frequency of crisis planning activities. From a C^3 systems perspective, another major difference is the near total dependence on deployable systems for the command and control (C^2) connectivity required.

Objective of the C^3 Planner

Whether it is deliberate or crisis planning, the C^3 planner's objectives are the same: to build a reliable C^3 system that supports the commander's preferred C^2 structure; or, if that is not possible, to precipitate agreeable changes in the C^2 structure to fit a C^3 system that can be built and operated reliably. Most C^3 planners work diligently at the first objective, sometimes succeeding and sometimes not. Fewer are skilled at the second objective, influencing change in the C^2 structure to match what can be built; in crisis planning, that seems to be a key ability.

Planners obtain information on developing crisis situations through countless means, including intelligence reports; CINC situation reports (SITREPS); the media; notifications from non-DOD government agencies, such as the Department of State; and our allies. Once the crisis situation has been identified (potential or actual), the crisis planning process begins.

The process for crisis action planning is flexible and may be followed as established in JOPS Volume IV or tailored by combining steps to meet the situation, whichever is necessary for proper and timely response to the nature of the event. Formal activity begins typically with a Joint Chiefs of Staff (JCS) warning order to a CINC, summarizing the situation, describing the military objectives desired and requesting his recommended courses of action. The CINC's response likely will contain a list of alternate actions and any additional factors he deemed appropriate for consideration by JCS. The courses of action are reviewed by the Joint Staff, and an alert order is issued designating the course of action selected. The alert order also gives guidance on such issues as rules of engagement, special constraints and public affairs considerations. A subsequent execute order is the authority to carry out the operation.

In today's high technology world, the challenge for a C^3 participant in crisis planning is easy to discuss—to plan and provide a secure, reliable C^3 system to support the mission—but it is not normally so easy to accomplish. Without extensive knowledge of the capabilities and availability of deployable C^3 systems and the interoperability characteristics of these systems—and any systems organic to the mission forces—The C^3 planner is destined for severe problems and possibly failure.

But even a thorough knowledge of the organic and deployable systems is not enough. A C^3 planner must perform well under stress and be credible to the operations planners. The importance of a good working relationship with the operations planners cannot be overstated, because that alone often has determined when the C^3 planner was brought into the activity and how much the planner could influence the C^2 structure. With time at a premium, the sooner he is involved, the better. Ideally, the planner is included at all levels when the process begins. If not, he will have to catch up in a time-compressed process.

Details of the C^3 planner's actions vary from echelon to echelon. The planner at the unified command responsible for the operation has the heaviest load—consolidating the requirements of the CINC and participating forces, determining the best workable solution, obtaining assets to meet shortfalls and issuing the directions for establishing the C^3 architecture, all the while building to ensure the required C^3 interoperability between the component forces. Since the mission force to be employed may vary between the alternate courses of action, knowledge of the range of organic C^3 capabilities within the forces, the types of communications equipment, the types and quantities of communications security (COMSEC) devices and keying systems is essential. Planning for interoperability between forces in the broadest sense should begin as the courses of action are being developed, and the C^3

planner must help ensure that each possible course of action considers the ability of combat forces to intercommunicate in a secure mode.

As the planning progresses, command relationships, lines of C² and intelligence and warning sources are determined. The C³ planner must track these as they develop to determine how best to provide deploying forces the supporting connectivity for each. Since the systems involved frequently must deploy with the mission forces and will not have time to "season" as in-place systems would, a key desire in this process is to provide redundant systems using different transmission media. As the C³ planner works the puzzle of how best to ensure that deploying combat forces will be able to interoperate and that the CINC or the Joint Task Force Commander will have the C² needed to control the operation, he may discover that available theater resources are not sufficient. When this happens, help is normally only a telephone call away.

At the Joint Chiefs of Staff level, the Contingency Support Division (C³SDO) within the Directorate for Command, Control and Communications Systems (C³S) is the focal point for JCS controlled communications assets and staffing the C³ portions of contingency OPLANs. C³SDO planners check that JCS controlled assets are ready to deploy if needed by the unified commander. The C³S planner at the Organization of the Joint Chiefs of Staff (OJCS) is in contact with the C³ planner at the unified command, helping when asked and telling the C³ planner about equipment and military satellite communications (MILSATCOM) resource availability. Meanwhile, the C³S planner is coordinating with the Defense Communications Agency (DCA) for any CINC requirements that can be provided over the Defense Communications Systems (DCS).

The C³ planner at the component level is busy too. Though possibly smaller in scope, the component C³ planner's problem is just as complex. He must determine the deploying forces' ability to operate in the desired configuration, with the proper COMSEC system and on the proper frequencies. Additionally, the component C³ planner determines whether unified command requirements can be met with component resources. All of this takes place concurrently with the unified command's planning.

C³ planning for crisis response has matured substantially with the increasing number of crisis incidents, the increasing complexity of crisis response options, the changing nature of crisis situations, the increasing compartmenting of planning actions and the increasing demands for higher quality secure voice and data connectivity. Only through the determination and personal efforts of the C³ planners at all levels tasked with making it happen does the process work.

Not only is it working, but it is improving as more people become proficient and as more assets become available. Through the CINC C² Initiatives Funding Program, significant capabilities have been fielded for secure voice and high-speed data using satellite communications (SATCOM) long-haul connectivity, all with smaller and more reliable equipment. While these successes are to be applauded, there is a danger of too much reliance on MILSATCOM systems as the preferred connectivity solution. The reluctance even to use high frequency or commercial compatible terminal equipment as a backup option is disturbing.

Institutional Efforts

The following section highlights some areas that are currently being improved to provide better C^3 planning capability.

Secure Voice Communications—Innumerable records show that there never has been enough secure telephone capability for planning and coordinating between unified commands, their components and the units that provide the combat forces for crisis response. Through the Secure Voice Improvement Program (SVIP), Secure Telephone Unit II (STU-II) telephones have been provided to many of the organizations involved in crisis reaction planning, and the STU-III will be available in even far greater quantities at lower costs.[3]

COMSEC—In 1980, following the Desert One experience, a program was established to define common COMSEC keying systems for forces that might be required to deploy into a theater and operate with in-place forces and other deploying forces. The Intertheater C^3 COMSEC Package (IC^3CP) does this by listing for C^3 planners the COMSEC systems and their distribution to organizations worldwide.[4] This package has proved to be an invaluable tool in planning crisis response.

JCS Controlled Tactical Communications Assets—In addition to resources that have been fielded through the CINC C^2 Initiatives Funding Program, the quantity of JCS controlled tactical communications assets is growing. The Joint Communications Support Element (JCSE), at MacDill AFB, Florida, will double in capability by 1989.[5] Other JCS controlled assets, operated and maintained by the Army and Air Force, will grow also, providing a quantum increase in the ability to support crisis responses.

MILSATCOM Management and Access—JCS Memorandum of Policy Number 178 (MOP 178), "Military Satellite Communications Systems," has been rewritten to provide a clearer general policy for the use of military communications satellites. MOP 178 now includes a new priority scheme for allocating this limited resource based on the required use and the urgency of the requirement and not just on the organizational level of the user.[6] Each crisis event has high priority. SATCOM access requirements necessitating preempting accesses assigned to operational forces. This preemption process frequently has illustrated to MILSATCOM users the wisdom of each operation having an alternate means of connectivity for each communications requirement.

Interoperability—Interoperability is an obvious ingredient of success in many crises. JCS Memorandum of Policy 160 (MOP 160), published last winter, institutionalizes new interoperability policies and procedures and outlines the management structure to attack interoperability problems throughout the life cycle of C^3 systems from statement of requirement to actual employment.[7]

Joint Tactical C^3 Agency (JTC^{3}A) Role—A new aspect of MOP 160 is the process of testing and certifying new C^3 equipment for interoperability. Conducted by JTC^{3}A, this program will allow the testing of new gear before it is fielded in large numbers and the polishing of joint interfaces before employment to validate that they work the way they are expected and needed to work. Interfaces that meet interoperability standards then will be certified by JCS for joint use. The program looks at the technical documentation supporting interface to see if it accurately describes how to make them work. Technical data deficiencies are corrected through a joint/combined interface procedure, the results of which go to system users and the publishers of the normal technical data as recommended changes.

Training for C^3 Planners—In the past, people assigned to C^3 planning positions received no special or qualifying training other than on-the-job experience. Success was often a result of luck, individual determination, exceptional leadership and enough time in place to study the situation before a serious incident occurred. To speed the learning process, a joint C^3I staff officers' course was developed and established by the Armed Forces Staff College (AFSC). It provides a good background in the JOPS process and the national and service C^3 systems available to commanders for planning and conducting operations. The course includes a three day practical exercise in crisis planning.

Crisis C^3 Review—An after-action program has been established to analyze specific C^3 plans and actions for crisis response. The constructive criticism approach identifies the positive and effective planning actions of unified commands and their components as well as real and potential problems that were not anticipated, and it exports the findings to the community at large. The program is credited with sensitizing the operational planners on the need to include C^3 planners from the start of any crisis activity.

Major Needs

An alternative training capability is needed for those C^3 planners who do not have the opportunity to attend the six week course at AFSC. A mobile training capability is being developed that is structured around the three day crisis planning exercise of the AFSC course. This training will be available to the C^3 staffs at unified and specified commands, at the various service signal schools and elsewhere, if desired, to provide an expedited course on the crisis action system; the tools available for planning; familiarization with the IC^{3}CP; satellite communications; and practical exercises in crisis response.

A major need of new crisis C^3 planners is practice, not the experience one obtains by solving real world problems in real time, but practice gained through some new realistic exercises. In fact, some unified and specified commands already are planning some surprise exercises to test current concepts on-line, to present a variety of crisis situations and to foster creative problem solving.

Only the next crisis will tell how well all of these improvements, this attention and those initiatives equip planners for C^3 planning in crisis response. If trends continue as they have, we should not have long to wait for the answer.

References

1. SM 502–85, Joint Operation Planning System, Volume IV, Crisis Action System, August 12, 1985.
2. Office Analysis of Crises Chronology (Informal) as performed by the Joint Operations Division, Directorate of Operations, Office of the Joint Chiefs of Staff.
3. MJCS 144–86, Department of Defense Secure Telephone Requirements, July 21, 1986.
4. Intertheater C^3 COMSEC Package Distribution Scheme, USCINCRED, June 1, 1986.
5. SM 344–85, Restructure of the JCS Controlled JTF Communications Capability, May 29, 1985.
6. Memorandum of Policy Number 178 (MOP 178), Military Satellite Communications Systems, September 4, 1986.
7. Memorandum of Policy Number 160 (MOP 160), Compatibility and Commonality of Equipment for Tactical Command and Control and Communications, 3rd Revision, January 7, 1986.

Developing the Technology for the SDI Battle Management/C³ System

C. Richard Whelan

The Strategic Defense Initiative (SDI) program is now in its second year. It is following a program broadly outlined by the presidentially directed Defensive Technologies Study Team (DTST). The DTST voiced a consensus among experts. It identified a number of challenging problems in developing the technology for an effective ballistic missile defense, but none that could be overcome. The DTST also identified the main features of such a system and outlined the needed technology development program. The nature of the battle management (BM) and command, control and communications (C^3) tasks now are becoming evident.

The total ballistic missile defense system would be based on a layered defense concept. An attacking missile force would be attacked by successive layers of defensive weapons through all phases of flight, from the boost phase, through midcourse flight and into the final terminal phase. The system would involve an array of space based, airborne and ground based surveillance and tracking systems and an array of defensive weapons suited to each phase of the engagement.

The battle management/C^3 system would provide overall system control. Each layer of the defensive system would function somewhat independently

**This article originally appeared in the July 1985 issue of SIGNAL.*

of the other layers. The BM/C^3 system also would operate in layers, with each layer having its own data processors and communications. These would provide command and control functions appropriate for that layer and provide for interface and handover functions between layers. The analysis and design of the BM/C^3 architecture, as well as the development and demonstration of critical technology to support the system, is now one of the major elements of the overall SDI program.

The BM/C^3 system for the SDI program is the main focus of the System Concepts and Battle Management program element in the DOD budget. Ninety-nine million dollars is allocated for this program in FY 85, with a significant increase to $243 million requested for FY 86. Although some reduction can be expected in this amount in the congressional appropriation process, it is clear that a substantial program in BM/C^3 for strategic defense is gaining momentum. This program will support continuing studies to determine the selected architecture and overall system design for the BM/C^3 system and the development of critical data processing and communications technology to support the system design.

The central effort in system design has been the work of 10 industry teams working on parallel competitive SDI architecture studies initiated in December 1984. While no specific results of these studies have been made public, some of the features of the system architecture are known, based on the earlier work of the DTST and subsequent published views by those involved in the program. Some of these features, coupled with essential attributes needed for an effective overall ballistic missile defense system, indicate where the emphasis will be placed in developing the necessary technology for the BM/C^3 system.

The system will require high performance. The greatest demand for high rate processing appears to be associated with the surveillance and tracking platforms. High resolution multispectral sensors needed for accurate detection, discrimination and tracking result in signal processing rates of billions of operations per second. These rates appear to be achievable using highly parallel signal processing arrays. These computers will each require thousands of processing elements.

Very Large Scale Integration (VLSI) technology will be widely used to minimize weight and power consumption, especially for the large number of computers operating on space platforms. The integration of low level processing functions with sensor arrays will tend to hold communication link requirements to a reasonable level in the large surveillance platforms.

Use of Computers

There will be hundreds of computers of a number of different types in the total system. The computers must be extremely reliable, using a fault tolerant architecture to achieve dependable operation without operator attention. Computers will not be accessible for ordinary adjustment and maintenance on space platforms. Fault tolerant operation for periods of 10 to 20 years is one kind of requirement being targeted for these computers. Moreover, these computers must be designed to withstand high levels of radiation.

The entire population of computers must be dispersed geographically. Subsets of the computer population will be dispersed by virtue of the geography of the several phases of the defensive engagement. Thus, computers associated with sensor and weapon platforms operating in boost and mid-course phase defense will be located on a number of orbital platforms. These will be deployed in an orbital cluster that will move relative to the engagement zone. As individual platforms move into, through and out of the potential or actual engagement zone, they will need to receive current data and pass along data to other platforms in the cluster. Computers associated with platforms in terminal phase defense will be located in clusters or zones allocated to target areas.

These clusters may be mobile, ground based or airborne platforms, with comparable needs for exchanging data between platforms. Thus, the computers required in the defensive system must be distributed geographically and operated in networks. The distribution of platforms and the allocation of functions between platforms will establish some of the basic requirements for communications in the system.

Communications Links

Several types of communications links will be required. Some will involve largely handover of data between sensor and weapon platforms operating in successive phases of the engagement. Others will involve transmission of status data and command and control information to widely dispersed platforms and locations. These networks will involve space to space and space to Earth links. The platforms on Earth will include fixed sites, mobile or movable sites and airborne platforms.

From a communications standpoint, the networks must be redundantly organized so that loss of individual nodes cannot disable total network operation.

Communication technology available today should be adequate to handle the data transfer rates required for various links in the BM/C^3 system. The DTST estimated that rates of 10 Mbps would be sufficient to handle maximum data transfer rates.

The communications links must also be highly resistant to jamming. Laser communications or 60 GHz links may provide acceptable approaches to achieving wideband links with a high degree of resistance to jamming for links between orbiting platforms. Techniques such as those currently under development in the Military Strategic/Tactical and Relay (MILSTAR) system satellite program could provide high resistance to jamming, at least for narrowband communications for space tc Earth communications.

Security of communications should not pose especially difficult problems for large portions of the communications traffic. Many links probably will be space to space links, associated with the transfer of highly volatile data. Channels carrying command and control communications will have to be given a high degree of protection in a multilevel secure system. Extraordinary security protection must be provided for the communication of weapon release authorizations. These security measures must be considered in all phases of the potential system evolution, from the design of hardware and software through future potential system implementation and operation.

The communications components and subsystems for the BM/C^3 system also must be designed for minimum weight and power consumption and must be highly resistant to severe radiation environments.

The attributes sought for the communication links of the BM/C^3 system are not unique. Technology for producing the desired properties, such as resistance to jamming, low weight and power consumption, redundancy, security and immunity to radiation environments, has been developed for a number of military communications systems. Perhaps the more challenging communications problems in the BM/C^3 system are associated with the organization and dynamic control of an extremely reliable and robust network of this size and complexity. The issues associated with this aspect of the communications system design are closely coupled with the details of the overall system architecture, which is now the subject of intensive study and analysis.

Strategic Computing Program Work

While the requirements for computers for BM/C^3 may be beyond present capabilities, they are not as demanding as some associated with the Defense Advanced Research Projects Agency's (DARPA) Strategic Computing Program (SCP). This program is supporting the development of extremely powerful new computers, based on unique highly parallel architectures. The development of hardware and software is being shaped by applications in artificial intelligence. Significant funding is also stimulating the underlying microelectronics technology in silicon and gallium arsenide VLSI systems.

In addition, significant funding is stimulating the research and development infrastructure, design tools, rapid prototyping facilities and computer facilities. The selected application thrusts of the DARPA SCP are not directly related to ballistic missile defense. The Pilot's Associate application is aimed at the tactical air mission; the Autonomous Land Vehicle is aimed at the future land battlefield environment; and the Battle Management System, at the carrier battle group mission. Nevertheless, SCP will provide a powerful base of technology support to the development of computers for the SDI BM/C^3 system.

The benefits to the SDI program flowing from SCP supported work are likely to be realized in several areas. In computer hardware development, SCP is supporting the development of highly parallel processors implemented in VLSI technology that will have the processing power needed for the most demanding high volume signal processing needs of SDI systems. In software technology, SCP is aiding the development of techniques and tools to increase software development productivity by at least one order of magnitude.

Further, SCP is investigating techniques to assure more rigid standards of verification and validation. Much of the basic work in microelectronics and enhancements in the supporting infrastructure will benefit the development of computer technology for the BM/C^3 system.

Applications of Artificial Intelligence

Although no one expects an intelligent computer to ever command a BM/C^3 system, a number of applications of artificial intelligence (AI) are likely to support the BM/C^3 system. These include the application of intelligent design tools in the development of hardware and software, expert systems for diagnosis and repair on the ground, autonomous platform operation in space, image analysis and understanding surveillance functions, situation assessment, battle management decision aids and many others.

Technology Development Work

The current funding for the Systems Concepts and Battle Management program represents about seven percent of the total SDI funding. Even with the sharp increase projected for SDI funding in FY 86, this percentage remains about the same. Currently, a little over one-half of these funds support system analysis and architecture studies; the rest are used to develop BM/C^3 technology.

The technology development work is focused on specific issues critical to the BM/C^3 system. Hardware and software are being developed for circuits that can withstand high radiation dose rates, as well as single event upsets. Methods and tools for software specification and verification are being developed. The objective is to develop efficient machine aided methods of generating software that can be formally specified and verified. Weapon release and command doctrines are being studied to establish a basis for designing and demonstrating effective command interfaces in the BM/C^3 system. Protocols for the distributed communications network are being developed. Battle management algorithms are being developed that will be suitable for use in a widely dispersed, loosely coupled, real time distributed computer system. These efforts are not extending the capability of computing and communication technology as much as they are tasks that effectively will apply advanced technology to the particular architectural and environmental requirements of the BM/C^3 system.

Studies and component technology developments, however, are not sufficient. A strong emphasis throughout the SDI program has been placed on demonstration of technology in all key areas. Several major simulation facilities are planned for development for the BM/C^3 system.

Both the Army and the Air Force are managing programs in BM/C^3 system and technology development. Th Army's Ballistic Missile Defense Systems Command in Huntsville, Alabama, is managing a number of projects in BM/C^3 system and technology development. Procurement is in progress for the development of an Advanced Fault Tolerant Computer System, a project that will include the development of the models, emulators, software and testing approaches needed to develop and to validate the processing concepts. Another project being initiated is BM/C^3 Algorithm Development, an effort that is expected to involve one approach based on improving and enhancing existing methods and a second approach using new methodologies drawn from applications of artificial intelligence. One objective is to develop unified methods of synthesizing algorithms suitable for application to specific architectural arrangements.

Another significant effort at the Missile Defense Systems Command is the planning and development of a BM/C^3 center. The focus of this activity is the development of a test bed that will permit the evaluation and demonstration of computer and software developments.

The Electronic Systems Division has the responsibility for the Air Force aspects of BM/C^3 for SDI, with the Rome Air Development Center (RADC) as the principal management center. RADC is initiating the development of a computer system architecture for space based battle management processing that will support ultra high performance, fault tolerance, multilevel security, 20 year survivability and autonomous operation. Another new task is the development of Protocols, Algorithms, Decision Aids (PAD) Applications Programs associated with battle management. This will develop the engineering environment used to produce conventional and AI based software, as well as the applications programs themselves. RADC also is initiating development of a battle management test bed that will be used to demonstrate and validate critical technology developments, like the fault tolerant computer and PAD applications programs. In addition to these computer oriented activities, RADC is initiating work on communications technology for a space based battle management communications network to complete the development task supporting the BM/C^3 system.

The division of SDI battle management activities among separate Army and Air Force management structures and development centers does not imply any duplication of effort, nor does it suggest that a competition is developing for control of this function. The division of activities reflects, in part, the practical reality that the management of the new SDI technology development projects is best accomplished by the in place service organizations that have the established background to carry out the tasks effectively. It also reflects that, within the general framework of established service roles and missions, the global scope of the layered ballistic missile defense concept is inherently a multiservice mission. The Air Force is pursuing the technology associated with space based surveillance and weapons platforms required for boost phase and midcourse defense. The Army is applying and extending the technology that it has been developing for over a quarter century for terminal and late midcourse defense.

The Navy, which has kept a low profile on SDI involvement, is now showing signs of increased participation. For example, the Navy Center for Applied Research in Artificial Intelligence (NCARAI) at the Naval Research Laboratory recently has shifted a major portion of its resources over to SDI supported work on the use of AI techniques in multisensor data fusion, decision aids and adaptive communications for battle management functions. Another positive indication was given in the testimony of Assistant Secretary of the Navy Melvyn Paisley before the House Subcommittee on Research and Development. He noted full Navy support of the SDI program, and added that: ". . . We continue to examine additional areas, such as battle management, where our expertise can be used to provide guidance. . . ."

The segmentation and geographical separation of several distinct development centers and test beds working on SDI battle management C^3I should not cause concern. After all, the eventual system is envisioned as distributed among dozens of internetted computers, dispersed on a global scale.

Communications for SDI

James Offutt and J. Bryan Danese

The Strategic Defense Initiative (SDI) concept presents a major challenge to the communications development community, both for accelerated development of existing technologies and for innovation in communications technologies. Although the challenges in communications technology may not be as formidable as those in SDI weapons technology (directed energy weapons for example), it is not clear that evolutionary development of existing communications technology will suffice. It is possible that basic performance requirements can be met in limited cases. But, at present, it appears questionable whether these same technologies can meet the rigorous size, weight and power requirements identified by the SDI research program or, most importantly, the cost, reliability and the goals of being able to manufacture it. In other cases, such as space communications networks with large numbers of platforms and changing topologies, we are dealing with a completely new and complex set of system requirements. Innovation will be required.

Communications provides the critical, indispensable means by which command, control and battle management (BM) functions are executed. Sufficiently capable, reliable, survivable and resilient communications systems are critical to the functioning of SDI systems. They must operate effectively in the face of anticipated intense enemy disruption efforts including many forms of physical attack, electronic countermeasures and nuclear disturbed environments. In any future conflict, efforts must be made to ensure that communications does not become the Achilles' heel of a potential SDI system, by which an enemy can nullify the system with relatively inexpensive electronic countermeasures at critical junctures.

If the United States eventually decides to develop and to produce an SDI type defense, an extensive array of communication systems would be required to interconnect a multitier, global, real time SDI system. The SDI commu-

**This article originally appeared in the July 1985 issue of SIGNAL.*

nications system would consist of large numbers of components on diverse space, airborne, fixed and ground mobile platforms.* Many types of links will be required ranging from low data rate, secure, high reliability command and control links to wideband, low probability of intercept data links. A major anticipated problem would be controlling the communications for a network of hundreds of space platforms with a constantly changing network topology. An SDI type system also would require integration with other national command and control systems, such as those for tactical warning and attack assessment, space defense and the strategic offensive forces.

The following sections briefly discuss the SDI research and technology program, the battle management command, control and communications (BM/C^3) architecture, communications requirements and communications technology base development areas.

The SDI Research Program

The SDI program has evolved rapidly, starting with President Reagan's speech in March 1983. This resulted in the Defensive Technologies Study (DTS) (also called the Fletcher Study). Based on the recommendations of DTS, the President formally started the SDI program and organization in April 1984. The goal of the SDI program (a five-year research and technology program) is to explore the critical technologies and to determine the feasibility of eliminating the threat posed by nuclear ballistic missiles to the United States and its allies.

The SDI System Architecture and Key Trade-offs Study, started in December 1984, is currently in Phase I with 10 contractor teams concurrently developing candidate SDI system architectures through trade-off analysis of the major components of sensors, weapons and BM/C^3. In Phase I, the contractors are examining the major system trade-offs to develop system architectures that are consistent with U.S. policy objectives, effective against the responsive threat, survivable and affordable. At the end of Phase I (summer 1985), the best four or five contractor teams will be selected to continue in Phase II.

In Phase II, the most promising architectures will be refined with more complete and detailed trade-offs and assessments of key technologies, including communications. Phase II activities also will include the development of plans for a BM/C^3 test bed to be developed as part of Phase III activities in FY 86.

SDI Multitiered Defense

The overall SDI system concept of operations is a multitiered defense designed to defend against ballistic missiles in all four phases of their flight:

**A number of SDI architecture types are under study, several of these include large numbers of space-based components.*

the boost, post-boost, midcourse and terminal phases. The four phases are illustrated in Figure 1.† This multitiered engagement of attacking missiles, which is fundamental to the SDI concept, results in a high degree of attrition of the attacking missiles. A defense with this capability would serve as an effective deterrent against enemy attack since it represents a very significant reduction in the overall attack size, and it destroys the structure of the attack. These features would prevent an enemy from achieving its strategic objectives.

The characteristics of the targets to be engaged and the weapons, sensors, engagement time lines and environment are different for each phase of missile flight. This leads to significantly different requirements on BM/C³ for each tier of the defense. We will survey briefly the weapon and sensor concepts in the following paragraphs and in the following section expand on the BM/C³ architecture.

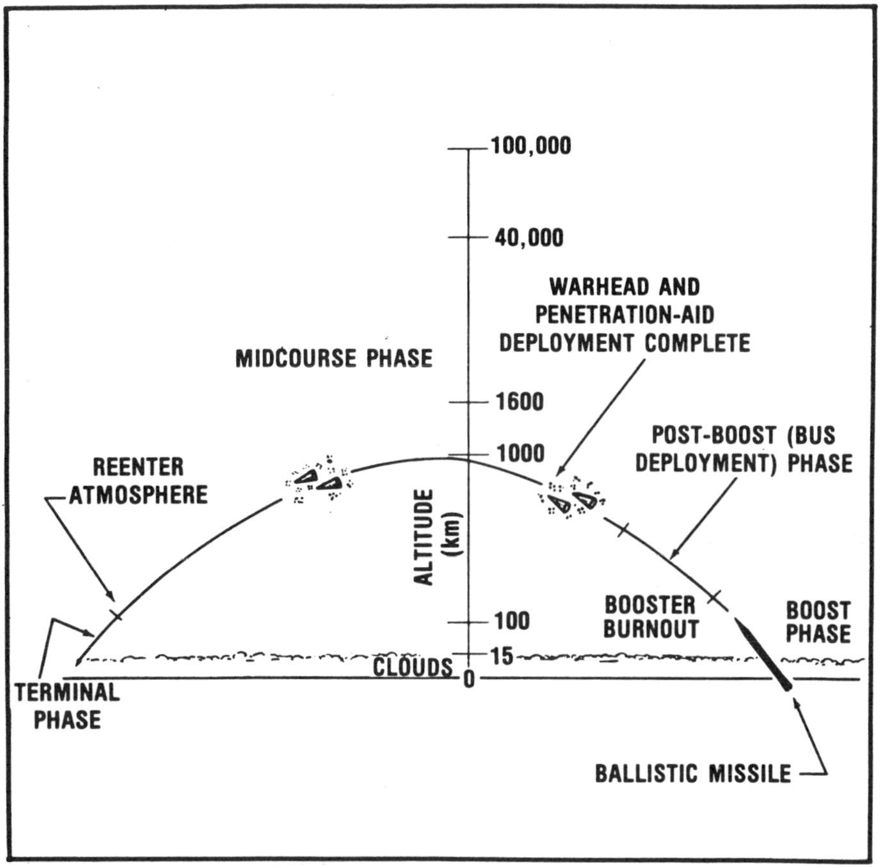

Figure 1. Phases of ballistic missile flight.

†*The Strategic Defense Initiative Defensive Technologies Study*, DOD publication, April 1984, p. 14.

A variety of weapons concepts are under study for SDI. These include space-based lasers, neutral partical beams and kinetic kill vehicles (KKV) (chemically propelled or shot from electromagnetic rail guns). Ground-based weapons include pulsed lasers with mirrors in space and ground-based interceptors for late midcourse and terminal defense. The space-based weapons typically are proliferated systems based in low earth orbits.

A number of sensor concepts are under consideration to support the SDI weapons concepts. A representative boost surveillance and tracking system would be an earth-looking shortwave infrared (SWIR) system based at high altitude or other survivable orbit. Surveillance of boost phase also might be performed as an additional function onboard the anticipated proliferated weapon platforms at low altitude. Above the horizon scanning longwave infrared (LWIR) sensors, based at lower altitudes, will be used for tracking the postboost vehicles and midcourse objects after deployment of the reentry vehicles (RVs). The sensors for terminal defense probably will consist of high performance, ground-based radars complemented by airborne LWIR systems. A critical problem in midcourse is discrimination of RVs and several sensor concepts have been proposed including imaging radar (60 GHz) and imaging lidar. It is assumed that all of the sensors perform onboard target track formation in order to reduce significantly the communications data rate requirements.

Battle Management and C³ Architecture for SDI

The BM/C³ system consists of the system components, interfaces, functions and procedures required to achieve and support the command, decision-making and control functions for an SDI type system. The battle management decision-making structure, with its supporting C³ system, is the ultimate determinant of system performance. The BM/C³ system implements the defense strategy and tactics and coordinates the functional elements in the order of battle. Communications provide the fundamental means for the execution of battle management, command and control. In addition to communications, the major technologies supporting BM/C³ include processor hardware, software and algorithms and man-machine interfaces (including C² facilities).

Specific BM/C³ architectures and their implementation depend strongly on the specific weapons and sensors employed. This subject currently is under active study in the System Architecture and Key Trade-offs Study. The first definitive results will come out of the study at the end of Phase I this summer. Study will continue in Phase II, going into 1986. For the purpose of generating BM/C³ architectures, general functional requirements and broad technical requirements still can be determined based on the character of the weapons, sensors, operational concepts, basing and key engagements time lines.

The multitiered SDI system, consisting of four tiers deployed worldwide in space, within the atmosphere and on the ground, is highly complex, requiring a very capable and sophisticated BM/C³ system to activate, assess, plan, direct and control it. As currently envisioned, it would consist of several hundred weapons, sensors and BM/C^3 platforms. A high degree of inter- and intratier coordination is required as well as intense regional and localized coordination in real or near-real time.

A notional schematic of BM/C^3 is shown in Figure 2. This schematic illustrates the major generic types of components, communications links and message categories in a BM/C^3 system. An actual system would contain many more components, more types of components and an accompanying proliferation of communications links and networks.

Note that control of the orbital weapons would be directly traceable to the National Command Authority (NCA)/Joint Chiefs of Staff (JCS) via the commander in chief (CINC) and spaced-based "battle manager." Weapons in space, like other strategic forces, would always remain under civilian political control. The role of human control (man-in-the-loop) is connected closely to the type of weapon and tier. This is currently under study, particularly for the boost phase intercept (BPI) system. It is clear that man will be in the loop and will have the power to activate or intervene and deactivate the system at any time. The structures and means by which this will be accomplished are under study.

The communications for an SDI type defensive system must interconnect hundreds of space-based platforms, airborne platforms, interceptors in flight and ground-based system elements. Because of the global reach of such an SDI type system and the basing of many of its elements, space communications play a central role. A number of functional characteristics or general requirements can be derived based on the nature of the systems involved. Those requirements discussed below represent only some of the major requirements that are beginning to emerge from ongoing systems studies.

Data Rate Requirements

Some representative data rates can be derived. (See Figure 2.) For example, the data rate from a single boost surveillance sensor at high altitude can be estimated.‡ Assume the sensor has a field of view that allows it to see 1,500 simultaneously launched missiles. With a track update rate of once per second and 1,000 bits per target state vector, the sensor output information rate (assuming onboard track formation) is 1.5 mbps. If we apply a minimum factor of two for encoding, encryption and overhead, the rate is 3.0 mbps. Applying an additional factor of 1,000 for signal spreading in an RF system (a moderate 30 dB processing gain) leads to a bandwidth of 3.0 GHz. Assuming the bandwidth is five percent of the carrier center frequency leads to a carrier frequency of about 60 GHz.

Going into the midcourse, the number of objects resulting from staging debris, RVs and decoys in the field of view may be much larger—a factor of

‡All numbers used here are representative estimates.

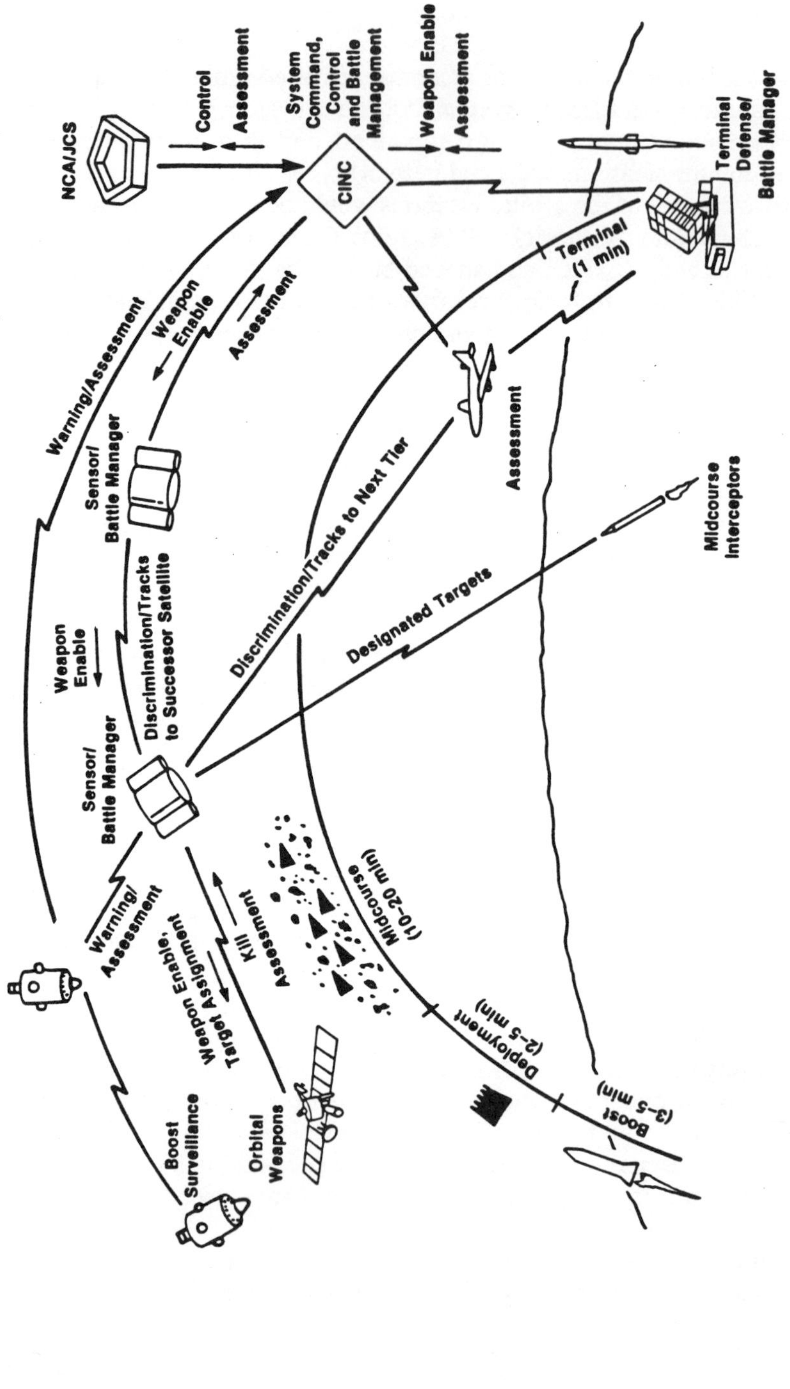

Figure 2. Battle management/C³ for multitier ballistic missile defense.

100 larger. However, the time to engage is somewhat longer—on the order of 20 minutes (1,200 seconds)—and the objects and in ballistic flight. Therefore, the track update rate may go from one per second to 10 per second. The net result of these competing effects is a sensor output on the order of 15 mbps rather than 1.5 mbps.

For the terminal defense system, the estimated data rate output for an airborne sensor is 100 to 200 kbps.

The above estimates provide a rough measure of the kinds of peak data rates to be expected for sensor outputs in the boost, midcourse and terminal phases. More precise estimates of sensor data rates require somewhat more precise system architecture definition and a corresponding level of sensor design definition. To extend the quantitative estimates to other links, such as those between battle managers in the boost phase tier, would be premature at this time since it is highly architecture specific. The results of Phase I of the System Architecture and Key Trade-offs Study should begin to shed light on these estimates for the several types of BM/C³ architectures under study by the contractors.

Command and Control Links

Command and control links are the critical links of the system, and their disruption or exploitation must be prevented. Command and control links usually are characterized by low data rates but often have stringent timeliness constraints imposed. The reliability and the probability of correct receipt of message on these links must be very high, particularly for critical activation or enabling messages. Very high security must be enforced to prevent exploitation, unauthorized use, alteration or damage to the system. Some critical command and control links will pass through the atmosphere, making the links vulnerable to uplink jamming and nuclear effects. High antijam performance will be required to prevent enemy disruption of C³ even if considerable resources are devoting to jamming these links.

Network Control

Many interconnected networks will be within the BM/C³ architecture. They may occur by region, tier or function. Functionally, they will include command and control, sensor, weapon and status reporting networks. Many of these BM/C³ networks probably will be disjoint because of the separate locations of their members or the nature, volume and timeliness requirements of their unique data traffic. The battle managers will serve as gateways for concentrating and distributing traffic, as switches for routing and interfacing and as controllers for running the networks. The battle managers are also sources, sinks and processors of information in these networks.

The changing network topology of satellites in many different orbits adds a unique dimension of complexity to the design of an SDI type defensive communications system. For a low altitude satellite, as it moves in its orbit, the relative locations of the other satellites in the system are changing continuously. After a period of time, some satellites will leave the line of sight while others will arrive. In battle, some satellites will be destroyed and others may malfunction. This introduces problems in the control of the changing system resources and coordination of joint and related activities. An SDI type communications network control system must be designed to cope with the changing network topology and the attrition of system resources. This is a unique requirement.

Antennas

The many space-based platforms of an SDI type system will be deployed at a number of altitudes (many being at low altitude) and possibly in several types of orbits. As a result of this, the line of sight between platforms typically will have time varying angular and radial rates. To communicate directly between these platforms would either require omnidirectional antennas, broad beam antennas or pointing and tracking of high gain antennas (we also include here laser telescopes). Since narrow beam systems will be preferred for reasons of security, performance margin and antijam capability, trade-offs in performance, weight, complexity and cost will be required.

Multiple Access

There would be many components of an SDI type communications system that require multiple access capability. Many of the space-based components will require simultaneous connectivity to several (or even a larger number) other satellites and ground locations, some very probably at widely separated angles. Some of these components will be within line of sight and others beyond the horizon, requiring relay. This multiple, simultaneous connectivity must be achieved further in the environment of a changing network topology with respect to the other satellites and ground locations.

Potential types of implementation of multiple access include spatial multiple access, time and frequency division multiplexing and combinations of all of these. Spatial multiple access, particularly for wide angle separations, will be a big drive on antenna design for RF systems and optics and detector systems for laser communications systems.

Aspects of the Communication Threats

A full discussion of the potential physical and electronic warfare threats and system countermeasures is beyond the scope of this article, which briefly covers some of the major features of the threat and system countermeasures. Because of the nature of an SDI type system, an enemy's strategy must include substantial efforts to destroy, degrade and/or nullify the system in order to carry out an attack with any assurance of achieving significant strategic objectives. The command and control system and, in particular, communications are lucrative targets. Further, attack of the C^3 system is emphasized in Soviet doctrine. Electronic countermeasures (ECM) are often a cost-effective solution to nullifying a system compared to outright destruction. For example, destruction of a low orbiting directed energy weapon with a multi-antisatellite (ASAT) attack may not be as easy, quick or cost-effective as jamming the command and control link at a critical juncture of the battle.

The potential physical threats to the space and ground segments of an SDI type of communications system and the required countermeasures are much the same as for other components of the system. Generic space segment threats include ASATs, space mines, directed energy weapons and prompt and delayed nuclear effects. Ground segment threats include damage and direct and collateral nuclear and conventional attacks, EMP, fallout, sabotage and enemy surveillance activities. The countermeasures to physical threats include hardening, proliferation, observable control and covert basing, maneuverability and mobility, safe basing and self-defense. All of these countermeasures add to system cost; detailed study will be required to assess the optimum balance of survivability countermeasures.

There will be many platforms in an SDI type system, all of which must communicate in some manner—spaced based, airborne and ground based. Because of the large number, it may be possible to design the communications system so that very few additional dedicated communications platforms will be required. In any case, the communications onboard other platforms will benefit from the survivability countermeasures implemented for the host platform. The communications must be as survivable as the rest of the platform and should not add to its vulnerability, for example, by a unique or targetable vulnerability.

Communications systems have some unique physical vulnerabilities such as antenna vulnerability both for space-based and ground-based systems. Communications front end low noise amplifiers may be subject to burnout by EMP or other high power electromagnetic sources. In addition, the integrated circuits employed may be susceptible to radiation effects. (They are generally less susceptible than high performance processors, however.)

Electronic warfare threats include signals collection and analysis, location and targeting, jamming, spoofing and deception, exploitation and nuclear degraded communications environments.

Two nuclear effects have the greatest effect on communications through the atmosphere: absorption and scintillation caused by the geomagnetically structured thin plasma remaining from high altitude nuclear detonations; and dust generated by nuclear ground bursts. Scintillation phenomena are complex, and they are most severe right after a high altitude burst in the localized region nearest the burst. The generated plasma spreads with time, decreasing the effects. Generally, the effects decrease with increasing frequency. For this reason, EHF has some significant advantages over SHF or UHF frequencies.

The suspended dust raised by ground bursts causes increased signal attenuation as the frequency is increased, becoming a significant problem in the EHF frequencies.

SDI Communications Technologies and Countermeasures

This section surveys some of the key technologies needed to meet the requirements of an SDI type defense and discusses some of the important potential survivability countermeasures. The technologies discussed here derive from a number of considerations including systems performance requirements, the threat and technology availability. Most of the technologies are near term. As systems studies continue, the technology requirements will become better defined as will the state of development of the requisite technologies. Then, the areas can be pinpointed where significant technology base development is required.

Space Cross-links

Substantial need for space cross-links exists in the SDI concept with baseband data rates perhaps as high as hundreds of megabits. The requirements for the types of cross-links vary, however, so it is doubtful that one solution will satisfy all cases.

For longer range, high data rate systems, either laser or mm-wave—probably 60 GHz—communications seem to be the solution. Both will be able to support data rates into the hundreds of megabits, are very low probability of intercept (LPI), and are very antijam (AJ) systems. (Although, laser communication is more LPI and AJ than 60 GHz). The state of the art is moving briskly in both areas.

For lasers, the recent development of solid-state (GaA1As) phased array lasers** provides a reliable, powerful pump source for Nd:YAG lasers. In the future, if problems with beam quality and thermal control can be overcome, solid-state phased array lasers might produce the main beam directly. For 60 GHz systems, recent laboratory developments in programs sponsored by National Aeronautics and Space Administration (NASA), the Air Force Wright Aeronautical Laboratories (AFWAL) and the Naval Research Laboratory (NRL) have demonstrated the feasibility of single 60 GHz Impatt device powers of 1.5 watts at 15 percent duty cycle. Therefore, in the near

**W. *Streifer et al., "Phased Array Diode Lasers," Laser Focus/Electro-Optics,* Vol. 19, June 1984, p. 100.

term, reliable solid-state amplifiers with power up to ten watts (employing power combining) will be available for space applications. For both laser and 60 GHz systems, realistic conceptual system designs exist for systems with megabit rates at ranges up to 40,000 Km.

Shorter range, low data rate, low duty cycle cross-link requirements might be met with omnidirectional SHF systems. To provide AJ protection, these systems might employ higher order encoding and signal spreading or operate in a burst mode. Having an omnidirectional antenna avoids potentially serious problems with antenna tracking and pointing and simplifies spacecraft integration.

Security Measures

Command and control links would interconnect the entire SDI type system. In order to ensure accurate and timely receipt of critical messages, multimedia or distinct redundant paths should be employed as in some present strategic offensive systems. Employing multimedia and redundant paths make enemy disruption efforts much more difficult.

Encryption and other security measures will need to be taken to ensure communications security. Security measures will need to be comprehensive since the SDI system would extensively involve many types of systems and basing modes and many levels of data classification.

Enemy jamming efforts can be defeated in a number of ways, including the use of spectrum spreading, antenna nulling, frequency and path diversity and LPI systems such as laser communications. Because of possible advances in the threat, antijamming systems may need the higher performance provided by combinations of measures, such as spread spectrum and antenna nulling. Path diversity offers another alternative. Using lasers for command uplinks is still another possibility. As a result of the laser's susceptibility to clouds, multiple ground sites connected, for example, by hardened fiber optic cable, would be required to ensure cloud-free line of sight and survivability. However, this approach still may be cost-effective due to the laser's relatively lower costs. The United States, Japan and the European Space Agency have studied the use of ground-based lasers for satellite uplinks. The United States is planning a laser uplink for the Galileo spacecraft. For NASA's Advanced Communications Technology Satellite (ACTS), an experimental 50 mbps laser uplink has been proposed and is being evaluated.

Atmospheric Links and Nuclear Effects

High altitude nuclear scintillation effects were described briefly in the previous section. As explained, going to higher frequencies generally reduces these effects. EHF frequencies, particularly around 44 GHz, represent a good compromise between decreasing nuclear scintillation effects and increasing attenuation effects due to dust and precipitation. (Frequencies around 22 and 60 GHz must be avoided because of strong atmospheric absorption bands.) Error correcting codes, interleaving and coherent phase shift key (PSK) waveforms offer additional protection.

Path diversity offers a way of overcoming high altitude scintillation effects from the localization of these effects. Because of the anticipated proliferation of SDI space assets (many with direct cross-links), path diversity is a countermeasure readily adapted to an SDI type communications system. Dispersion of communications transmitters plus appropriate interconnecting links will provide the needed diversity to circumvent localized scintillation effects. This is illustrated in Figure 3.

Another possibility for circumventing scintillation effects is laser uplinks. As discussed for command and control links above, if enough laser transmitters are dispersed and interconnected sufficiently, high probability of cloud-free line of sight can be ensured. Additional experimental data and proof of concept demonstrations are needed to assess this concept.

Multiple Access

One aspect of this subject is spatial multiple access. As discussed earlier, space-based SDI type systems would, in some cases, require simultaneous connectivity to multiple platforms in relative motion and at widely separated angles. It will be difficult to meet these requirements with present and near-term laser and RF cross-link technology without incurring possibly unac-

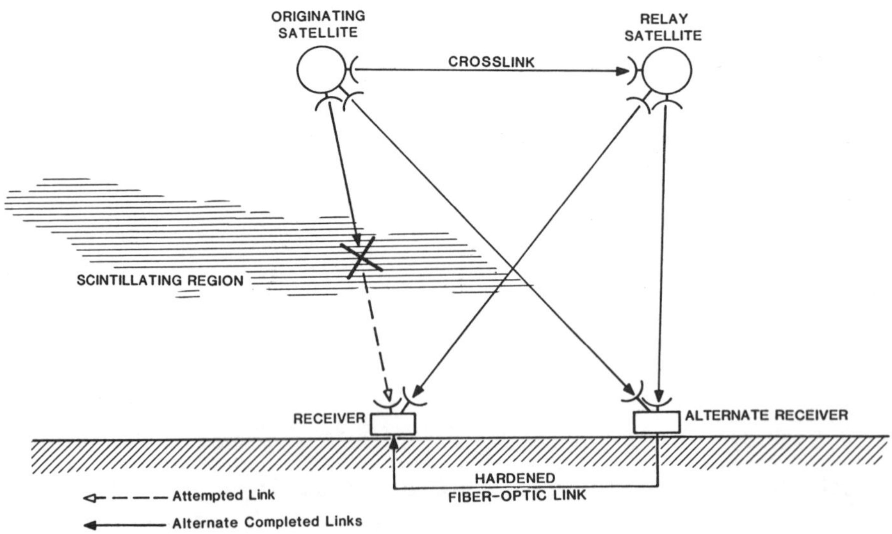

Figure 3. Path diversity to circumvent scintillation effects.

ceptable weight, complexity or cost penalties. New developments are needed for both technologies.

In many cases, narrow beams will be preferred for LPI and AJ as well as link performance. However, the transmitted and received beams will have to be very rapidly scanned to achieve connectivity with multiple platforms in near real time. For an RF system, this might be achieved with a phased array. However, beam scanning through 360 degrees is a substantially greater scan width requirement than is readily achievable with near-term technology. Using present technology and a large number of elements would result in a heavy and costly and, therefore, undesirable system.

Laser space cross-links typically require closed loop tracking and pointing. However, for shorter ranges on the order of 1,000 Km, a smaller aperture or defocusing might be used to spread the beam and remove the requirement for closed loop tracking. Spreading the beam reduces the power at the receiver, and this must be compensated for by increased beam power. Potentially, laser systems might be designed in this manner for shorter ranges, but more powerful, reliable lasers will be required. With less stringent beam tracking and beam stability requirements, mechanical scanning mirrors might be used to sweep the beam through 360° for multiple access. Proprietary designs for large field of view laser receivers have been advanced, but these are limited to less than ten degrees. As for RF, innovation is needed.

Networks

As discussed under requirements, there will be many interrelated networks within BM/C³ architecture. The nature of these networks depends on the BM/C³ architecture and the type of information traffic it carries. The requirements for these networks will be established in accordance with the architectural candidates for the SDI system. Potential BM/C³ architectures include centralized, distributed, federated and hybrid versions. It appears that circuit, message and packet-switched networks all will be required, with variable operating modes, such as broadcast and point-to-point for packet-switched networks. RF, IF and baseband switching and other onboard signal processing also will be required. It reasonably might be assumed that advances in space qualified, onboard processors for battle management functions will provide whatever processing power is needed for communications onboard processing. However, the major problem for space-based SDI networks appears to be developing the network system control functions and protocols for networks with changing topology and attrition of system hosts.

Conclusion

An SDI type concept presents major challenges and opportunities to the communications development community. Interconnecting an SDI type multitier, global, real-time system would require an extensive array of communications systems. It would require integration with strategic command and control systems as well as strategic communications systems. We have discussed some of the major communications requirements of an SDI type system and some of the key technologies, including high data rate space cross-links, high reliability command and control links, resilient atmospheric links, multiple access systems, advanced antennas and space-based networks.

Affordability is a key issue for communications no less than for the SDI system as a whole. Affordability for SDI communications will be achieved through the development of new higher performance and cheaper technologies, advances in manufacturing technology, and the exploitation, wherever possible, of existing and developing technologies. For example, for near-term deployments (mid-1990s), technologies derived from the MILSTAR program might be employed for antijam command and control links. And developments in several other programs for space cross-links also might be exploited.

In the case of SDI space communications networks for large numbers of low altitude platforms and changing topologies, we are dealing with new and complex system and technology requirements which require new system developments.

Soviet C^3: Theory and Practice

W. Robert Reeves

The Soviet concept of the control of troops or forces, *upravleniye voyskami* (silami), is in many ways different from the corresponding U.S. concept of command, control and communications (C^3). Troop control is a function and a product of Soviet military thought. It fits into a rigid and elaborate structural framework of concepts and definitions maintained by formal agreement of the highest Soviet military and political authorities. It is developed and coordinated by a large, centrally directed and highly controlled Soviet system of military research, publication and dissemination.

An understanding of concepts such as troop control can be gleaned to a great extent from the literature produced by this system. An excellent example of this literature is the major treatise published in 1977, *Fundamentals of the Control of Troops in Combat.*[1] Concepts like troop control undergo change as they are put into practice and as they are influenced by new methods and technologies. The development of these concepts and problems arising from their practice is often discernible through analysis of articles in professional Soviet periodicals such as the Soviet Navy's *Morskoy Sbornik (Naval Digest,)* The Ground Forces' *Voyennyy Vestnik (Military Herald),* the Air Force's *Aviatsiya I Kosmonavtika (Aviation and Cosmonautics)* or the Air Defense journal *Vestnik Pvo (Air Defense Herald).*

At some point, however, basic authoritative reference documents must be printed in order to disseminate the officially sanctioned definitions to those who must put them into practice. The often cited *Basic Dictionary of Military Terms,* published in 1965, is one such document,[2] and the more recent and comprehensive *Soviet Military Encyclopedia (SME)* is another.[3] *SME,* an eight volume work published in Moscow from 1976 to 1980 by Voyenizdat, the publishing house of the U.S.S.R. Ministry of Defense, is an unclassified official authoritative source of information with 13,000 articles. A more cur-

**This article originally appeared in the December 1985 issue of SIGNAL.*

rent authoritative source is the *Military Encyclopedic Dictionary (MED),* published in 1983, which provides a concise and updated version of *SME.* Both *MED* and *SME* are edited by the same Main Editorial Commission chaired by Marshal Ogarkov, the former Chief of the General Staff, who remains highly influential in Soviet military affairs.[4]

This article reviews the concept of troop control to provide a Soviet perspective as reflected in *SME* and *MED.* Other open-source Soviet literature is used to highlight aspects of the concept.

Troop Control: Concept and Management

Troop control as a concept occupies a unique place in the framework of Soviet military thought and a growing influence on all Soviet military activity. It is both a separate theory within the theory of military art and a function that permeates all aspects of the application of that art. In the Soviet theoretical view, the increased demand for control derives from the continuing "revolution in military affairs," the current stage of which some specialists have called the control phase.[5] Soviet emphasis is underscored by the definition of military science in *MED:* "In the postwar period, Soviet military science continues to develop. . . . Its main efforts are directed at [1] the investigation of the methods of preparing the country and the armed forces to repel possible aggression from imperialism, especially with the use of weapons of mass destruction, and [2] research of the problems of contemporary military art connected with the growing influence of scientific-technical progress on the development of military affairs [and research of problems of military art connected with] the influence of new means of armed struggle and of control."[6] A more current source is Col. Gen. Gareyev's book *M. V. Frunze—Military Theoretician: Views of M. V. Frunze and Contemporary Military Theory,* considered by many Western analysts to be the first major Soviet treatise on military strategy since the book of that name by Marshal Sokolovskiy.[7] Gen. Gareyev emphasizes the importance of the control process, pointing out its relevance "not only to the conduct of military operations, but also to realizing the development of the armed forces, the study and indoctrination of personnel, etc."[8] In other words, the control process is relevant in virtually every aspect of peacetime and wartime military affairs.

Control in the sense of management, another meaning of the Russian word *upravleniye,* is a watchword not only of Soviet military affairs, but of the entire Soviet society and economy. The application of systems analysis, cybernetics and computer technology to optimize the control or management of Soviet industry, agriculture and the entire economy has been a high priority of the Soviet leadership for years. Application of control in the military management sense has expanded through several levels, from Politburo level allocation of resources to the tactical commander in the field.

In the article "Troop Control," *SME* states that troop control is "activity by commanders, staffs, political organs, and support services to maintain the combat readiness and capability of troops (forces), to prepare for operations and combat, and to lead troops in the execution of assigned missions."[9] The Soviet concept places heavy emphasis on control, orchestration and management. Note that the "command" part of the Western concept command and control (C^2) is missing from Soviet troop control. "Troop Control' indicates that the commanders, staffs and political organs will conduct "control activity." But the Russian word for command, *komandovaniye,* is not used in the *SME* article or in its updated *MED* entry. Instead, the Soviets define a separate concept of *komandovaniye* and attribute three meanings to it: first, people who are endowed with authority to control subordinate forces and resources (commanders, deputies and chiefs of staff); secondly, military organizational entities, such as the Strategic Air Command; and thirdly, the execution of the functions for controlling troops (or forces) by command personnel.[10] This difference in terminology is a reflection of an important distinction in Soviet military thought: The authority to exercise control is a separate issue from the functions of controlling troops. Communications, so predominant in most discussions of U.S. C^3, also finds a different emphasis in the Soviet scheme.

SME and *MED* definitions of troop control itemize nine elements of the troop control "process":

- the continuous acquisition, collection, study and display and analysis of situation data;
- making a decision for an operation or an engagement;
- assigning missions to subordinate forces;
- planning of an operation;
- organization and maintaining coordination among troops and forces;
- preparing troops and staffs for military actions and conducting direct leadership of them;
- organization and conduct of Party political measures and all types of support;
- creation of a system of troop control; and
- organization of oversight and assistance to subordinate commanders, staffs and troops (forces).[11]

This list shows that the process is oriented toward troop control for a large-scale operation with little apparent time constraint. Indeed, a separate definition in *SME* of a commander's "decision for an operation" underscores this same orientation, and apparent contradiction, of planning well in advance and with a somewhat static representation of events. This belies the usual Soviet emphasis on rapid situational change. The commander's decision for an operation, according to *SME,* includes "the concept of the operation or engagement [*zamysel operatsii* (boya)], the missions for troops (forces), the organization of [troop] control, the fundamentals of coordination [among the branches and types of troops and forces] and the tasks in political work, all of which have been determined by the commander." The decision is "drawn up on a map; it is elaborated during the process of [further] planning the operation, and the operational directive and other documents are drawn up on the basis of the decision."[12]

It is clear from other open-source materials, however, that control is important at every level, not just the operational level, and at all times during the course of combat actions, not just during low threat conditions.[13] The process, as defined by *SME*, demonstrates that the Soviet struggle to adapt to the time sensitive, spatially extended C^2 (in Western parlance), and especially to the methods for exploitation of cybernetics and computers using automated systems or automated systems for troop control, in many ways just has begun. A review of individual elements of the process demonstrates how comprehensive the Soviets consider control to be.

The focal point of the elements or functions in the troop control process is the decision of the commander. All other elements of the process are aimed at or support this step. Collection, processing and analysis of situational data reflect the need for ample, valid data to enable the commander to make an informed decision. The increasing volume and complexity of data that are necessary for this to happen fuel the drive for automation.

Assigning missions to subordinate forces immediately after the commander makes his decision enables parallel rather than sequential planning of operations by the staffs of the commander's subordinates. This approach points out that time is the most critical factor to Soviet leaders in their implementation of troop control.

Planning an operation, as defined in a separate *SME* article of that name, is the "detailed elaboration of the content of combat missions, the sequence of their performance by troops (or forces), and the issues of allocating their efforts. . . ." It is accomplished on the basis of the decision of the commander, the directions of the staff of the immediate superior in command, a comprehensive situation estimate, the anticipation of possible developments of events, and accurate operational tactical calculations. The staff performs the primary work of planning.[14]

Organizing and maintaining coordination reflects one of the principles of military art and one of the fundamental tenets of Soviet military thought. *Yzaimodeystviye,* or coordination, is the coordinated actions of forces of the various services, branches and special troops. Success by Soviet definition is achieved only by the joint, synergistic efforts of all troops and forces in a combined arms sense. Assurance of that is achieved through its specific pursuit in the troop control process.[15]

As a separate function in the troop control process, preparation of troops and staffs and their leadership in combat seems to be an expression of the importance assigned to optimum combat readiness through training and careful orchestration of control over the troops during actual combat.

Organization and conduct of Party political measures and all types of support acknowledges the Party input and accounts for the pervasiveness of Party influence and power, even though a strict military chain of command is theoretically a fundamental principle of troop control.

Of great significance in this element of the troop control process are the important areas of endeavor hidden behind the phrase "all types of support." The Soviets have taken pains to define in detail what they consider "support of combat actions." They divide that support into four types: combat (or operational), special, rear and technical. Combat support includes reconnaissance (intelligence), defense against weapons of mass destruction, radioelectronic warfare, cover and deception, engineer support and security. Special support common to all branches of the armed forces include topogeodesic and hydrometeorological support. There are also types of support characteristic of each branch. Rear support includes a long list of topics, such as transportation, medical support and housing. Technical support includes such topics as rocket engineering, artillery engineering and motor vehicle engineering.[16] This list of support efforts, which must be organized and orchestrated as part of the troop control process, underscores the comprehensiveness of the troop control concept, and highlights the point made by Capt. W. J. Ruhe, USN (Ret.) (See "Soviet Submarine C^3" in this book), that Soviet troop control is not just C^3, but is more like command, control, communications, intelligence, navigation, electronic warfare and so on.[17] Of the lengthy list of support types, two bear special mention because of their increasing importance and effect on troop control.

One is radioelectronic warfare, *radioelektronnaya bor' ba* (REB), a Soviet concept close to the Western concept of electronic warfare (EW).[18] REB is a "set of measures conducted with the aim of identifying and suppressing enemy radioelectronic systems and equipment, and also [with the aim of] defending the radioelectronic systems and equipment of one's own troops (or forces); . . . [Electronic] measures for REB are conducted in combination with the destruction of enemy REB equipment."[19] Radioelectronic systems are a critical link in any system of control. The organization and conduct of suppression or destruction of enemy control systems and the protection of one's own is thus a critical portion of the troop control process.

The frequently discussed Soviet concept of *maskirovka,* roughly translated as cover and deception, is the second type of combat support of special interest in Soviet troop control. It directly influences the success of combat by confusing the enemy "concerning the number and distribution of troops (or forces), military targets, the content, capabilities and actions of troops, and also the plans of those in command."[20] Concealment of troop control is thus both a goal of cover and deception and a means of ensuring continuity and security of the troop control process.

Highlighting the creation of a troop control system as a separate element in the troop control process reflects a Soviet focus on the functioning of troop control via an integrated man-machine system. The Soviets long have been enamored with the systems approach and view troop control as an activity that must take place through a control system, and that must employ automation of that system wherever possible to improve information handling and reduce the time required for decision making. It is within the context of this element of the troop control process that the related and important but separate issue of communications and communication equipment is addressed. The overall "system of troop control" by *SME* definition is one in which "organs of control [such as staffs, departments], control posts, a communication system, automated troop control systems and equipment, as well as special systems to plot nuclear bursts, warnings, etc., are functionally connected."[21]

Automated troop control systems (for other services it is given the more cumbersome term "automated systems of control of forces") increasingly have become common throughout the Soviet military, and their use is one of the most frequently discussed topics in Soviet military literature. Automated troop control systems (ASUVs) are being developed "to increase the stability, continuity, efficiency and security of controlling troops."[22] Their need "stems from the growing complexity of the nature of combat operations and from an increase in pace, scope and intensity of warfare." The time factor clearly is the most important element in troop control. However, *SME* adds notably, "the importance of analytical soundness of decisions has risen sharply."[23]

Interestingly, the 1976 *SME* article, "Automation of Troop Control," is oriented toward the operational strategic level, like the 1980 article on troop control. For example, *SME* states, "The most important processes of computed activity are: prediction of the possible results of enemy use of weapons of mass destruction, evaluation of the possible consequences of using one's own conventional weapons and other weapons, as well as evaluation of the relative effectiveness of variants of decisions, estimates of the operational rear logistics and other estimates needed for making optimum decisions and planning troop operations."[24] Since 1976, however, with increasing Soviet accessibility, ASUVs are being stressed for decision making at the lower tactical levels.

Reaction to the ASUVs and to increasingly complex decision making has focused attention on how forcefully each commander or his staff oversees and assists subordinated commanders, staffs and troops—the final element listed in the troop control process. The issue is derived from the necessity to reconcile personal responsibility for decision making with a centralized culture where decisions have been kicked up the chain and actions have sometimes been based mindlessly on directions from above. This is aptly borne out in the Soviet principles of troop control.

As described in *SME* and updated in the *MED,* principles of troop control may distinguish the Soviet concept from Western C^3 better than any other characteristic. They are:

- unity of command;
- centralization of control in all links;
- firmness and steadiness in carrying out decisions;
- efficient response to a change in the situation;
- continuity and concealment; and

- the personal responsibility of commanders for their decisions, for the effective use of their troops and for the results of executing their assigned missions.[25]

Most of these principles embody a sense of exhortation to the commander. Unity of command, or one-man command *(yedinonachaliye),* is an excellent example. One of the most frequently cited concepts in Soviet military literature, one man command attempts to reconcile the need for a single person in complete charge of a given organization to the ideological requirement for collectivism and, in particular, to the practical requirement for Party control.

Undoubtedly, conditions for a commander are better now than in earlier times, when a political commissar had to countersign written orders. But the wording of unity of command issues in the Soviet press still treads a careful balance between explicit guarantees of unity of command on the one hand, and the acknowledged role of Party authorities and a certain level of collective decision making on the other hand. While commanders themselves report through the military chain of command, the commanders are clearly there in the end to implement Party policy. Obligations of military commanders and everyone in Soviet society always are listed as being to the Party and the State, never in the reverse order. Few individuals who are not themselves Party members are given positions of command, but the dual chains of authority in the culture at large and in the military apparently weigh significantly still. Impedance to genuine "one-man command" by the ever present alternate Party chain of command is complicated by the Soviet principle of the centralization of control. Even the "one-man commander" always is subject to orders from above in the centralized chain of command. Soviet efforts to counter the cultural tendency to wait for instructions and then to obey them unquestioningly clearly are reflected in the constant deluge of exhortation in the Soviet press for the military commander to be creative and to show initiative. The other side of that coin for the commander, however, is the further exhortation to know what the best course of action is or to know how to determine it, and, as the last principle states, he must take personal responsibility for his decisions, for the employment of his forces and for the successful execution of assigned missions.

In an interesting proposal made in 1978 regarding commanding officers (COs) of Soviet ships, VAdm. A. Mikhaylovskiy addressed the question of executing troop control properly and of possessing authority to control troops by virtue of holding command. Adm. Mikhaylovskiy, then CO of the Leningrad Naval Base, proposed a new system of graduated CO ratings based on skill and experience. This system formalized the stages of a CO's independence in controlling his ship. It motivated COs who, having arrived at command of a ship, began to experience a "dulling of his sense of responsibility" and provided a way to reward diligence and punish stagnation. The system consisted of six ratings, ranging in ascending order from fourth through first class, followed by "superior class" and "Captain-Instructor."[26] The rating system highlighted characteristics alien to the U.S. concept of commanding a ship. For example, controlling a ship independently while underway without the supervision of a senior officer would be reserved for COs with second class ratings, and who, by the time they receive that rating, would have served as the CO for one year. Controlling the ship in formation without supervision would require a first class rating; in other words, command of a ship for at least two years.[27]

Although this system was not adopted, its suggestion by a professional of Adm. Mikhaylovskiy's stature and its publication in *Morskoy Sbornik* suggest the fairly radical means, by U.S. understanding of command authority, of restricting that authority to improve knowledge and skills of military commanders (the Navy in this case).

The most important way to improve competency and enhance tactical decision making now is with the use of ASUVs to determine the best courses of action, judging from Soviet press reports. In one recent article, for example, a Soviet submarine CO lost an exercise engagement with another submarine because the CO of the latter submarine had used his ASUV even for making decisions that appeared obvious. The former CO, who did not rely on his ASUV, was caught in a simple situation that suddenly became too complex for him to judge correctly without automated help.[28]

Although *SME* defined troop control at a more operational than tactical level and with less of a sense of time constraint, Soviet press articles for several years increasingly focused on unit level decision making. An early indication of this emphasis in the Soviet Navy is the direction to each CO to treat his ship as a system and to depend more heavily on his assistants, watch officers and department heads in coordinating ship functions.[29] A more concrete indication of this was the creation in 1978 of a new shipboard control department, which performs the steps in the troop control process with the aid of ASUVs.[30]

A primary reason for having ASUVs, and one of the first functions they performed, is to manage combat data at a rate far greater than humans can perform. A step beyond that reason is Soviet attempts to program ASUVs with canned variants to various combat challenges. With such a capability, the commander can manipulate the continuously updated situational data and, on the basis of his judgment of that data, choose a canned response. This makes the troop control principle of efficient response to a change in the situation more manageable by commanders who cannot meet the standard in the troop control principle of firmness and steadiness in decision making.

The principle of continuity and concealment for troop control bears special attention because of the implications discussed above for REB and cover and deception. Together with stability and efficiency, continuity (also translated as secrecy or security) are achieved, according to *SME*, by several measures, many of which can be grouped under the rubric of protection of the overall system of control that links communication systems, ASUVs, control organs such as staffs, and commanders themselves.[31] A concept closely related to such protection is its antithesis, "disruption of control" *(narusheniye upravleniya)*.[32] While measures for disruption of control are aimed at the enemy, its provision in the Soviet scheme of armed conflict points out the requirement to protect against corresponding measures from the enemy, especially when that enemy is known to have the technological advantage.

Disruption of control is achieved mainly by destroying or neutralizing the enemy's electronic equipment and control systems (in the broad sense of that term). Nuclear weapons are considered the primary means to destroy these systems, while "aviation, artillery, airborne assault forces, airmobile troops and sabotage and scouting groups are also assigned [such] missions. . . ."[33] Protection and maintenance of control systems, through which troop control is implemented, logically must protect against a corresponding threat. Hardened, redundant communication systems and control posts, together with antijamming devices and other electronic protective measures are the main methods for such a defense. "The main efforts in disruption of control are concentrated on disrupting the control of the enemy's nuclear strike weapons in order to deprive him of the ability to employ them effectively."[34] The main efforts at protection, therefore, are likely to be corresponding control systems for Soviet nuclear strike forces and for the forces to be employed in disruption of enemy control.

Concluding Observations

The Soviet concept of troop control is a comprehensive set of critical measures or functions in a process aimed at orchestrating success in combat. The Soviets have developed this concept and have sustained an accurate conceptualization of activity while providing for its modification through lessons learned from practice and from new technology. The concept suggests a threat tightly woven both in terms of implementing troop control and disrupting that of the enemy.

Troop control seems to have three levels of focus, much like the Soviet breakdown of combat forms into strategic, operational and tactical. At the strategic level during peacetime, troop control approximates control in the sense of managing the development of forces and weapons and the allocation of resources as with other sectors of the society and economy. The Soviets look to the marshaling of countless data through automation for more efficient maintenance of combat readiness.

Soviet efforts to implement its concept of troop control appear to be heavily aimed at expanding availability of troop control equipment and improving methods for effecting control, particularly at the tactical level as more equipment becomes available. Definitions of concepts relating to troop control over the years show an increasing sophistication born of these efforts. Military professional literature underscores this trend.

At the same time, the paradox of a theoretical guarantee of one man command with dual authority in practice reducing command effectiveness especially at the tactical level apparently continues to plague the conceptualization and practice of Soviet troop control. Troop control, the process of what, how and when, is separate from the person in command and his level of authority and competence. The Soviets appear to be taking a two-pronged approach to deal with that separation. Along with a massive effort to encourage initiative and creativity, there is a simultaneous effort to build the competence of decision makers into the process. Authority then can be given to lower echelons with rigid, uncompromising rules programmed into the canned variants of automated control systems. While this can be seen as a serious weakness, it also should be viewed in a context that accounts for the massive effort to marry professional military competence and high technology with complete Party loyalty.

The Soviets see troop control as both a capability and a target. Attention to troop control in Soviet literature suggests that both aspects will be nurtured on a top priority basis.

Footnotes

[1]D. A. Ivanov, V. P. Savel' yev and P. V. Shemanskiy, *Osnovy Upravleniya Voyskami V Boyu* (Moscow: Voyenizdat, 1977), published in the United States under the auspices of the U.S. Air Force as *Fundamentals of Tactical Command and Control: A Soviet View* (Washington: U.S. Government Printing Office, 1980).

[2]*Slovar' Osnovnykh Voyennykh Terminov,* published under the auspices of the U.S. Air Force as *Dictionary of Basic Military Terms: A Soviet View* (Washington: U.S. Government Printing Office, 1976).

[3]*Sovetskaya Voyennaya Entsiklopediya (Soviet Military Encyclopedia)* (Moscoe: Voyenizdat, 1976–1980) in eight volumes. Hereafter referred to as *SME.*

[4]*Voyennyy Entsiklopedicheskiy Slovar' (Military Encyclopedia Dictionary)* (Moscow: Voyenizdat, 1983). Hereafter referred to as *MED.*

[5]Colonel-Engineer A. Volkov, "Effectiveness of Control of Fleet Forces," *Morskoy Sbornik* (December 1977) p. 21.

[6]"Voyennaya Nauka" (Military Science), *MED,* p. 137.

[7]M. A. Gareyev, *M. V. Frunze—Voyennyy Teoretik: Vzglyady I Sovremennaya Voyennaya Teoriya* (Moscow: Voyenizdat, 1985).

[8]Ibid., p. 406.

[9]P. K. Altukhov, "Upravleniye voyskami" (Control of Troops), *SME,* Vol. 8, 1980, p. 203.

[10]A. K. Slobodenko, "Komandovaniye" (Command), *SME,* Vol. 4, 1977, p. 259.

[11]"Control of Troops," *SME,* Vol. 8, 1980, p. 203.

[12]P. M. Petrus', "Resheniye na Operatsiyu" (Decision for an Operation), *SME,* Vol. 7, 1979, p. 115.

[13]Ivanov et al., p. 17.

[14]S. F. Begunov, "Planirovaniye Operatsii" (Operation Planning), *SME,* Vol. 6, 1978, p. 349.

[15]A. I. Radziyevskiy, "Vzaimodeystviye" (Coordination), *SME,* Vol. 2, 1976, pp. 123–125.

[16]V. S. Prokhorov and I. V. Shcherbakov, "Obespecheniye Boyevykh Deystviy" (Support of Combat Actions), *SME,* Vol. 5, 1978, pp. 651–653.

[17]Capt. W. J. Ruhe, USN (Ret.), "Soviet Submarine C^3," *SIGNAL,* December 1984, pp. 64–68.

[18]A. I. Paliy, "Radioelektronnaya Bor'ba" (Radioelectronic Warfare), *SME,* Vol. 7, 1979, pp. 29–30. For an in-depth analysis of radioelectronic warfare, see Floyd D. Kennedy, Jr., "The Radioelectronic Struggle: Soviet EW Doctrinal Development," *SIGNAL,* December 1984, pp. 59–63.

[19]"Radioelektronnaya Bor'ba" (Radioelectronic Warfare), *MED,* p. 615.

[20]"Maskirovka" (Cover and Deception), *MED,* p. 430.

[21]"Control of Troops," *SME,* Vol. 8, p. 203.

[22]"Avtomatizatsiya Upravleniya Voyskami (Silami)" (Automation of the Control of Troops (Forces)), *MED,* p. 17. The *SME* article is slightly different. See V. I. Vare, "Avtomatizatsiya Upravleniya Voyskami" (Automation of Troop Control), *SME,* Vol. 1, 1976, pp. 75–76.

[23]"Automation of Troop Control," *SME,* Vol. 1, p. 75.
[24]Ibid., pp. 75–76.
[25]"Upravleniye Voyskami" (Troop Control), *MED,* p. 766.
[26]VAdm. A. Mikhaylovskiy, "Klassnost' Komandira Korablya. Kakoy Yey Byt'?" (What Type of Rating Should a Ship Commanding Officer Have?), *Morskoy Sbornik* (November 1978), pp. 38–42.
[27]Ibid., pp. 40–41.
[28]Capt. First Rank V. Lushin, "The Commander and the ASU," *Krasnaya Zvezda,* April 11, 1985, p. 2.
[29]Admiral of the Fleet G. Yegorov, "The Effectiveness of Command," *Krasnaya Zvezda,* February 27, 1977, p. 2.
[30]RAdm. G. Kostev and Capt. First Rank K. Nikulin, "For Victory In Battle" (The Ship as a Control System), *Krasnaya Zvezda,* March 31, 1979, p. 2.
[31]"Control of Troops." *SME,* Vol. 8, 1980, p. 204.
[32]"Narusheniye Upravleniya" (Disruption of Control), *SME,* Vol. 5, 1978, pp. 515–516.
[33]Ibid., p. 516.
[34]Ibid.

Appendix

The following are articles from *Soviet Military Encyclopedia* that address control of troops (forces) and related topics, including command and communications. Most also are entries in the *Military Encyclopedic Dictionary.*

Airborne Command Post—Vozdushnyy komandnyy punkt
Automated Communication System—Avtomatizirovannaya sistema svyazi
Automated Troop Control System—Avtomatizirovannaya sistema upravleniya voyskami
Automated Weapons Control System—Avtomatizirovannaya sistema upravleniya boyevymi sredstvami
Automation of Control—Avtomatizatsiya upravleniya
Combat Effectiveness—Boyevaya effektivnost'
Combat Information Center—Boyevoy informatsionnyy post
Combat Organization of a Ship—Boyevaya organizatsiya korablya
Combat Readiness—Boyevaya gotovnost'
Command—Komandovaniye
Commander—Komandir
Commander in Chief—Glavnokomanduyushchiy
Commander in Chief of a Service of the Armed Forces—Glavnokomanduyushchiy vidom vooruzhennykh sil
Command Ship—Korabl' upravleniya
Communication System—Sistema svyazi
Control of Troops (Forces)—Upravleniye voyskami (silami)
Coordination—Vzaimodeystviye
Cover and Deception—Maskirovka
Cover and Deception Resources—Maskirovochnyye sredstva
Deputy Commander—Zamestitel' komandira
Disruption of Control—Narusheniye upravleniya
Flag Control Station—Flagmanskiy komandnyy punkt

Flag Officer—Flagman
Flagship—Flagmanskiy korabl'
Flag Specialist—Flagmanskiy spetsialist
General Staff—General' nyy shtab
Information Support for Automated Control Systems—Informatsionnoye obespecheniye v ASUV
Intelligence/Reconnaissance—Razvedka
Main Military Council—Glavnyy voyennyy sovet
Main Naval Staff—Glavnyy morskoy shtab
Main Staff—Glavnyy shtab
Mechanization of Troop Control—Mekhanizatsiya upravleniya voyskami
Military Art—Voyennoye iskusstvo
Military Discipline—Voyennaya distsiplina
Military Science—Voyennaya nauka
Naval General Staff—Morskoy general' nyy shtab
Network Planning—Setovoye planirovaniye
Nuclear Planning—Yadernoye planirovaniye
Operation Concept—Zamysel' operatsii
Operation Decision—Resheniye na operatsiyu
Operation Plan—Plan operatsii
Operation Planning—Planirovaniye operatsii
Operations Research—Issledovaniye operatsiy
Organs of Control—Organy upravleniya
Political Organization—Politicheskiy apparat
Political Organs—Politicheskiye organy
Primary Control Station of a Ship—Glavnyy komandnyy punkt korablya
Radioelectronic Equipment—Radioelektronnyye sredstva
Radioelectronic Warfare—Radioelektronnaya bor'ba
Secure Troop Control—Skrytoye upravleniye voyskami
Senior Commander—Komanduyushchiy
Shore Flag Control Center—Beregovoy flagmanskiy komandnyy punkt
Software—Matematicheskoye obespecheniye
Staff—Shtab
Systems Analysis—Sistemnyy analiz
Systems Engineering—Sistemotekhnika
Unity of Command—Yedinonachaliye

Soviet Submarine C^3

William J. Ruhe

Some general characteristics of Soviet command and control (C^2) of military forces appear applicable to submarines as well as to the command and control of other military forces. These common characteristics are well delineated by Soviet military writers, and they are evident in the Soviet Union's major military exercises as well as in its day to day military operations. These characteristics comprise: 1) a quick reaction flexibility for shifting command and control of forces from peacetime to wartime operations, for shifting from centralized to decentralized control as necessary and for maintaining firm, continuous control of forces despite situational changes and enemy efforts to disrupt communications and deny intelligence inputs; 2) the secure and deceptive means to provide offensive operations with a good element of surprise; 3) the means to absorb and efficiently make usable great amounts of information on the enemy; and 4) a combat stability (survivability) of C^2 systems under adverse conditions of enemy strikes and electronic warfare (EW). In addition, the Soviets emphasize the need to appreciate the enemy's policies, ideology, military doctrine, organization and the character of its commanders so that the Soviets can direct their own C^2 efforts more profitably—taking advantage of the psychological attitudes and the weaknesses of the enemy wherever possible. At the same time, the Soviets include under their C^2 efforts the importance of disrupting an enemy's control mechanisms in order to ensure the survivability and efficiency of the Soviet C^2 systems.

These generalities characterize Soviet submarine command and control, which mostly involves simple, unified command relationships and control of one or a few submarine units rather than control of groups of mixed military forces, including air and surface units. Nevertheless, control of Soviet submarine operations, which has appeared marginal in the past for the concept of employment being used, now seems to have matured to a point where

**This article originally appeared in the December 1984 issue of SIGNAL.*

Soviet submarines can apparently carry out operational concepts such as the first salvo attack on a fleet, or the discreet use of strategic weapons in an ongoing nuclear war. Determining the best way to control submarines in a variety of situations, in times of peace and war, has been a steadily developing capability. Landmarks such as the OKEAN exercises and this year's spring deployment of more than 90 submarines to sea for a grand exercise have provided evidence that the Soviet Navy has taken to heart the dictum that "skillful control of naval forces can bring victory."

Soviet Control of Forces

The Soviets place a major emphasis on the concept of "control of forces." Command, navigation, communications and intelligence appear to be only ancillary supporting elements for ensuring force control. Thus, in writing about Soviet command, control and communications, one should be hesitant in using the term C^3. Something such as CCCINEW (command, control, communications, intelligence, navigation, electronic warfare) more closely approximates the Soviet position on the C^3 function. Control is examined in great depth and is seen as a dynamic quality of military force employment that changes with different situations, different political objectives and different rules of engagement. A Soviet officer knowingly writes that, "the outcome of a naval action hinges on the status of force control." Adm. Sergei Gorshkov, head of the Soviet Navy, emphasizes the quality of control, which is being achieved "only by the use of automated control systems." He considers this to be a new weapon capable of reinforcing the intellect of the military commander. It is thus a quasi-collective intellect that now makes control decisions in the Soviet Navy. Hence, for submarines, decision making by a central commander or by a commanding officer (CO) of a unit is necessarily computer assisted. This application of computers, according to the Admiral, "concentrates theoretical knowledge and the experience of many military commanders in a form that assures its utilization by any CO with due regard to the specific situation that has arisen." Such a situation will be defined for the computer from the situational data inputted. The volume of this data is likely to be so great that only through the use of computers can it be synthesized to a form where it can be related to the mathematical algorithms and programs in the computer complex and can provide decision making suggestions. Yet, Adm. Gorshkov notes that the final control decision of the officer in charge "will always be subjective, with selection objectively substantiated by the computer's output." This does not rule out a creative approach by a CO for solving the problems at hand. Wisely, the Admiral also recognizes that, "even among indecisive officers in charge, an original (computer) search for the best decision cannot continue endlessly." A plan of action sooner or later must be determined by a hesitant CO. Thus control of submarines is not likely to result in a form of inaction and missed opportunity, as too often occurred in submarine operations during World War II. Adm.

Gorshkov's belief in the increased efficiency of control in modern day operations, despite "their large spatial scope and the shortening of time for decision making by an officer in charge," should be evaluated in light of the impact of a political commissar serving with each staff or submarine. It is clear that this Party representative, who is likely to be close at hand when major operational or tactical decisions are made by an officer in charge, will try to ensure the ideological purity and consistency of such decisions with Communist Party military policy. This additional factor acting on the control of a submarine is foreign to Western concepts but should be recognized when optimizing Western antisubmarine warfare C^2 functions. Although such a political officer has been made a watch stander on most submarines, it is unlikely that he would have adequate tactical knowledge either to approve or disapprove a CO's control decisions. But any departure by the CO from doctrine or the computer's suggestions, whose programs have been Party policy sanitized, would likely be viewed as a form of deviation and would tend to be suppressed. Thus, because the political officer can report the CO's deviation, he has a certain overriding effect aboard a submarine, and the CO's creative responses to tactical situations become far less likely. The isolation of the submarine in its operations effectively prevents a rapid resolution of such differences by higher authority. This is unlike the functioning of the political officer on submarine staffs, where his suspicions of deviation on the part of the commander can be resolved quickly.

As suggested earlier, in addition to controlling Soviet submarine forces with existing assets, the Soviets stress that control of enemy C^3 is of great importance in order to assure survivability and continuity of Soviet elements of control. Radioelectronic combat (or EW) is the essence of this form of control of the enemy's C^2 systems. The main two measures the Soviets employ are: 1) jamming enemy sensors; and 2) physically destroying them. Moreover, by gaining good intelligence on the enemy's control network, the Soviets believe they can deny up to 50 percent of the enemy's communication and intelligence collection capabilities. A capability to jam U.S. submarine broadcast frequencies except for the extremely low frequency (ELF) ones has been demonstrated by the Soviets over the past few years. In addition, the Soviets see the use of deception, such as false information, decoys, camouflage, etc., and electronic counter countermeasures as providing protection of Soviet control systems from enemy efforts.

Control in Peacetime Operations

Normal peacetime deployments of Soviet submarines are singly and comprise strategic deterrence patrols by nuclear powered fleet ballistic missile submarines (SSBNs) and high seas patrols by attack submarines to gain operating experience and to collect intelligence. In either case, the submarines are apparently controlled by decentralized land based submarine commands, which are responsible for the Atlantic, Pacific, Baltic and Black Sea theaters of operation. These commands collocate theater information and issue patrol orders to each deployed unit. Communications to the submarines are normally ultra high frequency (UHF) relayed from one of the 30 land transmitting stations within Soviet Bloc territory (five of these stations have a listed power of 500 kw). These transmissions are received by the submarine through towed antennas close to or on the surface of the ocean. Submarine transmissions back to the central command are high frequency (HF) or VHF via satellites and are normally held to a minimum during a submarine patrol.

An accurate geographic position for each submarine is maintained on board, derived mainly from Glonass, a satellite navigation system. This makes transition to war efficient, since a central command can develop its plans and orders knowing the approximate geographic positions of its submarines on a world plot of all potential enemy naval targets. The decentralized, shore based submarine commanders then can translate these plans and orders into a control of submarines for carrying out a global plan of action. Thus, with an approximate position for each deployed submarine, the central command can designate geographically to each appropriate CO of an attack submarine, the enemy targets that should be taken under fire at the initiation of a war. In addition, the central commander can give, via the theater submarine commands, positions of other surface ships in the area of the targets, the time to commence firing, the probable movement of the designated targets and possible submarine activity in the area of operations. From this data, the CO of each submarine can adjust the course and speed of his submarine to arrive at a good firing position, consistent with the established time for launching of weapons.

While it is unlikely that the Soviets keep a comprehensive peacetime plot of potential targets on board each submarine, U.S. submarines maintain such a plot with their systems such as Outlaw Shark. Coordinated operations of Soviet submarines during peacetime tend to occur only for occasional exercises that test wartime concepts of operations and for the development of peacetime capability for trailing of Western ballistic missile submarines.

The peacetime trailing of Western submarines is part of the Soviet effort to enable its submarines to stay close to an enemy strategic submarine. The Soviets want their submarines close enough so that at the initiation of strategic war, Soviet submarines can put an enemy SSBN out of commission so rapidly that few, if any, of its ballistic missiles can be launched. Several Soviet submarines usually are involved in such an operation, since holding a trail with one submarine is seemingly too difficult. Thus, several submarines would probably be under a coordinating commander located on board one of the submarines, who would control the operation. The submarines would try to detect a SSBN, usually when it sorties from its base, and then maintain continuous contact on it by at least one submarine of the group. Much of the tactical operation would be controlled by predetermined doctrine, but occasional reorientations, additional coordinating information and navigational checks of submarine positions will require underwater telephone communications between Soviet submarines, which can risk the overtness of this means of communication because of the peacetime environment. Similarly, *the overtness* of active sonar is of little consequence in peacetime operations.

Control of Strategic Submarines

It is becoming apparent (especially within the last two years) that the Soviets are bringing together their intercontinental strategic nuclear forces into one organization, a Strategic Nuclear Forces (SNF) command. Under the SNF command, decisions on the employment of strategic nuclear weapons would be vested in a Supreme High Command (SHC) chaired by the Supreme Commander in Chief and comprised of select members from the Party and professional military apparatus. SHC thus should include the Commander in Chief of the Navy, who would lend SSBN operational expertise to the SHC's deliberations. However, it is likely that operational control of SSBNs would be vested in a Navy Nuclear Commander in Chief, who would report directly to the SNF Commander in Chief. With centralization of command indicated for all SSBNs as well as for submarines carrying nuclear tipped land attack cruise missiles, the operational control of strategic missilery on submarines is the function of the Navy's Nuclear Commander in Chief. The CO of a submarine employing strategic weapons merely carries out the orders of his higher command. Whereas Soviet SSBNs were formerly expected to play a role in an initial massive strategic nuclear exchange, today they are planned for use as a blackmailing force for late-war bargaining and as an influencing force for peace terms. SSBN and its communications for control of its operations are least affected by nuclear blasts. Hence, in a nuclear war, the survivability and viability of this strategic system tend to be assured.

The latest class of SSBNs to be built, the Typhoon, has been designed to be extremely tough and difficult to sink. In addition, the Soviets have indicated a strategy for deployment of their SSBNs in protected bastions close to home or under the Arctic ice cap. Communications for operations under the ice are satisfactory when VLF and ELF broadcasts are used. These low frequency transmissions easily penetrate the surface ice and can be received down to depths well below the deepest ridges of the ice covering. Communications between Soviet SSBNs and protecting nuclear powered attack submarines (SSN) are apparently available for close range tactical coordinated actions, using underwater sonar telephones, or possibly explosive signals.

Control of Coordinated Operations in War

In 1970, the Soviets conducted an OKEAN sea exercise mainly to iron out the control problems connected with a coordinated attack against a major enemy fleet—a large attack carrier task group. The concept of a shore based centralized control of what was intended to be a first salvo initiation of a war at sea, using long-range cruise missiles, was tested. Later, off Lebanon, Adm. Gorshkov apparently tested the control of a similar exercise from a fleet command ship. But OKEAN '70, '75 and a most recent massive fleet exercise, suggest that it is shore based centralization that is used "especially when controlling forces employing guided missile weaponry," Adm. Gorshkov has indicated. In this type of preemptive attack, control would be accomplished while retaining a high element of surprise. The enemy battle group would be located geographically through surveillance by airborne means (satellites and aircraft). The contact information then would be sent by radio via satellite to the shore based command center. In fact, all information concerning the combat theater, such as enemy intentions, weather, location of other naval and commercial units including own forces in the theater and constraints on the operations of own and enemy forces, would be similarly made available to the command center. A great portion is made available from satellite readouts. Meteor satellites in a three satellite constellation provide ocean weather. "The Soviet ocean surveillance network appears to be one of the Soviet Union's front line tactical space systems. In 1983 alone there were four launches of nuclear powered radar surveillance satellites and three of the conventionally powered ELINT variety," according to Nicholas Johnson (see references).

In the case of first salvo operation, data would be collected from several oceans of the world, making possible coordinated attacks by Soviet forces in several ocean areas, nearly simultaneously, employing a mix of submarines, land based long-range aircraft and possibly surface warships. The central command would promulgate the overall plan, with the decentralized submarine commands issuing control directions to submerged submarines by VLF broadcasts, backed-up by ELF transmissions (to alert deep submarines to come shallow and receive the VLF broadcasts). Missile carrying attack submarines within a thousand miles of an enemy battle group are likely to be mustered for a massive strike in accordance with first salvo doctrine. Since firing of missiles can efficiently be carried out from widely dispersed positions, the functions of land based control are: to designate the geographic position of the battle group and its movement; to update information, continuously if possible, on enemy targets to the firing units; and to designate the time for missiles to arrive on target to assure a nearly simultaneous attack. The CO of a missile carrying submarine needs only to control the speed and the course of his submarine so as to arrive at a point within missile range of the battle group targets. From this point he can launch his antiship missiles so they will arrive on target at the designated time. The coordinated firing plan may be corrected by the central commander due to changes in the situation, with new orders transmitted to individual submarines to achieve the corrected plan. The Soviets see that targeting information will be supplied mainly by airborne means, and that target information derived organically by submarines is not essential to the control of weapon fire. By using a NAVSTAR type of global geographic positioning, each tracking unit can know the target's geographic position with high accuracy. Therefore, missiles can be directed accurately to long ranges with the terminal homing capability on each missile allowing for trajectory errors. Additionally, many Soviet antiship misiles can downlink video target information to the launching submarine or to an aircraft much closer to the target, and corrections then can be uplinked to the missile to ensure proper target selection. After launch of missiles, each attack submarine CO would take over control of his unit's operations, either for purposes of evasion or for mopping up enemy forces by means of torpedo attacks. Shifting from land based to on board control would be a matter of doctrine. However, at any instant the central commander can order all submarines to desist from their efforts and commence some other operation. It is also possible that during mop up operations several submarines would coordinate their torpedo attacks through underwater telephone means, with the senior commander in a group of submarines assuming tactical control over the group. (The Soviets have developed underwater communications that appear to be reliable, though not secure, and that have sufficient range to carry out coordinated torpedo attacks.)

It should be noted that the coordination of submarines with other naval forces in a first salvo attack requires but few communications between submarines and other forces. However, where submarines are used in coordinated operations in antisubmarine warfare (ASW) protection of surface ships, control would likely be vested in a surface command. Communications between surface, air and submarine units would frequently be used, and overall they should be reliable. Transit of Soviet submarines through enemy submarine barriers is likely to be with several submarines in coordinated operations, or with an escort group of surface ships and aircraft with the operations controlled by the surface escort command. Where coordination of submarine operations is indicated for an attack on groups of ships, even when land based aircraft are being used, a central, shore based submarine commander will probably control the submarine operations, similar to the manner in which Adm. Doenitz controlled his wolf packs of German submarines in World War II. Adm. Gorshkov has indicated, however, that the fatal mistakes of World War II are not likely to be repeated by his command. A minimum of submarine communications would be called for, with detection and tracking of shipping expected from Soviet satellites and long-range aircraft. Even damage assessment should be made by airborne means. Use of tactical doctrine apparently is expected to minimize the chances of collision between submarines or hits from friendly torpedo fire.

The wrestling match over command and control of submarine forces for various U.S. Navy operations is not evident in the Soviet Navy. Adm. Gorshkov has identified submarines as his main fighting forces. By using global communications, satellite surveillance, accurate geographic navigation and submarines of high performance, the function of command can be kept relatively simple, and control of submarines can be made possible from a remote and secure command center. At the same time, Soviet efforts to reduce an enemy's command and control efficiency are likely to be effective in a piecemeal fashion; but they will be crippling enough to ensure that their own command and control systems have sufficient survivability to produce a continuity and reliability that must be reckoned with.

References

1. Mario de Arcangelis, "The Case for C^3 Countermeasures," *Military Technology,* February 1984.
2. V. Chernavin, Admiral of the Fleet, "On Naval Theory," *Morskoi Sbornik,* No. 6 June 1982.
3. V. Charkin, "Controlling a Force in a Naval Battle," *Morskoi Sbornik,* No. 1, 1978.
4. Michael Deane, Llana Klass, Andrew Porth, "The Soviet Command Structure in Transformation," *Strategic Review,* Spring 1984.
5. S. Gorshkov, CinC of the Soviet Navy, "Problems with Respect to Control of Naval Forces," *Morskoi Sbornik,* No. 5, 1980.
6. S. Gorshkov, CinC of the Soviet Navy, "Problems with Respect to Control of Naval Forces," *Morskoi Sbornik,* No. 6, 1980.
7. John Hemsley, *Soviet Troop Control, The Role of Command Technology in The Soviet Military System.* (Oxford/New York Brassey's Publishers Ltd., 1982).
8. *Jane's Defense Weekly* May 12, 1984, "Soviet Intelligence."
9. *Jane's Defense Weekly* June 2, 1984, "Soviet Intelligence."
10. Nicholas Johnson, "Soviet Strides in Space," *Air Force Magazine,* March 1983.
11. S. Kritnitskiy, "Aircraft Against Submarines," *Morskoi Sbornik,* December 1971.
12. D. B. Lawrence, LTC., USAF, "Soviet Radioelectronic Combat," *Air Force Magazine,* March 1982.
13. W. F. Scott, Col., USAF (Ret.), "Continuity and Change in Soviet Military Organization and Concepts," *Air Force,* March 1982.
14. M. P. Skirdo, Col., Soviet Army, *The People, the Army, the Commander, A Soviet View,* Moscow 1970, Published under the auspices of the United States Air Force.
15. Milan Vego, "Submarines in Soviet ASW, Doctrine and Tactics," *Naval War College Review,* March–April 1983.
16. Milan Vego, "The Soviet Navy's C^3I: Doctrine and Capabilities," *Janet's Defense Review,* Vol. 4 No. 8 1983.
17. A. Volkov, Colonel-Engineer, "Effectiveness of Fleet Forces," *Morskoi Sbornik,* No. 12, 1977.

Making the C^3 Pieces Fit In Central Europe

Glenn K. Otis and Robert F. Driscoll

Command and control (C^2) is the exercise of authority and direction by a properly designated commander over assigned forces in the accomplishment of a mission; communications is the method by which the C^2 information is disseminated. Taken together, it is obvious that command, control and communications (C^3) is central to the conduct of battle. Effective C^3 assists the commander in acting at the most advantageous time and place with the most suitable means. Most battlefield automation systems are designed either to help the staffs do their jobs better and faster or to make tactical weapons systems more efficient on the battlefield. Staffs need automation to handle tremendous volumes of information and to analyze, integrate and package the information better to support the commander's decision process.

Computers are invaluable components of warfare. From the individual weapons system level, where, for example, the computer on the Abrams tank has helped to revolutionize tank weapons firing, to other networks, such as the tactical fire and direction (TACFIRE) system, automation has established itself and is here to stay. Nonetheless, commanders will not be running a push-button war: War remains a human activity, and the commander's intuition, savvy, daring and bravery will remain essential in recognizing and exploiting the decisive moments in battle. Technology will not replace the value of battle drills, nor the value of the initiative of subordinates in adhering to the intent in the absence of specific guidance.

It is not surprising, given the indisputable value of automation, that managers and commanders have used their initiative to automate their own functional areas and commands. Everyone is familiar with the negative aspects

**This article originally appeared in the November 1986 issue of SIGNAL.*

of such developments when they are nonstandard and uncoordinated: fairly sophisticated vertical (functional) systems, but poor interoperability and compatibility with other systems. The good news is that there is a growing urgency behind spending money across the spectrum of design objectives, so that the Army now procures systems that work together. We clearly are in an era when almost everyone recognizes the imperatives for standardized planning and design, not only in Army systems, but also in joint and combined arenas. This will become evident in the subsequent discussion of U.S. Army Europe's (USAREUR) Tactical Command and Control System (UTACC).

The NATO Environment

C^3 in the alliance environment is our greatest challenge. Secretary of Defense Caspar W. Weinberger has noted that, although NATO nations have more than twice the gross national product (GNP) of the Warsaw Pact countries, one of the causes of a widening military capabilities gap is the failure of NATO both to standardize equipment adequately and to rationalize resource allocation. He indicated, for instance, that C^2 tactical switching systems have suffered from parallel, redundant and nonstandardized development; that technical incompatibilities are found in cryptologic equipment, protocols, multiplexing equipment and disparate standards of mechanical, electrical and signal processing design. The result is a lack of netting capability, which is particularly crucial when NATO commanders attempt to coordinate their efforts. The Supreme Allied Commander Europe, Gen. Bernard W. Rogers, USA, agrees. He stated recently that NATO must collaborate more closely on research and development; coordinate procurement efforts; and use its financial, industrial and technical resources more efficiently. Once again, the good news is that the Alliance in general, and Central Army Group (CENTAG) in particular, are making significant strides to do just that. The unabated Soviet modernization programs, the high cost of modern weaponry, the Nunn Amendment monies and the recognized logistical and combat burdens of redundant, nonstandardized NATO systems contribute to pressures for increased coordination.

Interoperability

Within the Central Region, older existing national systems use national hardware not designed to interface. However, there are short- and long-term interoperability programs to remedy this problem. Currently, national army headquarters interact with Headquarters (HQ) CENTAG/4ATAF through communications links and liaison officers. The links are largely of the "swivel chair" type, in which either a national liaison officer brings national automation equipment to HQ CENTAG, or a NATO liaison officer with a terminal goes to the national army headquarters. In addition, of course, the corps across the CENTAG front establish their own links. NATO HQs in the Central Region are connected to a common C^2 system called the Central Region Automated Command and Control Information System (ACCIS), while the air forces are additionally connected to a different C^2 system, the German Electronic Information Command and Control System called EIFEL. The USAREUR Command and Control Information system (UCCIS) also provides information to HQ CENTAG via appropriate interfaces. Other existing C^2 systems are the German HEROS and the British WAVELL.

The Central Region ACCIS is the current NATO C^2 system used by CENTAG. It is an example of a system CENTAG is modernizing with an eye to interoperability. It currently uses U.S. worldwide military command and control system (WWMCCS) type equipment. With this equipment, commanders and staffs can pass reports and files and have access to personnel and enemy data bases. Meanwhile, more automated data processing (ADP) equipment is being bought, a remote dual-homed processor is being installed and CENTAG liaison officers will have more terminals. In the future, Central Region ACCIS will be upgraded with hardened sites and mobile alternate sites, with all stand-alone and internetted elements. In addition, the Central Region ACCIS ultimately will interface with other NATO and national automated C^2 systems.

Another example of interoperability development is the Land Air Campaign Initiative (LACI), a program that was tested during the CRESTED EAGLE 86 exercise. Its purpose is to interface dissimilar systems through a communications gateway. In the test, CENTAG and USAREUR gained insight on how to go forward in the introduction of both exercise and wartime automatic data handling. This helps in defining how the various C^3 pieces can best be made to fit in an environment that allows for the analysis of various system options without commitment to a specific hardware solution.

In an alliance, it is necessary to give equal weight to the fusion of information. Parallel to the development of automated systems, we have to modify our staff organizations and standing operating procedures to make better use of the improved technology. Moreover, CENTAG's perspective on interoperability has to be broadened because its area of operations contains German, Canadian, French and U.S. forces. For its part, USAREUR provides major combat elements to CENTAG and also is responsible for the combat support and combat service support of U.S. forces. Current initiatives in USAREUR are designed to lend coherence and interoperability to soon-to-be-introduced Army Standard Systems and then their successors. USAREUR is refining the functional networks (for example, TACFIRE, TSQ-73), which support staff's information needs and provide the data for targeting. Meanwhile, it is stressing the interlinking of the networks that aid the battle commander (for example, Maneuver Control Systems).

Tactical Command and Control

The major USAREUR initiative for the theaterwide automation of C^3 is called the USAREUR UTACC. Experiences gained in exercises (for instance, in NATO exercise CRESTED EAGLE 84) and analytical work performed in the Battlefield Automation Architecture Study provided the input that produced UTACC. CRESTED EAGLE 84 tested electronic briefing and analysis decision aids, primarily in the logistics area. The Battlefield Automation Architecture Study was a year long investigation of automation issues. The study succeeded in defining a composite and complementary structure for automation that coordinates theater needs with proposed Army developments—especially at echelons above the corps (EAC) and in joint and combined operations with the tactical corps level systems. UTACC ties together the USAREUR major commands with USAREUR headquarters, and it provides essential automation for these headquarters until implementation in the mid-1990s of the Army Command and Control System (ACCS) EAC structure, which will run the SIGMA STAR systems (that is, Maneuver Control System, Combat Service Support Control System). The study of the overall theater architecture is more complicated in that it encompasses all functional combat support and combat service support linkages, as well as the combat C^2 picture down to battalion level.

UTACC is a major step forward. Its capabilities include shared data base management and graphics. It is designed to evolve in a controlled fashion from existing capabilities. Users throughout the theater currently are supported by WWMCCS hardware and locally developed UCCIS software. With the availability of UTACC equipment, the UTACC system will provide automated support for non-WWMCCS functions. Nondevelopmental items, Maneuver Control System hardware and software, as well as Maneuver Control System-developed software, are the near-term base for UTACC. UCCIS software, which currently uses the WWMCCS hardware for support, will be integrated into the UTACC application software.

UTACC in effect will be a transportable, theaterwide system supporting HQ USAREUR, USAREUR major and subordinate commands and other theater users. It will interface with the U.S. Army, Air Force, NATO and host nations' C^2 systems, although not all of these interfaces are as yet clearly defined. It will provide USAREUR with essential data management ability until ACCS is deployed. In so doing, it will limit proliferation to Army Standard Systems or their Department of the Army (DA)-approved surrogates.

Conclusion

Clearly, these descriptions of selected C^2 modernization programs and issues are far from complete. Nothing has been said, for instance, about the major question of the transmission means that are required to support these initiatives or of their survivability. Since the United States is but one member of NATO, USAREUR is not the only interested player in the automation game. Our attempts to standardize in preparation for ACCS and to interoperate, as justified as these are from the USAREUR perspective, must be supported by other key organizations.

The needs of the staffs determine automation requirements. The staffs know the mission essential tasks and the critical items of information they require to allow them to sustain the force and to brief the commander on courses of action. The commander has to be comfortable with the software but divorced from systems. The commander must taste and experience the battle with the troops who fight it.

Comparing the C^3 Systems of NATO and the Warsaw Pact

Charles H. Jundt

In comparing the command, control and communications (C^3) systems of the Soviet Union and of NATO, the pushing of strategic requirements and the pulling of technology are responsible for the shape of C^3 today. From the trip wire days of MC-14/2 and NATO forces protected by their nuclear shield, to today's Soviet Operational Maneuver Groups standing in the elongated shadow of their theater nuclear capabilities, each side has developed its own C^3 system to meet the demands of today's dynamic battlefield.

As parity in strategic nuclear weapons was reached and nonstrategic nuclear weapons were developed and deployed, changes were made in the way many people thought about the unthinkable. The capability for nuclear weapons to alter drastically the course of war in moments was recognized by both sides. The forces have become more mobile and the battlefield concepts more dynamic. These dynamics, plus the complexities of nuclear war, placed new and demanding requirements upon C^3 systems. Attempting to meet these requirements is the basis for today's C^3 systems. Developing effective C^3 systems transcends all previous levels of complexity, driving new applications of sensor, computer and communication technology and setting new levels for cost.

The C^3 problems are similar, but the solutions are almost diametrically opposed and reflect differing NATO and Soviet C^3 philosophies. There are some similarities in the concepts for the Warsaw Pact and NATO organiza-

**This article originally appeared in the December 1984 issue of SIGNAL.*

tions. Both organizations are comprised of nations manifesting their concern for security in pledges of mutual defense, and each organization has as its military leader a representative of its major nuclear power. However, in taking a deeper look, the differences are far greater. NATO is a voluntary and cooperative organization. The Warsaw Pact has a hierarchical command structure with its control coming from Moscow. Therefore, while some may be concerned that NATO's consultative process will subvert rapid and effective C^2 in response to the dynamics of modern combat, the Soviet commanders do not have that problem. It is conceivable in a NATO/Warsaw Pact conflict scenario, that at the highest political levels, lateral coordination among Warsaw Pact leaders could terminate with the initiation of hostilities just when high-level political and military talks are beginning among the NATO leaders.

Should war break out, the Soviet leadership can relocate to hardened or mobile facilities for survival. The NATO national leaders must face their country's crisis or conflict problems from a range of secure resources that reflect that nation's security priorities. However, SACEUR (Supreme Allied Commander Europe) must maintain contact with these civilian and military leaders in whatever facilities are provided.

Comparing the Systems

NATO's communications problems will range from setting message priorities in a time of unprecedented traffic levels to enemy actions involving exploitation, jamming, physical destruction and perhaps nuclear effects. However, NATO has taken many steps to improve its communications capability. Survivability through redundancy is provided for every level of C^2 from the U.S. National Command Authority to SACEUR, to national forces using the national communications infrastructures and dedicated communication links. NATO Integrated Communications Systems (NICS) is designed to meet the needs of civil, political and military users. ACE-HIGH, the primary subsystem of NICS from Norway to Eastern Turkey, uses both line of sight microwave links and troposcatter links. The Satellite Ground Terminal program is progressing. The NATO Status, Control, Alerting and Reporting System (SCARS) elements will be linked through the Initial Voice Switched Network (IVSN).

The Commander in Chief of the Warsaw Pact will be talking to his military theater commanders in chief over the integrated communications systems of the Soviet Ministry of Communications, using the national communications infrastructure with underground landlines providing secure and survivable communications. It can be assumed that command posts in fixed facilities will have landlines plus the redundant radios associated with that command echelon. It should be noted that the reorganization of the Soviet military has elevated the central command level from the front to that of the Theater of Military Operations (TVD). Elevating the C^2 level provides more resources to achieve the desired correlation of forces, supports their concept of centralized control and permits planning on a larger unified scale. Their command and control system supports their hierarchical organization and focuses upon supporting the decision makers and implementing their decisions and directions.

In supporting the decision maker's staff, the Soviet's consider automation a key element in developing detailed plans for the integrated employment of all the forces in the geographic area under the commander's control. Integral to this planning process is the detailed task of attempting to foresee how the plan will work using detailed employment algorithms developed as part of their cybernetics approach to planning. When the decision is made and promulgated to the forces, it becomes each "sub-commander's" responsibility to develop those plans that will implement their portion of the operation. The key to the Soviet concept is the ability of the commanders (decision makers) to assimilate the information available to them, decide upon a course of action and implement that plan before the enemy can react to the environment. Therefore, to support their hierarchical structure and their planning process, Soviet military communications that disseminate these plans are primarily vertical and redundant. The Soviets state that the commander cannot permit loss of communications, which would interrupt the leadership of the troops.

NATO's command and control is a decentralized system that is beginning to use microcomputers to support corps, division and regimental size units. Air Force units also need computer support in maintaining a dynamic data base of those Soviet second echelon targets that provide the sustainability to the Warsaw Pact thrust. Both Army and Air Force commanders need information about their areas of influence and interest. To provide that deeper look, new systems are available that are capable of covering large areas. The TR-1 has a Side Looking Radar (SLAR), which has a range in excess of 300 nm enabling the TR-1 to cover more than 130,000 square miles each hour. Multiple TR-1s, using the Precision Emitter Locator System, will intercept and locate radar signals from deep in hostile territory. Sensors covering large areas generate vast amounts of intelligence data on the enemy threat, and operations and logistics reports on the status of friendly forces all inundate the decision makers with data.

The collection and correlation of data from monitoring and reconnaissance resources is one of the most difficult tasks to emerge in the evolution of tactical C^3. The Joint Tactical Fusion-Limited Operational Capability Europe (JTF-LOCE) is a prototype system that is designed automatically to distribute sensor data to all interconnected parts of the system, correlate sensor reports and develop and distribute battlefield situation displays and target nominations. Automation at many levels throughout the national forces is adding a new dimension of support to the decision maker.

The Warsaw Pact also has a heavy investment in intelligence collection. Using all the sources shown in Figure 1, the data are made available to Warsaw Pact commanders. Two productive intelligence sources for the Soviet

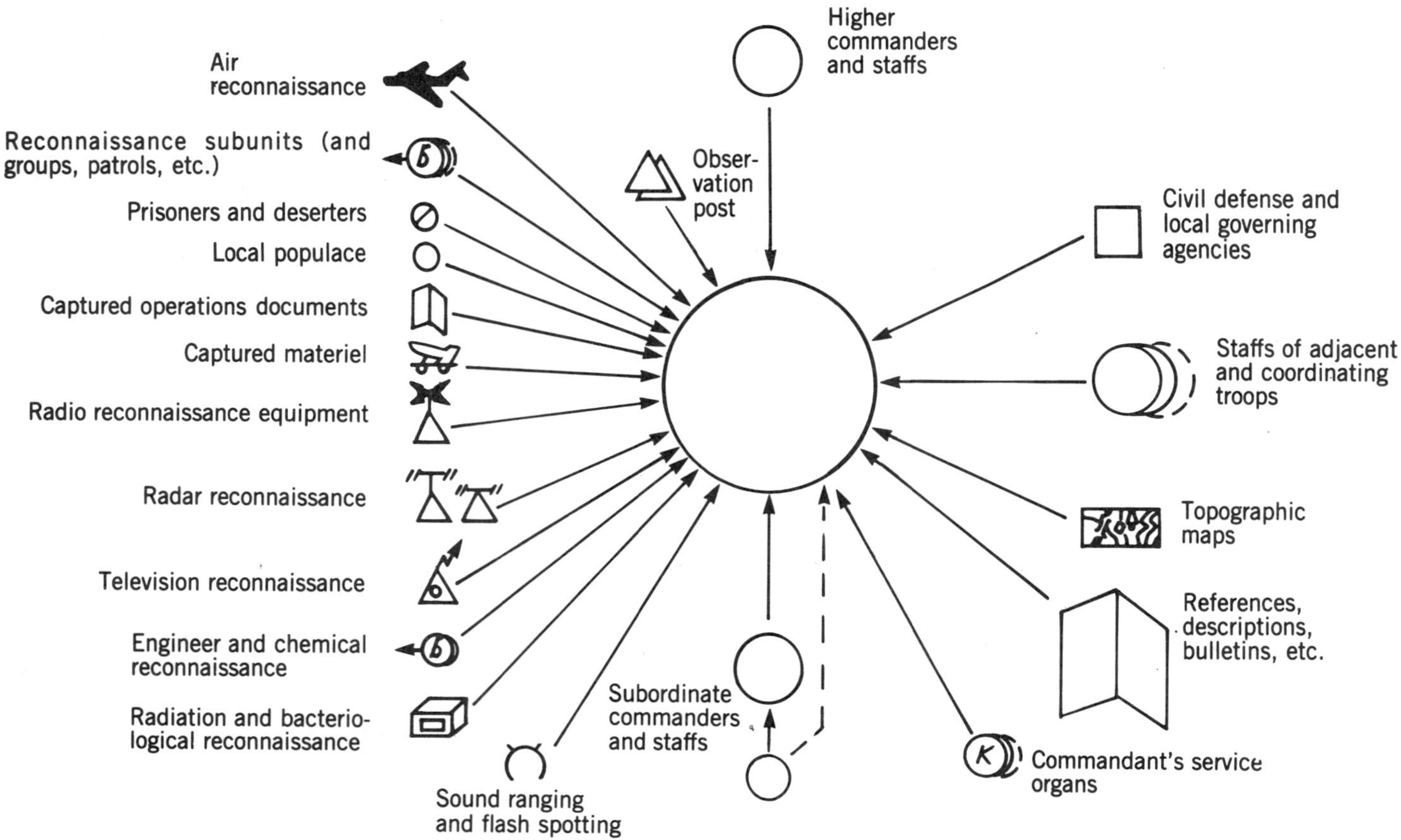

Figure 1. Warsaw Pact intelligence collection.

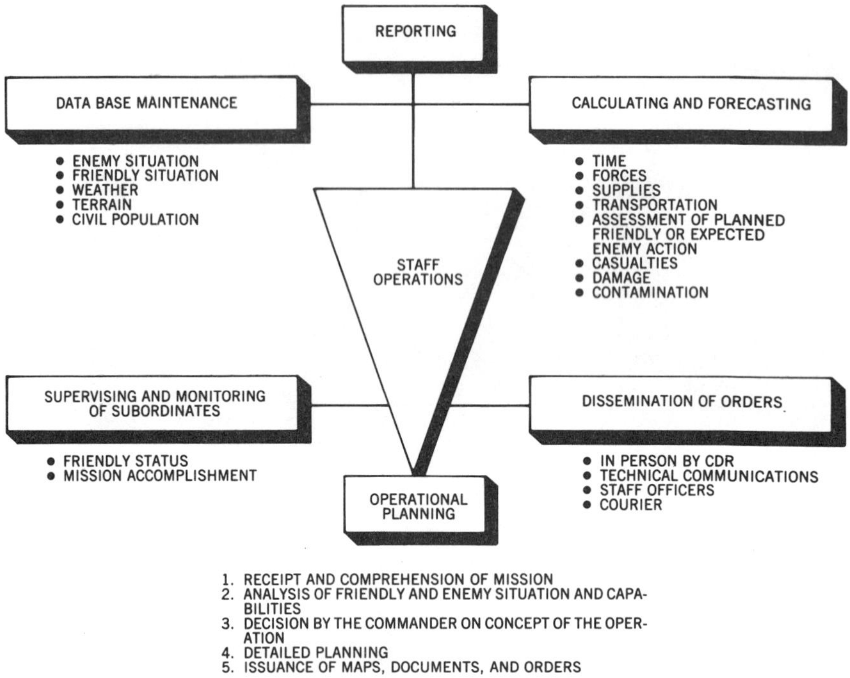

Figure 2. Staff functions reporting to final operational planning.

Chief of Reconnaissance are tactical reconnaissance aircraft and ground signals intelligence (SIGINT) units. The Soviets describe the combat requirement for tactical air reconnaissance as being primary and continuous. Fighter type reconnaissance aircraft would support missions along the lines of battle and to the rear of the NATO ground forces. Other reconnaissance aircraft would be assigned missions of longer range, such as locating assembly, bivouac and supply areas and detecting activities on communication or supply routes. These aircraft have visual, photo, infrared, radar and SIGINT capabilities. On a priority mission, targets can be identified, located and targeted in about two hours. Ground based SIGINT focuses upon detection of nuclear weapons systems and the ground forces and their associated C^2. Information collected by these means is used for both immediate targeting and as input for staff planning actions. The staff functions from intelligence and force status reporting to the final operational planning are shown in Figure 2.

With the development of its combat plan, each side must then promulgate that decision down to the forces under their command. In the NATO forces, this is over the communications links of each national force. The multilingual nature of the NATO forces may not be as big a hinderance to communications as the lack of interoperability among the NATO systems.

The preponderance of Warsaw Pact communications systems are of Soviet design and so reduce their interoperability problems. Figure 3 shows the radios connecting the Soviet forces.

COMMAND Echelon	Radio	Radio	Radio
REGIMENT	R-123	R-107	R-311
BATTALION	R-123	R-107	R-311
COMPANY	R-123	R-107	R-126
PLATOON	R-123		R-126
SQUAD	R-123		R-126
	Telephone & telegraph are not listed		

Figure 3. Soviet Army tactical communications equipment (Adapted from *Defense Electronics,* June 1983, p. 106).

The Soviet C^3 communications structure is vertical and therefore necessitates relaying communications from echelon to echelon. Therefore, message brevity and formatting are essential in minimizing the use of radios. When on the march and maintaining strict listening silence, radios are only for passing warnings of an attack or chemical, biological, radiological warfare (CBR) data, and there is no message receipt acknowledgement.

NATO and Warsaw Pact Air Defense

NATO air defense is perhaps one of its finest examples of multinational cooperation. However, its modernization has been ignored. Finally, after 30 years, major efforts are underway to provide an air defense capability to meet the threat. NATO's air defense C^3 must be able to do three things: survive; identify and track targets; and communicate instructions to the weapons systems. Fixed ground radar sites, the current eyes of the system, are subject to both physical and electronic attack. Attacking aircraft with antiradiation missiles and low-level penetrators able to penetrate below the radar line of sight pose a major threat to a ground based system. Airborne Early Warning Aircraft (18 E-3As and 11 NIMROD aircraft) provide supplemental coverage to ground sites and provide low altitude coverage, which would have required an excessive number of ground based radars. The existing Identification Friend or Foe (IFF) system is old, nearly obsolete and can be jammed. An integrated NATO identification system is essential to enable air battle commanders to see their "battlefield." Without communications to direct the air battle, the other two efforts are without meaning. Communications no longer mean just voice direction of the intercept, but rather digitized commands, which can conduct all functions of the intercept as long as both the target and interceptor are tracked and their spatial relationship can be supplied for an intercept computer.

Major technological and organizational changes have been occurring in Soviet air defense to meet the threats as they have perceived them. Soviet air defenses have reacted to the development of the B-70, the use of low-level penetration tactics, the use of electronic countermeasures, the current deployment of the cruise missile and the future threat posed by the B-1B and eventually a Stealth bomber. In weapons systems, the FOXBAT and SA-5 were developed as a counter to the B-70. The demonstrated ability of attacking aircraft to avoid or attack fixed surface-to-air missile (SAM) sites, coupled with the increased mobility of the ground forces needing air cover, brought about the development of mobile SAM systems. The development of a Soviet Airborne Early Warning System using the IL-76 can be seen as a counter to the cruise missile and as a means to provide survivable radar coverage. The new generation and modifications to existing air defense systems have improved the Soviets' low altitude and electronic counter-countermeasures (ECCM) capabilities.

The Soviet national air defense forces (PVO Strany) were reorganized in the late 1970s or early 1980s. PVO Strany consisted of the dedicated air defense interceptor aircraft, the fixed position SAM systems (SA-1, 2, 3 and 5) and the fixed radar sites which performed the early warning/ground controlled interception (EW/GCI) functions. Administratively, the 10 air defense districts have changed to five and come under the operational control of the Military District commander in peacetime, but under the TVD commander in wartime. This expansion of the defended area is an attempt to cope better with the threat of cruise missiles and low flying bombers. Perhaps it was the netting of air defense data and the use of computers in the Soviet Air Defense C^3 process that permitted the expansion of these air defense areas. An article by Marshall of the Soviet Union P. F. Batitskiy, who was Commander of Soviet Air Defense, described the role and importance of automated control systems in modern battle. He emphasized that it was not coincidental that the Soviet Air Defense forces pioneered advances in the cybernetic approach. John Erickson in his article "Computers and Commanders in the Soviet Military System: A General Survey," states that "manual or even semiautomated air defense systems can no longer cope with high density coordinated attacks designed to saturate the defense and to overwhelm both surveillance and C^3 systems. Fully automated systems are mandatory from initial target detection through fire control and interceptor vectoring." Newer Soviet radars with improved tracking will improve the control capabilities of all weapons systems and particularly SAMs. Automation will facilitate the solution of the multiple engagement problem by displaying information to each SAM battery of targets engaged by other batteries.

Initial attempts at improving the air defense capabilities included the use of the ARLS VHF data link system using commands generated by the VP-1 system. This system was introduced 20 years ago and is currently used in MIG-23 Flogger, late model MIG-21 Fishbed, Su-15 Flagon, Su-11 Fishpot and probably most other interceptor aircraft.[1] Considering the obvious C^3 requirements and the technological advancements, it is most reasonable to assume that the Soviets have continued their efforts in this area to provide automated control of their air defense resources over an extended area.

Conclusion

Testimony to the need for effective C^3 under modern conditions of combat is the emphasis placed by both sides on defeating the other's C^3 capabilities while protecting their own. The Soviet doctrine of Radio Electronic Combat (REC) emphasizes disrupting NATO C^3 by jamming or direct attack. U.S. C^3 countermeasures (C^3CM) strategy is the inclusion of the requisite destructive or jamming resources against key elements of the enemy C^3 system in attaining the overall mission objectives.

Today, the commanders of both NATO and Warsaw Pact forces are faced with C^3 problems that result from weapons systems capable of causing destruction to the opposing forces. The speed and mobility of today's forces necessitate that commanders be able to view extended areas of interest and influence, assimilate vast amounts of data, determine the appropriate actions of combined forces and disseminate their commands rapidly over extended distances. The challenge is great. The approaches to solving the problems differ between NATO and the Soviets. But one cannot compare the capabilities of NATO or Soviet C^3 in isolation. The side with the better C^3 capabilities cannot win unless it has the necessary forces and weapons, and the side with the necessary forces and weapons cannot win unless it has the better C^3 capabilities.

[1]Jane's Defense Review, Vol. 4, No. 2, 1983.

Developments in Integrated Circuit Technologies

James D. Palmer

Introduction

Command and control support functions for activities such as battlefield management in the 1990s will feature dispersed systems with distributed functions characterized by fully automated decision support systems. These systems will be hardened for survivability, secure for both up and down message management with common user communications systems that will rely on distributed data bases with a hierarchical interrupt capability. The systems will function on a multiplicity of hardware systems supported by interoperable software. The technologies that make this all feasible are the advances brought about by very large scale integrated circuits (VLSIC) and very high speed integrated circuits (VHSIC). The significant advances in graphics technology, which impacts displays and human-machine interfaces, will provide the boost to take these technologies to the field as well as at centralized remote locations. The major effect of distributed systems is the placement of responsibility for the gathering of data at the closest point of data collection. This implies that field command and control functions will be re-defined to include responsibility for entering information into the local data base, confirming and verifying this information, performing preliminary analysis and evaluation of the threat and making the information available for access at the earliest possible time to higher command authority. The quality of information and the capability to improve the accuracy of decision support tools will be much enhanced by technology advances that take the best of technology to all levels of the command and control structure and

This article was written especially for **Principles of Command and Control.*

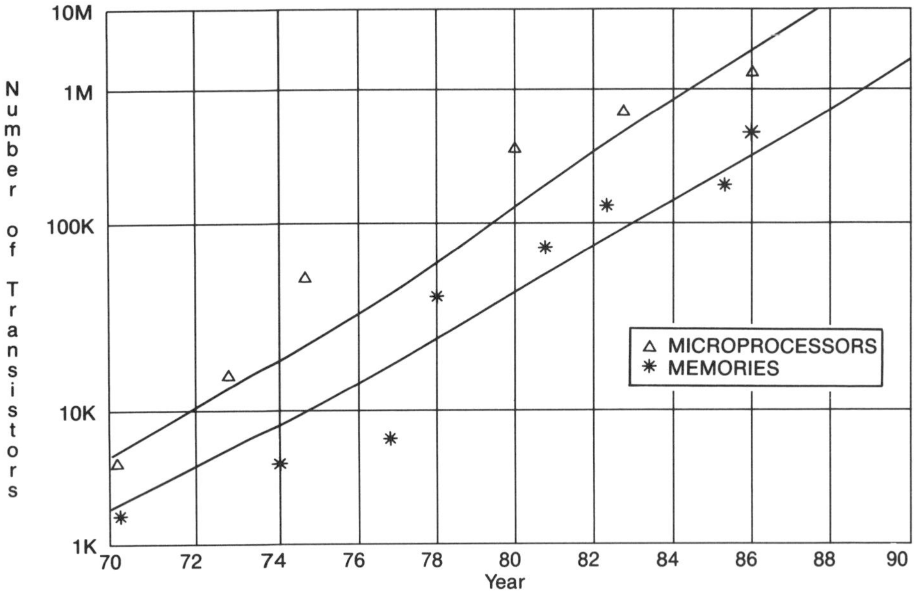

Figure 1. Number of Transistors per Chip per Year for Both Memories and Microprocessors

provide the ability to utilize displays, interact with various scenarios and bring the highest quality information to the decision makers.

Advances in Chip Technology and the Implications for Command and Control

The rapid pace of growth of microprocessor chip technology continues unabated, production costs continue to decline and the number of devices on a single chip proceeds on its exponential growth curve versus time, as is shown in Figure 1 and Figure 2.[1] We see the continued revolution in silicon microprocessor technology leading the way for more power, more flexibility, more storage capacity and greater ease of use. New generations of chips are being introduced at a rate that makes it extremely difficult for the software that drives these engines to keep pace. VLSI technology is having a tremendous impact, as the cost/performance ratios continue to decline. For computer graphics this means the development of high performance very high speed graphic chips that will enable multi-layer high resolution color graphic systems to be brought to the field.

The increasing sophistication of microprocessor architectures places them in a technology leading position, rather than following the lead of the mainframes and mini-computers. For command and control this forecasts a revolution in the way in which display presentations are made and in the way information is generated, stored, processed, distributed, and presented. The

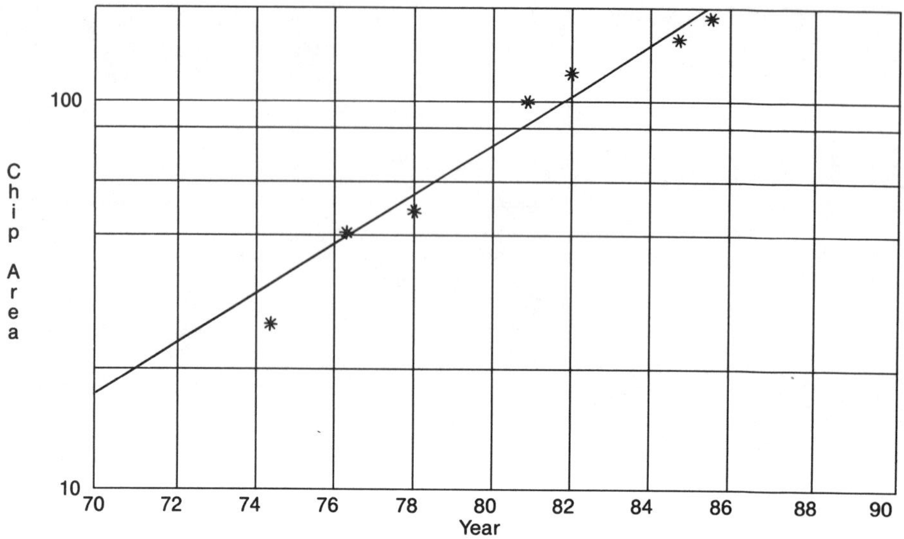

Figure 2. Chip Area Trend (in square km)

way is clear for low cost battlefield management architecture advances based on microprocessor technology to be made to an extent not visualized as achievable less than three years ago. Truly distributed data bases with intelligent data base management systems (DBMS) interfaces, coupled with architectures that place the power of mainframes in a microprocessor in the field at the closest point of contact, will enable command structure visibility into the battle scene now possible only by being there.

Emerging Architectures That Will Impact Command and Control

Emerging computer architectures are bringing a revolution in computing with microprocessors. Parallel systems that will increase the speed of execution of algorithms far beyond that achievable by a single processor, be it mainframe, minicomputer or microcomputer, are under development and test and will be available in the 1990s. The development of parallel computer systems does present the software designer with difficult problems that must be solved to take advantage of the explosive growth of hardware capability. Algorithms that are able to utilize parallel architectures are being developed and are forthcoming. Software tools that respond to the multitasking requirements of parallel computing include compiling technologies to partition global applications written in current high level languages and management of concurrent applications by the programmer coding in highly structured languages. A new concept presented by Intel proposes a functional language model bus oriented multiprocessor architecture with the information distributed among the multiprocessors.[1] The authors term this architecture "global memory system" because each microprocessor can access the memory of any other microprocessor board.

There are three basic multiprocessor design architectures:

- parallel processing with identical processors using one of many possible configurations; such as hyper-cube, pyramid structure and trees, that participate in the execution of a single application
- loosely coupled multiprocessing with each processor possessing its own private memory and performing one or more specific task
- tightly coupled multiprocessing in which processes can be migrated from one processor to another and processors share memory

This latter category is the most important for microprocessors and is essential to the development of advanced graphics processing requirements such as multi-tasking.

Another area of great progress is in the instruction-set with the introduction of the Reduced Instruction-Set Computers (RISC). RISC machines are based on the following concepts: single cycle execution, minimization of the cycle time, maximum use of all the space on the chip, optimization of the pipeline, reduced functions in the architecture and placement of more requirements on the compilers. The future of RISC architectures as a viable constituent in command and control functions such as battle management depends upon the development of capabilities to handle floating point arithmetic, virtual memory and multiprocessing. In addition, the systems must be compatible with other processors if any gains are to be made in operating software developed on earlier versions of any of the architectures.

Compact Disc (CD) Technology

Perhaps no other technology will give rise to a revolution in C^2 battle management computer architecture as will the realization of the CD. CDs are being developed that permit the user to write and store information and then access this data. The capability for massive information storage provided by this technology will certainly change the way in which we perceive and utilize optical disc drives such as the CD-ROM or the CD-I (interactive). Present laser driven CDs combine storage capability of 550 Mbytes of information, or about 1500 PC disks, with virtually instantaneous access to that information. Current CD technology formats for computers are all-text; however, work has commenced on more interactive forms of CDs that will serve not only text, but audio and video as well, functioning as a computer peripheral. Coupled with advances in graphics chips, this technology will provide 3-D graphic capability in the field with the capability of altering scenes or providing multiple windows with extremely high resolution and ruggedized construction. The ability of the CD to provide audio, video and computer information means that it will be possible to provide C^2 information at all levels with accurate and timely information derived from the entire spectrum of sensors, and mapped on stored scenarios prepared earlier to simulate a specific situation.

VLSI Graphics

Computer graphics is one of the fastest growing segments of a high growth industry. The development of new VLSI graphics chips is in demand in order to support the productivity improvements made possible by the application of high resolution graphics in the presentation of scientific and engineering information. High resolution graphics employed in command and control will be of inestimable value when deployed through distributed multi-tasking architectures made possible by the advances noted above in microprocessor technology. Coupled with CD-I compact disc technology advances, the future for human-to-computer interface simplification, data presentation and interpretation will see enhanced graphics stations with integrated audio, video and text presentations with multiple windowing and 3-D capability replacing the text terminals of today's technology.

VLSI Graphics Architecture

The issue of architecture is pervasive relative to the ability to take advantage of the advanced VLSI technology in microprocessors and computer graphics. Hardware architectures must be developed to support the increasing capability of display resolution that comes from improved VLSI chips. One of the most desirable forms for advances in hardware architecture will be in the compatible VLSI technology currently being developed. Graphic capabilities include display resolutions of 1024 × 1024 pixels or more with very high speed screen manipulations and color. Data should be able to be formatted on disks (for the near term) and be CD-I compatible (for the long term) for displays. This will involve database interpretation and translation to and from various standard graphics formats in order to take advantage of the multiplicity of equipment that will undoubtedly be in the command and control inventory (now and in the future).

The advantages in the VLSI graphics technology advances become immediately apparent in the application of generating realistic images. One of the most challenging problems confronting command and control displays is realism of the image in depicting the actual scene under observation. To be effective in the evaluation and assessment of the battle management situation and in the subsequent allocation of resources, the situation assessment must be realistic, in three dimensions and information rich. Advances in raster technology have made it possible to fill a screen with a tremendous range of color, shading and intensity. This is what makes it possible to display images possessing realism in three dimensions. Among the most difficult objects to model with any degree of realism are natural terrain and clouds. Stochastic processes are being utilized to generate the required detail for realistic depiction of natural scenes. Amanatides presents an excellent survey of the current state of development of algorithms that provide realism in computer graphics.[2] The importance of this is that present developments provide adequate realism for the observer to make judgements concerning the information displayed relative to the specific situation. Future advances using VLSI technology will generate more realistic displays that will be available for command and control decisions at all levels of the battle management hierarchy. These advances coupled with distributed architecture, CD-I massive data storage with audio and video potential, provide enormously greater capability at the disposal of command and control operators.

Chip Level VLSI Advances in Graphics

Graphics presentations bring such increased capabilities to problem solving to a wide variety of fields, including command and control, that the demand for improvements makes this one of the fastest growing segments of VLSI. Historically, VLSI graphics controllers have not been satisfactory for high-resolution graphics requirements of graphics terminals. As a result, much of the graphics hardware implementation defaulted to bipolar bit-slice processors and random logic rather than bitmapped terminals. The bipolar bit-slice processors are transistor-transistor logic (TTL) or emitter-coupled logic (ECL) technology, which results in high cost solutions compared to the potential savings derived from VLSI, and this logic is extremely difficult to map into future high resolution multiple-plane applications. Bitmapped terminals are not without problems related to performance, and new generations of chips are being introduced that address these high-performance displays.

To support and reinforce the growth of VLSI in computer graphics, it is necessary that hardware architectures be developed to provide for the rapid growth potential in display resolution that will surely occur over the next several years. These hardware architectures must provide for a high-degree of independence in the applications area without requiring any significant increases in overhead for the software interfaces. It is necessary that the hardware architecture be expandable to include the ability to take advantage of improvements surely to come in VLSI technology and not to require changes in the software each time a new chip architecture is introduced. The hardware architecture also must be in VLSI in order to control costs and to assure that the speed of response time is able to keep up with graphics implementations.

One of the most exciting opportunities in computer architecture brought on through VLSI is parallel architecture. Parallel computer systems make it possible to increase the throughput speed via the fast execution of algorithms designed for this application. As noted earlier, multi-processing and multi-tasking through multiple microprocessor boards provide true supercomputer capability at a fraction of the cost and complexity of present supercomputers.

The most interesting facets for command and control include the use of broad bandwidth buses with the concommitant computing powers of tens or more million instructions per second (MIPS) and throughputs of 100 MIPS, which is much higher than mainframes, but at a fraction of the cost and complexity. This means that it will be possible to handle large amounts of data from local databases on disks, CD-I's or received via communications networks directed to a graphics terminal capable of providing high-resolution three dimensional displays with realistic scenarios that include windowing and zooming capability. Further, this implies that the graphics processor is capable of handling the creation and manipulation of the massive amounts of data required for the development of actual bitmaps of the frame buffer, which is critical to the production of large bitmaps.

VLSI components provide significantly higher levels of functional integration when compared to current technology. This means that there is a substantial reduction in the number of IC packages required in a graphics system. It also means that it is possible to provide such silicon intensive operations as pixel support including video generation, the setting of bits in a bitmap structure and the actual movement of bits in a bitmap. These are accomplished at very high speed and offer the highest performance potential for systems with over 4 bits per pixel. With regard to architecture, packed pixel systems have individual picture elements packed into computer words in a linear-addressing structure with 16-bit words. One of these words will hold 16 1-bit pixels or two 8-bit pixels, much like the storage of ASCII characters as 8-bit bytes. The advantage of this architecture is that a clean view in the graphics bitmap is provided for software when computation involving individual pixels is needed. Consequently, this leads to a smaller number of pins in the graphics system and this means lower cost and higher reliability, while providing adequate performance.

Implementation of VLSI for bitmapped graphics terminals paves the way for graphics co-processors providing graphics capabilities at tremendous speeds. When connected via LANs and wide area networks (WANs), systems network architecture (SNA) gateways, X.25 or TCP/IP protocols and utilizing the OSI/ISO protocols, high speed graphics representations will become generally available via distributed architecture networks.[4]

In a recent announcement Intel introduced the 82786 graphics processor chip implemented in VLSI silicon technology. The 87286 provides for the utilization of high-level commands that in turn simplify the software and provide the hardware that will allow for acceleration for frequent operations.[4] According to Shires of Intel Corporation:

> Commands have been optimized to make implementation of graphics standards such as CGI, GKS, NAPLPS, Windows and GEM small and fast. The display works well with window manager systems and allows window manager software to concentrate on the organization of the display area and not on the need for creating and updating the window contents. The processor has the potential

> of the display-processor functions to become the silicon support for display managers as the graphics processor has become the silicon support for standardized graphics software. Many fast and flexible graphics primitives are provided. However, application always requires some graphics operations that are not contained in the primitives as some graphics engines do not allow the CPU to access the bitmap.

The Intel 87286 chip combines in a single processor nearly all that is required for the total operation of a graphics subsystem. It includes a complete dynamic random access memory (DRAM) controller that is built in, thus requiring no logic between the chip and the DRAMS that are used for the graphics/display memory. It can also provide up to 40 Mbytes per second memory bandwidth, and dual-ported DRAMS used for video may be utilized to increase the bandwidth for the graphics processor. In addition, the video logic is integrated so that no external logic chip is required to implement such characteristics as scrolling, panning, windowing or displaying bitmaps.

A new graphics processor has been introduced by Texas Instruments that provides for the entire graphics processor as a system. These new generations that are being introduced with ICs have been developed to address the performance problems that are associated with high-performance bitmapped displays that results from VLSI implementation. To assure that the massive amounts of data collected from a variety of sensors are readily available to the terminal, piplined CPUs built on VLSI technology will be capable of executing 128K bytes of ROM-based firmware and 8 VLSI gate arrays will be able to process raster operations at 13 million pixels per second. This means that it is feasible to draw 20,000 vectors or characters per second in 256 colors from a palette of 4096 colors using a TI 34010 graphics system.

The TI 34010 graphics system processor is a 32-bit graphics microprocessor possessing the ability to execute high-level languages.

> The 34010 combines a full general-purpose instruction set with a powerful set of graphics instructions that includes arithmetic as well as boolean PIXBLTS (pixel block transfers). Because it is completely programmable the 34010 can be used in many different graphics and nongraphics applications. The chip includes such system features as an on-board instruction cache, full interrupt capability, wait and hold functions and display timing control, as well as test and emulation support. It also addresses all memory down to the bit level with variable sized fields.[5]

One of the major attributes of the TI 34010 chip is its ability to provide field PCs machine linkages with facsimile interface boards and software with its support of a facsimile transmission instruction set. It is possible to connect this system via a facsimile device controller and an internal 9600 baud modem to handle ASCII to facsimile conversion and provide image handling software with all user friendly menus. The TI 34010 also provides 32-bit RISC capability with special silicon added to accelerate the graphics functions. It features a set of arithmetic operations on pixels that is innovative and adds new high-speed computation ability for colored-pixel values.

Conclusions

The capabilities noted in the review of advances in VLSI technology in microprocessors and graphics processors clearly demonstrate that the ability to provide the necessary graphics support to enhance presentations in battlefield management systems will be a factor in the manner in which these architectures are implemented in the future. As determined earlier in this article, distributed architectures based on microcomputing technology will be as capable as mainframe technology. These advances depend on what happens in VLSI technology, particularly in graphics applications. The opportunities for significant, substantive improvements in distributed systems architectures employing microprocessors that are as capable as mainframes will lead the way for the introduction of this technology in command and control of battlefield management systems. The potential of deploying hardened, survivable, secure equipment with outstanding capabilities of realistic three-dimensional displays and hierarchically controlled distributed databases with intelligent interfaces on microcomputer hardware represents a significant breakthrough for VLSI technology in the future.

References

1. Myers, Glenford J., Albert Y.C. Yu, and David L. House, "Microprocessor Technology Trends," *Proceedings of the IEEE,* Vol. 74, No. 12, December 1986, New York.
2. Amanatides, John, "Realism in Computer Graphics: A Survey," *IEEE Computer Graphics and applications,* Vol. 7, No. 1, January 1987, New York.
3. Palmer, James D. and Yi Qing Liang, "Distributed Systems Architecture and Decision Support Systems in Computer Integrated Manufacturing," *Proceedings,* 2nd International Conference on Computer Applications, IEEE Computer Society, June 1987, Bejing, China.
4. Shires, Glen, "A New VLSI Graphics Coprocessor—The Intel 82786," *IEEE Computer Graphics and Applications,* Vol. 6, No. 10, October 1986, New York.
5. Asal, Mike, Graham Short, Tom Preston, Richard Simpson, Derek Roskell and Karl Guttag, "The Texas Instruments 34010 Graphics Systems Processor," *IEEE Computer Graphics and applications,* Vol. 6, No. 10, October 1986, New York.
6. Litwin, Witold and Abdelaziz Abdellatif, "Multibatabase Interoperability," *Computer,* Vol. 19, No. 12, December 1986, IEEE Computer Society, New York.
7. Hayes, John P., Trevor Mudge, Quentin F. Stout, Stephen Coller and John Palmer, "A Microprocessor-based Hypercube Supercomputer, *IEEE Micro,* Vol. 6, No. 5, October 1986, New York.
8. Gaudiot, Jean-Luc, Michel Dubois, Liang-Teh Lee and Nadim G. Tohme, "The TX16: A Highly Programmable Multi-Microprocessor Architecture, " *IEEE Micro,* Vol. 6, No. 5, October 1986, New York.
9. Hwang, Kai, Joydeep Ghosh and Raymond Chowkwanyun, "Computer Architectures for Artificial Intelligence Processing," *Computer,* Vol. 20, No. 1, January 1987, IEEE Computer Society, New York.

Artificial Intelligence in Military Applications

Randall P. Shumaker and Jude Franklin

The term artificial intelligence is seen everywhere these days, from news magazines to technical journals. The subject would seem to be new judging by the sudden burst of public interest, but, in fact, AI has been around since the early 1950s. Why the sudden expansion of interest? The answer lies in the great potential of the methods rather than in any specific demonstration, although some significant success stories exist. Advances in computing hardware, and in particular the availability of inexpensive large-scale memories and powerful processors, have set the stage for practical applications, but the real story is in the software. After an early flush of high hopes, AI researchers found that intelligence, natural or otherwise, is not easy to understand, let alone to build. In the last decade, steady progress has been made, and in a number of areas, modest successes have been achieved, with the promise of much more to follow.

Relevant AI Technologies

Several technologies are commonly included under the umbrella of AI. For the purposes of this article, AI will be considered the branch of computer science dealing with the development of methods that solve problems previously considered solvable only by humans. These problems include rea-

**This article originally appeared in the June 1986 issue of SIGNAL*

soning, understanding natural language, planning and problem solving, learning, visual processing and manipulation of the physical world. Of the various AI specialities, two are relevant particularly to near-term military applications: expert systems and natural language understanding. Significant advances have been made in these areas, and a host of practical military applications are expected. While other potentially valuable efforts are underway in AI, most are not developed sufficiently well to be applied in the field within the next five years.

Expert Systems

Expert systems are computer programs that duplicate to some degree the kind of results achieved by human experts. These systems are able to solve problems, to predict, to provide a rationale and to give advice in a narrow area of consideration. Some impressive results have been demonstrated, and several commercially successful examples exist. The most well known expert systems have been developed for medical diagnosis, mineral prospecting, chemical analysis and computer configuration. The developers of these systems make no pretext of duplicating the reasoning process that experts use in problem solving; instead, they concentrate on achieving similar results. An understanding of the way in which these systems operate, their limitations and how they are constructed is important in being able to see why these systems are being considered for so many applications. The skill shown by an expert is the result of accumulated knowledge and experience in interpreting information. One theory behind expert systems is that expertise, the skill shown by experts, is the result of the accumulation of a set of rules for interpreting facts to reach a conclusion. The general idea is that if these rules could be collected and put into a form usable by a computer, then some of the skill of the expert could be shown by the computer. With some qualification, this has been demonstrated to be the case. The qualification is necessary because success has been obtained only in a limited context, and no pretense is made that experts actually reason in the same way that the expert system reasons. Even with these limitations, many problems can benefit from application of such systems, and many projects, both commercial and military, already are underway.

Figure 1 shows a common form of expert system based on rules. In actual implementation, there often are variations in the details, and other forms of representation of knowledge also are used. However, the general principles are the same. The three components of the system are a factual data base, a set of rules and an inference engine. The data base consists of facts, which may be simple *a priori* assertions, responses to requests for information or the results of deduction from applying the rules. The rules are in the form "if A then B," where A is a fact from the data base, or some logical statement concerning the facts (e.g., if A and B, then C). The "then" portion of the rule is a new fact that can be inferred from the "if" portion of the rule. The

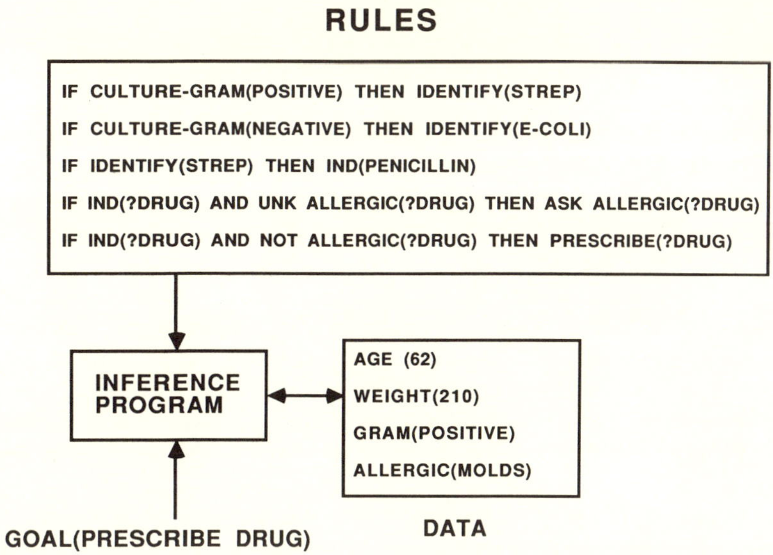

Figure 1. An example of a rule based expert system.

inference engine is a computer program that is capable of examining the factual data base and of finding rules that can be applied to deduce new facts. In operation, rules are applied repeatedly in an attempt to derive some desired result. Each application of a rule generates some new data, which, it is hoped, will be useful in reaching a conclusion by making additional rules applicable.

In practice, the process is not as simple as it might seem initially. For instance, with a given set of facts, there usually will be many rules that could apply. The strategy for choosing rules to apply is critical to the efficiency of the system. In many instances, there may not be a single conclusion from application of a rule, but there may be many with varying degrees of confidence or certainty for each. The growth of new facts, with an associated degree of confidence in their accuracy, complicates choosing rules to apply and interpreting the degree of confidence in their result. Various methods have been developed to cope with these problems, but often a method for choosing rules and for interpreting certainty that works well in one expert system will not be applicable in a different context.

The method of search of the rules is a design feature of expert systems. Three schemes commonly are used: data driven, goal driven and mixed. The data driven strategy also is known as forward chaining. This method uses information in the data base with rules to generate new facts. The process continues either until the desired result is achieved or until no more inferences can be made with the data or rules available. The addition of more data by an operator or through some other source allows the process to continue. A characteristic of this strategy is what seems to be aimless search, first pursuing one line of reasoning, then pursuing another in an apparently undirected manner. While this method often is successful, it is not satisfying to human operators. This is a disadvantage where operator interaction and cooperation is important.

The goal driven technique also is known as backward chaining. In this type of strategy, the rule search starts with some desired conclusion and looks for rules that could be applied to reach it. That is, it looks for rules that have a "then" part that contains the desired fact as a conclusion. It then determines whether the "if" portion is true according to the data base, and, if not, it looks for rules whose application would make the statement true. This backward search continues until the assumed result is proved or until no more rules can be applied. At this point, an interactive system may query the user for applicable information to allow the search to continue. If no additional data are available, another line of reasoning may be followed.

An important advantage of backward chaining is that selection of relevant paths is inherent in the method. Irrelevant rules and data never are invoked. This is in contrast to forward chaining where special care must be used in rule selection to avoid the generation of correct, but irrelevant, data. Operators also find this method more satisfying because it more closely follows the human strategy of developing a hypothesis and attempting to prove it. One problem with this technique is that it does not take advantage effectively of new information that may become available. This information may not support the path currently being investigated. Data driven systems would consider applicable data immediately.

A mixed strategy uses elements of both data and goal driven methods to attempt to speed the process of convergence to a conclusion, while taking advantage of the best of both techniques. Selecting an optimum combination is an important area of research, one with great impact on the success of the system.

No matter which strategy is used, an important concern is the efficient selection of rules. In the usual case, many rules may be applied in a given situation. The wrong selection will not prevent the correct conclusion from being reached eventually, but it will cause some irrelevant and time consuming rule chaining to be done. To help in this selection, there may be rules in the rule data base that deal not with facts but with the application of the other rules. These rules about rules, or meta-rules, are important in making an efficient expert system and are an important constituent of human expertise as well.

Expert systems typically have an explanation facility; they can provide rationale for the conclusion they have reached. In simplistic terms, this can be viewed just as an unraveling of the rules that were "fired" in making the inference, but the better systems also provide some "why" information as well. Expert systems generally are constructed by querying some subject expert to develop the rule base. At the time of rule collection, the "why" information also can be collected to permit a reasonable explanation to be provided. Having an efficient explanation facility is important for several reasons. For development, the facility allows the system to be tested in a convenient way, and it provides feedback during the development process.

Often, human experts do not know explicitly how they apply the rules that they know. Seeing a logical train played back by the expert system often will elicit additional rules or meta-rules that improve the system performance from the human expert. An explanation facility also makes possible the use of expert systems to train people by showing how conclusions were reached.

Most systems currently in use were built by using only a single area expert. Experts do not always agree on the ways they reached conclusions, even if they have reached the same conclusion. Because of differences in methodology among experts, if rules from several sources are mixed, the performance of the system may be poor or the results may be incorrect. The effectiveness of expert systems should improve as data and rules are added, just as for human experts. Often this is the case, but not always. The efficiency of the search process becomes more important as rules and data are added. A poor search method may work in a limited case, but it may run excessively long and follow irrelevant paths when more data or rules are provided. Furthermore, the addition of rules may weaken the performance of systems by modifying the application of existing rules. For these reasons and others, the task of building expert systems has been highly labor intensive, using a lot of the expert's time and requiring the use of a highly trained knowledge engineer to build the system.

Natural Language

Natural language is the term computer scientists use to refer to human communications. Because it tends to be highly variable, having computers automatically understand natural language has been an elusive problem. The difficulty lies in the variety of ways in which people correctly can express the same thought. Understanding what is said often depends considerably on implied information and context, neither of which usually is available to the computer. Speech understanding is even more complex because of problems with grammatical errors, accents, noise, separation of words in continuous speech and even the reliance on nonverbal cues in normal human conversation. Despite the problems, some significant advances have been made, and the payoff potential is high. However, in speech generation and text generation, the situation is a little more advanced. Very understandable speech now can be created, and of course, well formed sentences can be generated by a computer. The problem here is in giving the computers something worthwhile to say or write. In limited contexts, fixed format messages and output from expert systems can be presented effectively.

In the military, as in any technologically sophisticated and complex organization, people must be able to communicate with computers. Unfortunately, with the computers in current use, natural language cannot be used for communications. Instead, stylized, formal computer languages and limited key words must be used to convey commands to the computer. Frequently, even in this constrained environment, errors are introduced. What is needed most is the capability for the computer to understand native language and to be able to act on that information reliably. Three-year-old children can speak and understand spoken language, but currently no computer system can rival even this level of linguistic performance. The thrust of research and development in natural language is to improve this situation by developing techniques to enable the computer to understand written text and relatively unconstrained spoken commands. The availability of computers with natural language capability would help in many ways, including the following immediate capabilities: voice command of weapons systems settings, computer understanding of messages to facilitate dissemination, automatic information extraction from messages, interactive troubleshooting aids and training.

There are numerous examples of problems that could or would have been eliminated if computers had natural language capability. Mistakes in routing messages to the USS *Liberty* off the coast of Israel in 1967 could have been corrected automatically with an artificial intelligence based distribution system. It is likely that these messages would have eliminated the potential for harm to the ship and to her personnel. In wartime, such instances would be expected to occur regularly.

In message transmission, overdissemination often is found on the order of 95 percent. This not only causes wasted man-hours and manpower overload, but also delays reception of the message. Many messages are summarized manually prior to entry into a data base, causing further delays and potential for error. This summarization task may be suitable for execution by machines with natural language capability. Modern aircraft increasingly have complex interaction requirements with pilots, and an already heavy workload aggravates this problem. Advances in display technology have helped, but the increase in the number and complexity of avionics systems seems to have exceeded the pace. Reliable voice recognition with more than simple key word capability offers the potential to increase cockpit efficiency and to reduce the workload, particularly under combat conditions.

For a computer to understand natural language is difficult because the process used by humans is not understood completely. Formal techniques that use lexical, syntactic and semantic knowledge of the language have been developed, but they have not been completely satisfactory. It appears that at least some real-world knowledge is required to understand fully natural language. Grammar based techniques are, however, much more effective than the standard key word technique currently in use. Some significant progress has been made within the context of more formal natural language transactions.

Key word systems, while useful in limited situations, cause problems because they cannot discriminate meaning adequately. For example, key word systems often break down where abbreviations are used. Consider a message concerning a transmitter failure. The abbreviation for intermediate frequency often is written as "if," and no key word system possibly could differentiate this from the conjunction "if," which might also appear in the

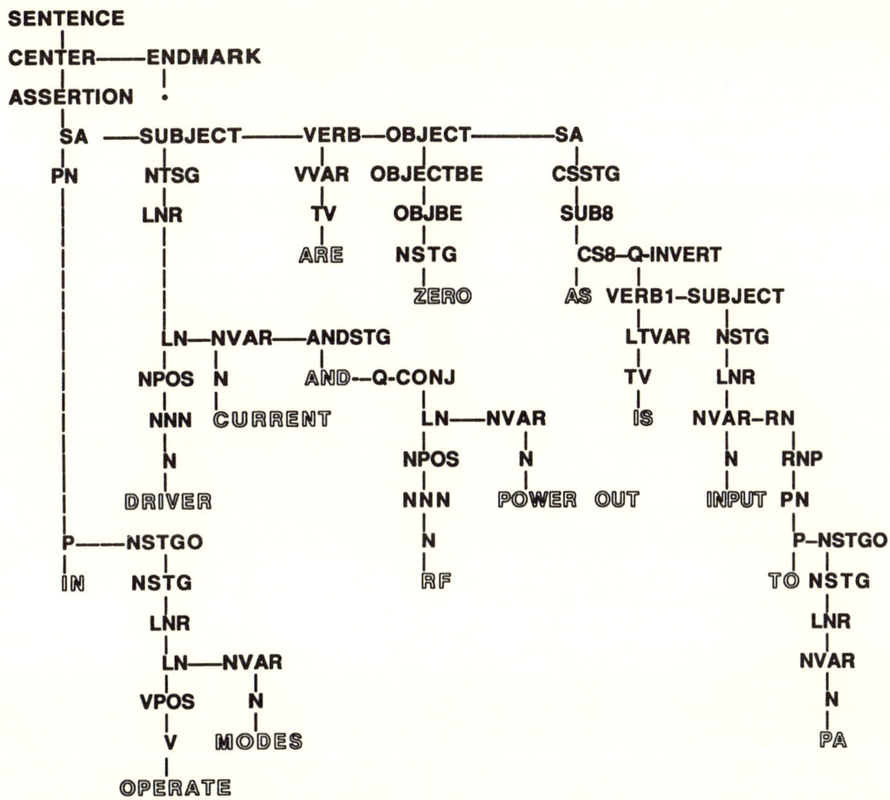

Figure 2. Computer parse of a sentence.

text. A grammar based system can resolve the problem by analyzing how each instance of "if" is used in the sentence, and in most cases, the system at least would be able to identify the potential use of an abbreviation.

We can compare the computer using grammar based natural language interpretation to understand written text with the way humans understand sentence structure. The process of diagramming sentences into their component parts (e.g., subject, predicate, object, prepositional phrases, etc.) is called parsing and is analogous to the way in which computers can understand a message. To provide a more specific example, consider a typical sentence from a message concerning an equipment failure. "In operate modes the driver current and the RF power out are zero as is input to PA." A portion of the computer parse for this sentence is shown in Figure 2.

Note that the computer parse found the compound subject of the sentence to be "driver current" and "RF power out." It found the verb "are" and the object "zero." The computer handles each sentence or major clause in the same way until the whole message is parsed. After the initial parse is performed, the sentence is regularized. For instance, a passive verb is changed to an active verb and an object. For the example shown, it breaks the original sentence into multiple assertions such as: driver current is zero; RF power out is zero; and input to PA is zero. The standardized format facilitates further computer interpretation.

Near Term Applications of AI

To make the development of an expert system possible and worthwhile, certain circumstances must exist. One basic requirement is that an expert must be available in the application area to help build the rule base. To be useful, a significant need must exist for the expertise either because of a limited number of experts or because of the difficulty in transporting experts to the problem site. An important corollary to the limited availability of expertise is that the requirement for expert analysis is time critical in some way. In the military, there are many applications that fulfill these requirements, resulting in a number of projects to develop expert systems. The following discussion focuses on some key applications.

Maintenance and Troubleshooting

Because the sophistication of military equipment has increased, problems in keeping the equipment operational also have increased. Today, a major cost of operations is fault diagnosis and repair. A number of expert systems have been produced by industry and by the military to aid in the diagnosis of equipment faults. Perhaps the best known of these is the Delta system from the General Electric Company. The Delta system was designed to diagnose faults in diesel-electric locomotives.

Each of the services has some effort underway to produce similar diagnostic aids for technicians to use. In any maintenance application where a limited pool of human experts is available, the application of an expert system maintenance aid becomes an attractive option. However, in electronics maintenance, the possibilities for immediate payoff are even more apparent. This is true particularly in avionics maintenance because of the heavy reliance of aircraft on electronics, the large number of different systems involved and the premium placed on rapid turnaround.

For the U.S. Navy, the problem of avionics maintenance is aggravated by the need to operate from ships where space, logistics support and limited manpower place heavy emphasis on speedy and accurate diagnosis and repair. The Navy relies heavily on automatic test equipment (ATE) for diagnosis of faults. In fact, all of the services use automated testing to some degree. Even though many avionics items have sophisticated, built-in test capability and even though testing is automated, high false removal rates, excessive levels of fault ambiguity and the continued need for human intervention still are problems. (False removal rates as high as 85 percent are found, and ambiguities of three to five circuit cards are fairly common.) The approach taken by the Navy in developing diagnostic systems for ATE illustrates the general approach for such systems.

ATE makes use of test program sets (TPSs), which consist of an interface between the avionics and the ATE and software for fault diagnosis. For each different avionics item, a separate TPS must be provided. The present method of test program generation is highly manpower intensive, and results are somewhat variable. A typical test program set to diagnose system faults may take 18 months to procure, cost between $100,000 and $10 million and may take between 20 minutes and 12 hours. On a typical aircraft carrier, more than 600 different TPSs support the avionics on various aircraft. In many cases, TPSs do not perform optimally; they either fail to identify causes of faults in a reasonable time or, more generally, produce a large, ambiguous group of suspected faults. High procurement costs, long acquisition times, a limited number of human experts and the critical importance of automatic testing make TPS generation an especially good application for expert systems.

For electronics fault diagnosis, the data base of the expert system consists of two kinds of information: detailed specifications for the equipment to be diagnosed and results of measurements. The specifications consist of such information as functional description, interconnections, nominal values for normal operating parameters and component values and tolerances. This kind of information must be available for each piece of equipment and is equivalent to the manuals and performance specifications that a technician would use. The additional data information in the data base consists of symptoms and results of measurements. The rule base of the system consists of general diagnostic methods, rules associated with particular classes of equipment and, finally, rules peculiar to the specific equipment being tested.

In expert systems developed for other applications, the data and rules are generated manually. This particular bottleneck may be bypassed to a large degree for electronic equipment because often the design data already are available in computer usable form from computer aided design/computer aided engineering (CAD/CAM) data bases. In addition, it is expected that at least some of the rules automatically can be captured from analysis of the system functional descriptions and circuit topology. The possibility for such automatic knowledge compilation is an important advantage in applying expert systems to electronic diagnoses.

In operation, the diagnostic expert system examines the data base, looking for rules to apply. In the early stages of diagnosis, the application of rules primarily is for the purpose of making key measurements to be added to the data base. In later steps, many inferences are possible with only limited additional measurements. One objective of using an expert system for diagnosis is to minimize the total testing time by reducing the number of measurements necessary to diagnose the fault. At each step, the expert system looks for a final diagnosis, and where that is not yet possible with the data available, it determines the next best test to make in order to rule out the most possibilities. The economic payoffs potentially are very high for such diagnostic expert systems. It is estimated that TPS and ATE procurement costs could be reduced by 25 percent to 50 percent. This is no small matter since the aviation community spends more than $1 billion a year on these items. Improvement in readiness, though harder to quantify, certainly would result from the faster turnaround and the reduced false removal rates for avionics components.

One final selling point for the application of expert systems in ATE is the already high level of commitment to automation within the avionics community. Most of the measurement and stimulus equipment needed for immediate application already exists and is designed for computer control. Significant benefits can be achieved in nonavionics applications as well, but much more attention must be given to interfacing with technicians and providing them with the information that they require in a readily usable form. Comment on the need for work on the man-machine interface problem already has been made. For maintenance aids, the problem is not so much free form communication as it is providing access to large amounts of information in a convenient way. Video disks under computer control are being explored as one potential solution to this aspect of the problem.

Decision Support

In any high stress, high information rate situation, human decision makers can be overloaded and can become unable to assimilate and to use all of the data available. This is true particularly in sensor interpretation and in combat management where data must be evaluated and used effectively in managing weapons and making tactical decisions under strict time constraints. Expert systems offer the potential for application in decision support by providing interpretation aids, automatic analysis or situation assessment. In these applications, expert systems automatically would assimilate the latest data from such sources as sensors, intelligence reports and human decision makers, and they would provide plausible hypotheses about the current situation to operators, planners and other human decision makers.

Operator decision aids for sensors are under development by all of the services, particularly by the Air Force and by the Navy, for such applications as radars, sonars, infrared and electronic warfare equipment. These generally are based on the previously described classical expert systems with the addition of links to more conventional computer aids and with heavy emphasis on the considerations of human factors. These sorts of aids are likely to be available in the relatively near future for some applications, and the work being done now certainly will have a strong influence on new designs for equipment. Both the Navy and the Air Force are developing prototype systems for air strike planning with a mix of conventional and AI techniques, and early work in battle management and operational planning aids is underway in all of the services. The Defense Advanced Research Projects Agency (DARPA) and the Strategic Defense Initiative Organization (SDIO) are supporting work in these areas, and the DARPA Pilot's Associate program is an excellent example of a project designed to integrate advanced techniques in computing and in AI to produce an advanced operating aid.

The highest future payoff area for AI technology probably lies in integrating information from a variety of sources and in providing high level interpretation and advising capability. The concept of multisensor integration has been explored for several platforms, including the P-3 ASW aircraft, and DARPA is sponsoring just such a project for fleet battle management in cooperation with the Navy. The DARPA effort is the most ambitious currently underway. It uses a variety of AI techniques including natural language, expert systems and advanced human interface methods. The system implements for five key management functions at the battle group level: force requirements, capability assessment, campaign simulation, operational planning and strategy assessment. These tasks are complex and labor intensive as well as time constrained—they make excellent candidates for the AI approach. Distributed cooperating systems using people as the primary data processor frequently are found in command, control, communications and intelligence (C^3I) applications, and the results of this and similar efforts are likely to have a significant effect on future designs. A current thrust of research interest in AI is how best to implement systems and the development of techniques to make them practical in real time applications.

In the simplest form, such a system acts as an information filter and compiler, forming reasonable estimates about the overall meaning of the information available. To be useful, it must be able to attach some figure of reliability, probability and importance to information in providing plausible scenarios to decision makers. This type of system is data driven; that is, changes in the current information library trigger updates of the situation assessment. Such a system must have competency in multiple areas of expertise to be able to produce plausible and useful assessments. It also must be flexible enough to allow the simultaneous pursuit of multiple competing possibilities. The arrival of a new piece of data, such as a sonar contact, triggers an update of the current situation assessment. This may allow an update of an estimated target position or identification, or may suggest some new possibility in combination with existing data.

A decision aid such as this must have expertise in a variety of areas, and within the current state-of-the-art, no single expert system would be likely to produce adequate results. Current work is focused on the use of multiple expert systems working in parallel; each system not only works as a specialist in some aspect of assessment but also is able to share results with the other systems. A method used to allow the sharing of data between component expert systems is a "blackboard," which serves as an interchange medium and current status summary device. In effect, this blackboard is a higher level data base that is readable by each component expert system, even if each component's internal organization is quite different from the others in the system. Proponents of such multiple expert system decision aids believe that the ability to use intermediate results from multiple sources is a key to powerful and general systems. At the very least, it is a way to make use of knowledge from multiple subject area experts without trying to integrate possibly conflicting approaches within a single expert system.

A closer examination of the way in which such a system would operate will help clarify both its mode of operation and the problems that must be solved. Individual component expert systems, such as a sensor analyzer, are driven by the availability of new data within their area of expertise, and may use hypotheses presented by other expert systems in reaching conclusions. Such cooperative problem solving is analogous to a human's taking several approaches to a problem to support or to refute proposed solutions, or perhaps is analogous to teams of experts working on the same problem from different perspectives. Critical to success is the management of the cooperative process to reduce the clutter of irrelevant information, to deduce information from the results of individual processes and to prevent a sort of electronic fantasy from developing. These overall management functions are needed to allow understanding in a more global way than the simple summation of results from individual processes. One such function is an analyzer for deducing data from existing results. These deduced results are a form of an educated guess that can serve as goals for a subsidiary analyzer either to confirm or to refute from observed data. Another required function is a "data pruner," which cleans up the current common data base by merging data and hypotheses that actually are the same or are subsumed by another hypothesis. An additional function is the deletion of marginal or low confidence hypotheses to increase the efficiency of the search process.

Once again, a key element for success will be an effective means for man-machine interface. One source for hypotheses to be tested is the human operators of the system. No matter how efficient the internal process may be, if it is too difficult to insert or to extract information, the system will be ineffective. The blackboard methodology allows a means of sharing data between operators and allows the use of underlying methods that may differ significantly in internal processes. Figure 3 shows a possible configuration for a decision support aid based on distributed cooperating systems.

Autonomous Vehicles

An attractive use of AI technology is in the implementation of autonomous vehicles for reconnaissance, intelligence gathering or smart weapons. Such vehicles are different conceptually from remotely piloted vehicles (RPVs) in that they would be capable of at least some degree of independent decision making and planning, whereas RPVs rely almost entirely on a human operator. The ideal autonomous vehicle would have the ability to accept mission requirements in a very high level form, perhaps in a natural language. From the requirements, the vehicle would be able to plan its own approach to mission execution and would have the ability to revise the plan when unexpected conditions arise. These capabilities are a long way from full imple-

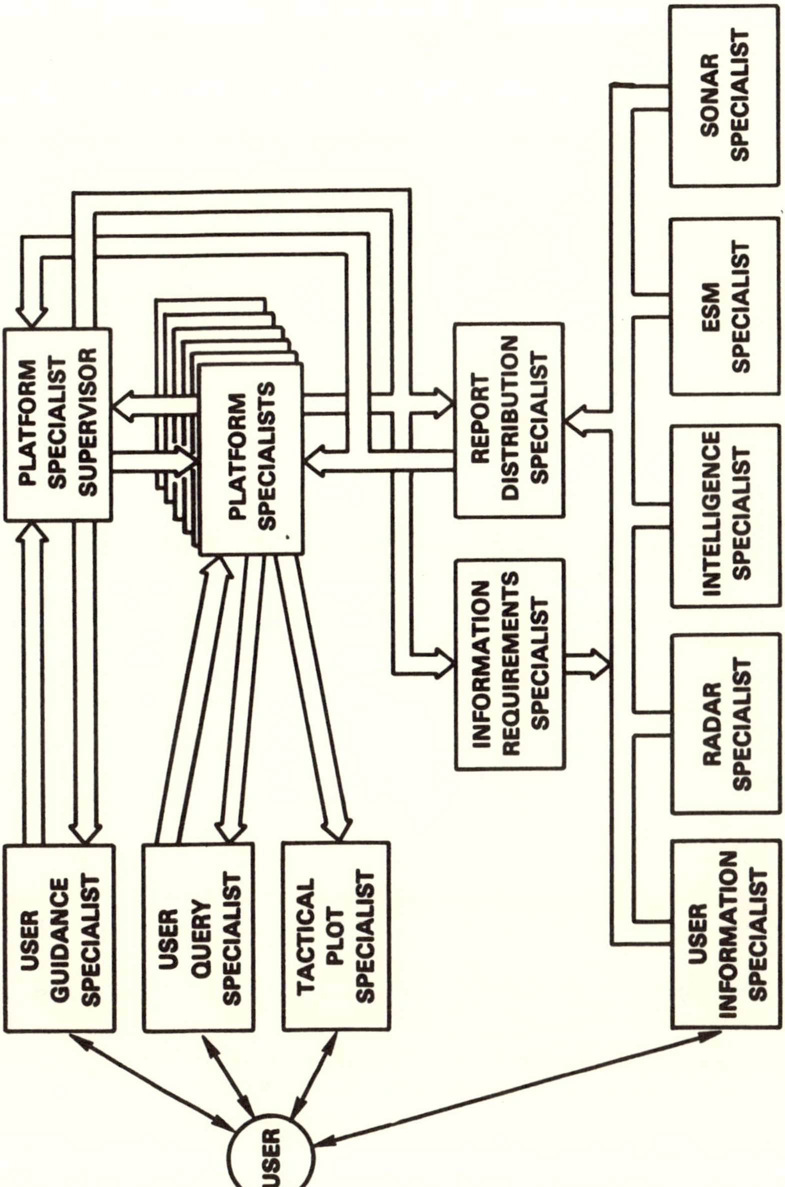

Figure 3. Functional organization of knowledge based specialist for platform centered MSI.

mentation, but project work underway is aimed at solving basic problems of navigation, sensor interpretation and control in a real time environment.

DARPA and the U.S Army are developing an autonomous land vehicle with the ultimate aim of producing a vehicle capable of traversing independently complex terrain to carry out some mission. The vehicle would be capable of solving problems of navigation and of automatically avoiding local obstacles such as trees, rocks and man-made objects. Initial work is focused on providing real time image interpretation for navigation on a roadway. Later work will include image understanding, adaptive planning and obstacle avoidance.

The Navy actively is investigating the development of autonomous sensor platforms (ASPs) with onboard intelligence as an alternative to manned and remotely piloted vehicles in certain situations. A key requirement is the ability to function effectively without a continuous data link. In addition to the ability to fly the vehicle safely, it must be able to identify friendly and enemy forces, to deploy countermeasures, to employ its sensors effectively and to operate in concert with other vehicles.

Message Automation

Text handling is a major problem for all of the services and currently is very manpower intensive. Message dissemination, summarization and information extraction from text now are handled either entirely manually or with limited automation. This is a particular problem when dealing with messages under time critical conditions. Message automation and natural language understanding projects are being sponsored by several agencies with the goal of automating much of this work. Figure 4 shows a system developed to automate message processing in a naval application. After parsing, regularization and linguistic analysis, the system uses real world knowledge stored within an expert system to understand the references, the relationships and the importance of various statements within the message. This analysis is used in a process of promotion of statements within the text to determine the most significant part for the purpose of providing a summary and of determining who would be interested in the content.

This process requires not only information about the relationship between systems and their relative importance, but also some knowledge of the motivation of potential recipients, for instance why they would be interested in this information. Initial results have been good; the summaries generated by the machine have agreed with the manual ones about 85 percent of the time, and in a few cases, the machine generated summaries were more explicit and were more useful than the ones generated manually. In other cases, the results have not been as satisfactory, and further work is required before such systems are ready for practical use.

An early application of message understanding systems is expected to be in message preparation, where the system would be able to generate a sum-

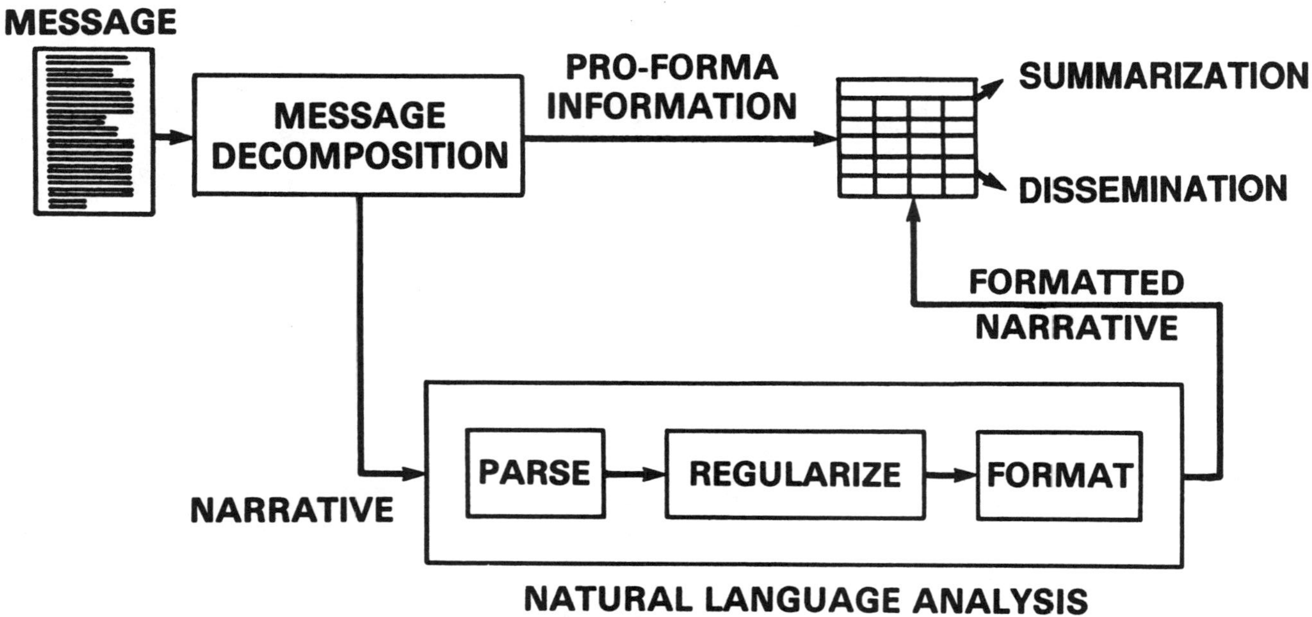

Figure 4. Message processing system.

mary for the approval of the sender before the message is transmitted. Automatic data base update, decision maker alerting and search of large volumes of text for relevant information, such as intelligence, are part of this work.

Training

A very important use of AI technology is in training. In addition to helping in the decision process as previously described, the same methods can be used to generate advice and to provide guidance to students during training. This is becoming increasingly important in helping operators and maintainers deal with the complexity of high technology products found on today's ships and aircraft.

One of the first AI efforts in this application was the Steamer Project carried out at the Naval Personnel Research and Development Center (NPRDC) in San Diego, California. The Steamer Project was a research effort concerned with the implementation of intelligent computer based training systems. The project focused on the construction of a system to assist in propulsion engineering instruction. A steam propulsion system is an exceedingly complex physical system, and the propulsion spaces account for about one-third of the space in most Navy ships. There are thousands of components interconnected by miles of pipes, not to mention a significant number of monitoring instruments and controls. The operation of the plant is supervised by an engineering officer of the watch and controlled by a team of 16 to 25 individuals who operate in the most trying of circumstances. They often work long hours in a hot, dirty and dangerous environment. It takes years of instruction and experience to become competent in the operation of a propulsion plant.

The Steamer Project has been used as a training aid in the Great Lakes Training Center and on Coronado Island, San Diego, California. The results to date are encouraging. Preliminary results indicate that personnel respond positively and that they can learn the same material in a shorter period. One interesting aspect of the project is that during a simulation run on the computer, the operator can activate the keyboard and can see what happens when certain things are done. In the simulation, the operator can "blow up" the steam plant.

The Air Force, the Army and the Navy actively are applying AI to training, with practical benefits available near term. A very attractive potential result of the introduction of AI into fielded systems is the concept of embedded training made possible by the existence of sophisticated knowledge based processing within the equipment. The possibility of reducing the training cycle and the maintenance of proficiency through realistic training modes on actual equipment is very attractive, and may be an important benefit of embedded intelligence within equipment.

Developmental Issues

The AI field is moving ahead in many areas, with expert systems and natural language processing nearing practical application. Even so, a number of important unsolved problems must be addressed before even these techniques can be used widely or easily. The most important of these issues is in knowledge acquisition, that is, the process of data base generation needed for expert systems. No clear alternative to the present labor intensive methods is currently available, but some promising work is being done. One project for the development of an expert system for maintenance is attempting to generate a significant portion of the needed knowledge from computer based design data. While this may be successful, more general methods must be found to speed the development process in other applications. Similarly, improved voice recognition methods and speech processing will make systems easier to use and more available to operators and to decision makers.

Many applications where AI could be applied successfully are not being considered fully due to a lack of understanding of the potential, limitations and requirements for use of AI methods. One important task of AI researchers, particularly those in military laboratories, is to familiarize officers with the potential, the use and the need for AI. Closely related to this is a wider acceptance of AI by acquisition managers; the industrial research community must be educated in the use of this important new tool in military projects. Successful completion of the pioneering projects that now are underway will go a long way toward gaining the recognition and acceptance needed to permit the wider acceptance and the use of AI methods.

Glossary of Common AI Terms

Algorithm—A procedure with a specific beginning and end that is guaranteed to solve a specific problem.

Artificial intelligence—a branch of computer science concerned with developing methods for carrying out tasks with the use of a computer, which would be considered intelligent if done by humans.

Backward chaining—A search technique that starts from a desired goal or solution and attempts to reason backward toward known facts to prove the goal true.

Blackboard architecture—A software technique used to facilitate sharing of information between multiple expert systems operating cooperatively.

Expert system—A computer program using facts and procedures to emulate the reasoning process of human experts in solving particular types of problems.

Explanation facility—A feature of expert systems that provides a rationale for the conclusions reached by the system. It usually provides an unraveling of the sequence of steps in the search process along with additional information designed to improve human understanding.

Forward chaining—A search technique that begins with known facts or data and attempts to reason toward a goal or solution.

Heuristic—A rule of thumb that applies in most cases, but is not guaranteed to lead to a solution. Such information usually is supplied by an expert based on experience in solving particular problems.

Inference engine—The part of an expert system that controls the selection of rules and their application to data in searching for a solution to a problem.

KBS—An abbreviation for knowledge based systems. It often is used as a synonym for expert system, but with a more general connotation.

Knowledge base—The parts of an expert system that contain the factual and procedural knowledge (facts and rules).

Knowledge engineer—A specialist in extracting and in encoding the knowledge required to implement an expert system.

Meta-rules—Rules about the use of other rules. They are used to improve the efficiency of application of rules in rule based systems.

Natural language processing—Computer techniques for processing text or spoken language to allow convenient communication with computers. Includes parsing as well as domain knowledge processing capability.

Parser—A computer program for analyzing the syntactical structure of a text or a speech.

Rule based system—A form of expert system in which the procedural knowledge is encoded using "if-then" rules (production rules).

Fighting Machines: Significance and Issues of the Soldier-Machine Interface for Tactical Decision Aids

Harlan H. Black

Significance of the Soldier-Machine Interface

Assets that cannot be accessed are worthless. When their lure becomes distracting, they can become dangerous. The bankrupt speculator, the lost treasure seeker, the strained athelete and the burned-out executive are unfortunate examples.

Battles are fought by decisions. Weapons, forces, terrain, weather and the like are factors, and constraints. Without decisions there is no command and no control; communication and intelligence lose their value.

Battles are fought by people. We cannot relate to, nor can we rely upon, a battle whose decisions are to be made only by machine. People use machines; we cannot allow machines to use people.

Battlefield decision aids are vital. Current tactical optimism for the free world relies upon aids that are supported or enhanced by high technology computer systems. We have the technological edge. This is our asset.

This article was written especially for* ***Principles of Command and Control.*

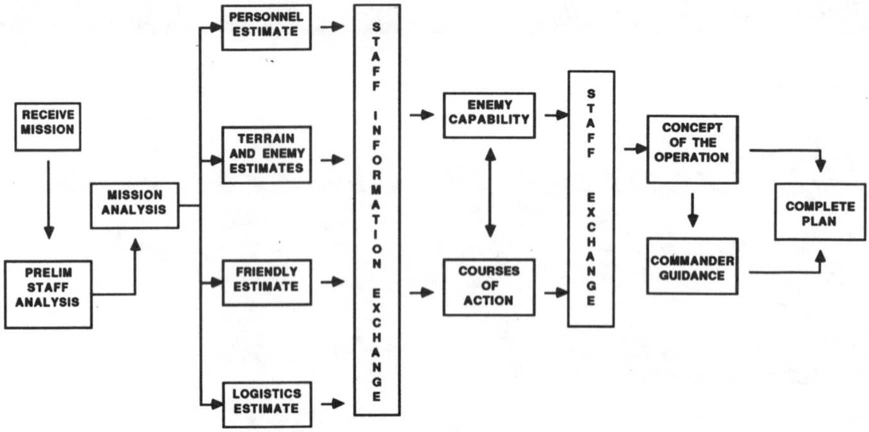

Figure 1. The doctrinal planning process.

The soldier-machine interface is the means by which this asset is accessed. The degree of this interface's effectiveness is the degree of access. This, in turn, provides the user with the capabilities of the system. Therefore, the system's value, regardless of its capabilities, relies upon the interface.

With a good soldier-machine interface, the soldier can better fight the enemy and with a poor interface, the soldier must resort to fighting the machine. A distraction of this asset can be fatal.

An Illustrative Context

The U.S. Army Communications/Electronics Command at Fort Monmouth, New Jersey, is performing internal research in artificial intelligence and is constructing a tactical decision aid prototype for maneuver control planning for a corps commander. The project's name is ARES, and it is being sponsored by the Center for Communications/Automated Data Processing. One branch of research is devoted to C^3I issues which relate to soldier-machine interface. Lessons learned by providing ARES with an optimal interface hopefully will provide additional insight and experience and will contribute to the outcome on the battlefield.

The maneuver control planning problem is crucial and complex. Figure 1 portrays the doctrinal planning process. Upon receipt of new or modified orders, the commander's staff first must analyze the mission to clarify the full intent of higher headquarters and the mission's implications. This usually results in the breaking down of stated goals into a hierarchy of realistic subgoals. Mission analysis is complete when realistic subgoals to achieve the goal have been generated.

This developing plan must match the actual battlefield situation and context. A detailed and accurate assessment is needed to provide estimates of

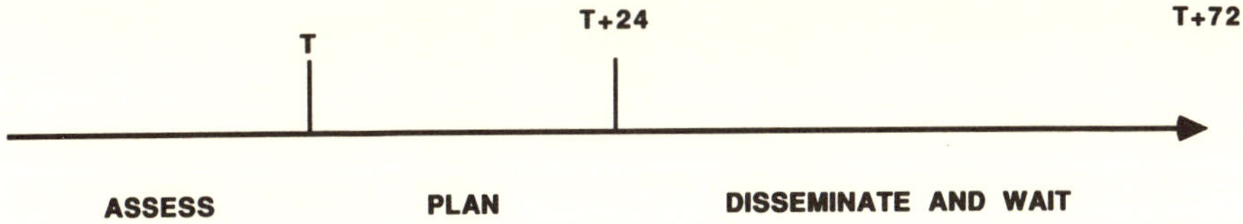

Figure 2. The Planning Time Line.

the friendly and enemy forces and assets, and of the terrain and logistical factors.

From the estimates and staff interaction, two classes of conceptual models emerge. The first consists of one or more models of the enemy's capability, based on such factors as his forces and constraints. The second holds one or more courses of action, one of which will be chosen as the plan to follow. The number of elements in these two classes and the degree to which there is any dependency between them will increase the complexity of the overall problem.

After evaluating the alternatives, one course of action is selected and the plan is complete and ready for implementation. Herein lies a major source of problem complexity. The corps echelon primarily deals with the allocation and management of resources. Levels of command, much lower than corps, do the actual combat. From the time that the plan is completed, the corps must wait up to forty-eight hours for the lower levels to do their own planning and plan dissemination. This means that the corps commander must construct his plan with only a projection of the situation assessment, not the actual assessment. Apart from the dissemination lapse, the planning process itself takes up to twenty-four hours. Thus, the commander and his staff must project up to seventy-two hours into the battle. Figures 2 and 3 illustrate these timing constraints. The further the projection, the lower is the quality of the assessment and the plan upon which it is based.

A requirement of the Army-21 doctrine is for our forces to react within the decision cycle of the enemy. Automated decision support systems can shorten the planning time of the commander and of his subordinates. This, in turn, can shorten the assessment lead time and increase the value of the assess-

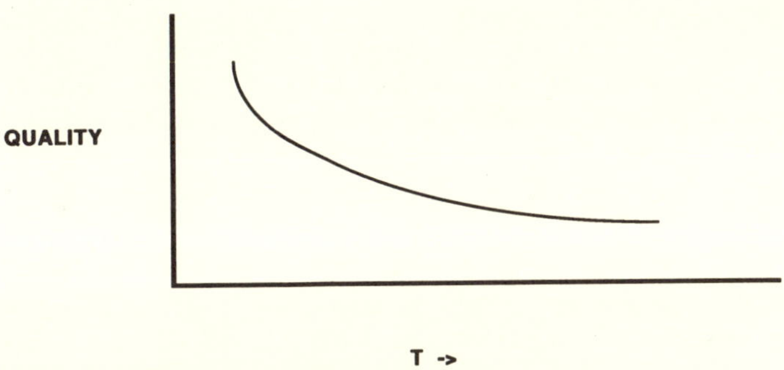

Figure 3. Plan quality deteriorates with time.

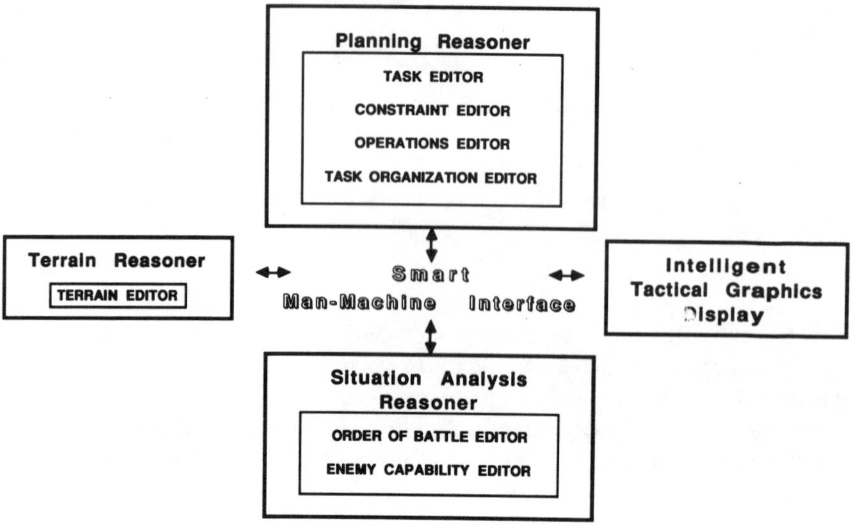

Figure 4. Architecture of the ARES prototype, version 1.

ments and the quality of the plan. Problems and opportunities can be recognized earlier, and decision support systems can provide more alternate courses of action to the commander.

These benefits can be derived only from a system whose assets are accessable through the soldier-machine interface. Figure 4 depicts the ARES project's architecture, emphasizing the centrality of the interface.

Selected Issues of the Soldier-Machine Interface

There is really one basic issue for interfaces: the bottom line. Simply stated, it must work for the soldier in the field. This has hardware and design implications.

Interface Hardware. Apart from hardening concerns, there is a significant gap between state-of-the-art commercially available hardware that is practical for lab prototypes and that which is practical for fielded systems. Taking the large-scale tactical display as an example, planners typically utilize detailed tactical maps that are printed in sections and then taped together as needed to provide a global view of the battlefield. The resulting map or set of maps may each be six-foot square or larger. It is impossible to display, much less see, this field of view on a computer monitor without greatly reducing the map detail. Using paper maps, the commander sometimes steps back to view the entire field of view. He sometimes moves in close to read significant information. This natural movement causes no cognitive break. With computer controlled displays, the system provides a mechanism to allow the user to pan and zoom in on the map. The actions required to invoke the functionality and the time lapse for the redisplay are a distraction to the user. When

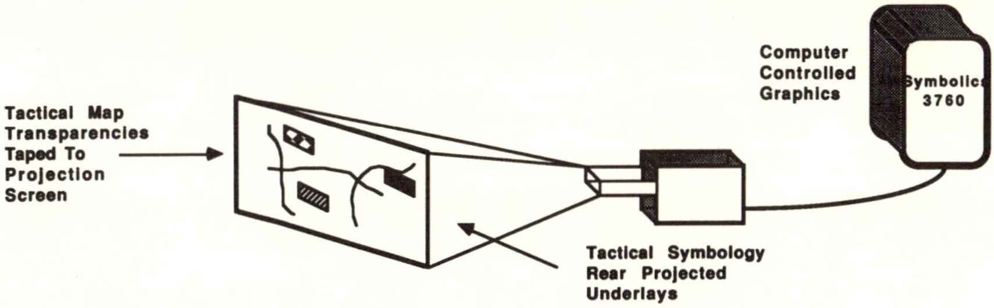

Figure 5. ARES large screen tactical display configuration.

the user zooms in, he is restricted by tunnel vision as he pans the screen to find the needed information. Reasoning is hampered. To resolve this concern, research is being performed in the ARES project to identify the available and optimal tactical display hardware configuration. It was proposed to utilize and evaluate large-size map transparencies and to project computer generated and managed tactical symbology behind them. The transparencies are photographs of tactical maps, in full detail. By projecting the symbology behind the map, the commander still can walk in close and then step back, with no cognitive break. Figure 5 portrays the ARES large screen tactical display configuration.

While the evaluation and refinement of interactive techniques can be done in the lab, fielding this system will require additional research and development. For one, the corps command post has no room for the required throw distance of a rear screen projector. A possible resolution is the projected availability of large screen computer displays. The Magnavox MX1208 large screen plasma display is forty-two inches square. With a density of fifty lines per inch, it is capable of handling symbology, albeit in one color.

Data input hardware is another example. People in life-threatening situations are rough on their equipment. A light pen which dangles from a cable may not last very long. The cable itself may get in the way of a user's abrupt movement. New technologies offer connection-free devices and must be considered for the field.

Interface Design The value of interface technology is constrained by fielding requirements. ARES project research has emphasized a number of design considerations for tactical decision aids.

The soldier-machine interface must be easy to use, and it must be designed to minimize the amount of specialized operator training. We must design with the awareness that there will be a shortage of technical support personnel, especially at the outbreak of hostilities.

System response to the user must be timely and prioritized. The user is working under stress. Undoubtedly, there will be delays as the system is processing information. However, when they occur, the user must be informed by the system that he must wait. Research points out that significant and unanticipated delays can cause operator panic.

The interface must provide multiple levels of warnings and alarms. Risk taking, a part of tactical life, is dependent upon one's ability to perceive the risk. This can be provided through careful interface design.

The interface must support and enhance the learning process as the commander must learn the battlefield reality. Quality feedback is a key to learning, and its design and incorporation into the interface are crucial. The facts must be organized and displayed so as to maximize their effect.

Another major issue is information overload. Technology provides an overwhelming amount of data which must be analyzed and filtered. Research points out that the extent of usefulness of computer-managed tactical displays is limited by the amount of information decluttering which the system can provide. Both the interface and the underlying reasoning subsystem must be designed to emphasize the significant, and to de-emphasize the nonsignificant. An open research issue is whether, how, and to what extent a system should eliminate nonsignificant information.

Another open issue is the extent to which the user should control the reasoning process. The user can do so only through the interface. The more he is aware of the underlying system architecture, the more complex the interface must be. To rephrase the issue, to what degree can or should the reasoning components be hidden from the user? The more they are hidden, the easier it is to control the overall system, but the less is the user's control of the reasoning process. What is minimally required from the interface is a means by which the user can understand how the system arrived at a given recommendation. This is vital if the user is to learn to trust the system.

One final and perhaps most important issue is the extent to which the interface supports human creativity. Creativity cannot be replicated by a machine. Creativity is stimulated by need. The user must never be allowed to rely fully upon a machine for problem solving. Reliable decision support must be provided; yet, the commander must be made to rely upon his own resources.

Implementation. Interface design and development are vital. It is a serious error to leave interface testing until the end of implementation. The preferred technique is iterative development. Design iteration is not expensive fine tuning; it is an essential part of the development. Prototypes such as ARES provide a means of defining user requirements. They also provide a means for iterative development of a simulated working system, and they can find design flaws before expensive implementation and fielding are begun.

To the soldier in the field, the machine is the interface. It must work for him; it must work with him. The more assets he has, the more he becomes an asset.

An Assessment of the State-of-the-Art of Advanced Analytical Methodologies for C^3 Decision Support

Gerald W. Hopple

Other chapters in this volume address the nature of command and control (C^2). The purpose of this chapter is to link command and control with the array of tools and techniques in the armamentarium of the specialist in analytical methodology. Computer based problem solving systems of all shapes, sizes and varieties have proliferated in the realm of C^2. Dozens of such systems—featuring a wide range of analytical methodologies—have appeared since 1970, and this trend is constantly accelerating. The very large universe of such aiding systems is canvassed systematically in a number of sources.[1]

Definition: Command and Control

There is extensive literature on the nature and characteristics of military command and control.[2] Among the more popular definitions of C^2 are the following:

This article was written especially for* ***Principles of Command and Control.*

- weapons control and unit maneuvering via sophisticated communications technology;
- relays, switches, displays and information;
- sensor subsystems, fusion centers and communications links between the sensors and fusion centers, and between the fusion centers and the forces.

More generally, a command and control system is defined "as an arrangement of personnel, facilities and systems for information acquisition, processing and dissemination employed . . . in planning, directing and controlling operations [Teates, 1982: 46]."

This definition highlights the fact that C^2 is coupled with the process of decision making for force management, which in turn implies that the sequence of perceptions and choices which constitute the phenomenon consists of a decision process. Many alternative typologies and frameworks of decision processes have been developed (See, for example, Hopple [1986c]), but the simplest isolates two key elements which feed into action by a command and control system: information inputs (perceptions of the enemy and other aspects of the external environment, own resources and other limiting factors, commander's guidance, and so forth) and choice (selection of a specific option from an array of alternative options or choice alternatives). A commander confronts two distinct forms of uncertainty in command and control decision making: information input uncertainty and consequences of action uncertainty.

Some decision situations allow the decision maker to ignore both variants of uncertainty; simple stimulus-response contexts exemplify this class of decisions, where the appearance of cues (data or a stimulus) automatically elicits action (a response). In other cases, the options are fairly clear, but the input data are of low quality; this is the classic military intelligence situation (for example, knowledge of where the enemy is and how his forces are distributed often provide the basis for acting quickly and effectively, but "where is he?" is typically elusive). Here, the emphasis must be on creating and testing hypotheses on where the enemy is and what he is doing. A third logically possible situation arises when the input data are certain or of high quality, but the options are open-ended, as in strategic or political decision dilemmas like the Cuban missile crisis, where the decision maker must expend energy on the creation and analysis of options and their potential consequences. Tactical decision making in support of combined air-land operations yields the fourth and most difficult decision making situation; in this case, as Figure 1 shows (cell C), the input data are generally of low quality and the options open-ended.

The four cells in Figure 1 concern the stages of hypothesis-testing and option assessment, the two phases which intervene between the initial data (the stimulus for the decision) and the action (or response). Very simple (automatic) decisions feature only a stimulus (S) and a response (R); most C^2 decision occasions, in contrast, inject all four, leading to what Wohl (1981)

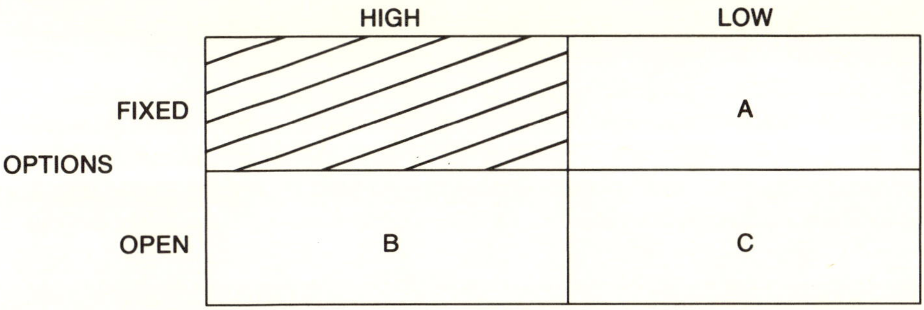

Figure 1. Input data quality.

refers to as the SHOR paradigm of the command and control decision process:

- S or stimulus (data);
- H or hypotheses (perception of alternatives); create/evaluate/select hypotheses;
- O or options (response alternatives); create/evaluate/select options; and
- R or response (action); plan/organize/execute the response.

Another way of identifying the elements of the command and control function is the following (from Harris [1979]):

- Receipt of necessary information;
- Review and analysis of information;
- Formulation of orders;
- Issuance of command; and
- Oversight of implementation of orders.

These and a variety of other typologies illustrate the view of command and control as a *process*. There are a number of C^2 oriented decision aiding taxonomies which identify phases or core functions of a command and control system.[3] A second and equally useful definitional perspective looks at command and control as a *structure,* a set of identifiable and Service specific organizations and procedures; for example, the Air Force structure is portrayed in Hopple (1986b) and Wohl (1981). The importance of this distinction is that command and control is comprised of a generic military process (of decision making and its antecedents and consequences) produced by a specific structure or structural configuration. The process and structure are interdependent but quite distinct.

Finally, experts on C^2 have identified a number of salient attributes of such systems. From the lengthy lists compiled in the literature, the following are the most pertinent. First, command and control is a multilevel phenomenon; almost invariably, there are flows of information across the levels of a command hierarchy (both up and down), command takes place at all levels, decisions are made at all levels and decisions at a given level entail coordination with those above and below. This suggests that the analytical process must be sensitive to command cycles and communications processes within *and* across levels and that information input requirements—and aiding paradigms and potentially useful methodologies—will vary appreciably as one ascends or descends the hierarchy.

Second, C^2 serves multiple and manifold functions, including perception and information management, decision making and direction giving and implementation. Both planning (preliminary and detailed) and operations are central to the process, and intelligence, fire and air support, and personnel and logistics all feed into the C^2 analytical and decision processees. In a more generic sense, central to command and control are both inference making and option identification and selection, which correspond at least in a very general way with intelligence and operations; both require a wide range of aiding systems and methodology bases to yield accurate, useful inferences and comprehensive but manageable and easily communicated option sets.

Third, C^2 applies to multiple military decision occasions. Included under this rubric are peacetime, crisis and wartime (conventional and nuclear) situations (operating at multiple levels, such as the national military command system, the unified and specified commands—and the subordinate unified commands—joint and individual service task forces, the service component commands, and the field commander and below). Overlaid on these tiers are strategic C^2, theater C^2, allied C^2 and tactical C^2, which injects even more complexity into the overall command and control milieu, reinforcing the realization that, by its very nature, C^2 decision making is both complex and "distributed."

Finally, the key shared characteristic of all C^2 decision processes (regardless of level, function or context) is the commander's *definition of the situation,* underlining the importance of psychology to the C^2 process—psychology, not in the sense of Freud and the labyrinthine dynamics and pathways of depth, but in the sense of perception, cognitive processing and the interface between artificial intelligence and the study of the mind. This emphasizes that C^2 is a human decision maker (usually with staff support) *plus* procedures, organization and equipment. The sensors, processes, displays and technologies *per se* do not constitute the core of the process; the commander, and his definition of the situation, are the essence of command and control.

Methodologies

The functional/decision process, structural and attribute related facets of command and control all have manifold implications for decision aiding and incarnating C^2 relevant tools and techniques (methodologies) in computer based problem solving systems for command and control support. This chapter cannot explore all of these linkages and ramifications, but it will address the issue of methodologies in relation to C^2. Initially, however, the universe of tools and techniques—grouped into clusters—will be surveyed.

To aid or not to aid is the first and most fundamental question. It is surprising how often this question is glossed over or not even confronted.

Further, in attempting to answer this question for a given command and control function and decision context, initially, it is necessary to underline the fact that the decision aid movement has spawned an army (or cadre) of decision aiders who have made significant professional and ego investments in designing and developing decision aids as ends in and of themselves. Give a child a hammer, and he or she will use it on everything encountered, as Kaplan (1964) and others have observed. This is the law of the hammer, and it constitutes the essence of "methidolatry." Decision analysts were the methidolatrists of the 1970s; those in artificial intelligence play a corresponding role now.

Understandably, methodologists tend to be guilty of methidolatry, which manifests itself at the most general level as the fetishistic belief that computer based decision aids are appropriate solutions to, and potential panaceas for, every analytical and decision problem or task that confronts a command and control problem solver. Rouse (1985) and others have pointed out, however, that computer based approaches are not invariably or inherently "better" than unaided decision making, and that rigorous evaluations should be used to demonstrate the superiority of a computer based problem solving system. To combat methidolatry, it is necessary to use a framework for decision aid design and development which distinguishes clearly and consistently among decision "types." The availability of a theoretically validated typology of decisions (and a coherent framework for guiding the design process) would enable potential systems designers to assign a candidate problem to a specific cell of the typology, thereby providing an early answer to the question "Is this problem susceptible to, or suitable for, computerization?"

Necessity, apparently, is not always the mother of invention; there is no viable theoretical framework or decision typology which could be used to resolve this first requirements analysis question. Wohl (1981) emphasizes that a theory of C^2 must start with a theory of decision making for command and control, which of course does not exist. Analogously, a theory of command and control decision aid design and development presumes the existence of such a theory of command and control decision making. This is the central question of aidability determination (an admittedly cumbersome term which simply asks: Is this problem susceptible to resolution via a computer based problem solving system?).

There are at least several families of analytical methodologies which can reside in computer based decision aids. Whereas methidolatry surfaces at the front end of the requirements analysis process in the form of suspiciously enthusiastic calls for computerization, an equally pernicious variant of the disease can threaten the method assessment/selection process. In this context, the law of the hammer refers to the overgeneralization of a particular tool (family) across the universes of decision process stages or tasks and functional command and control problems. The syndrome results from a number of causes, including ego investments on the part of those susceptible to the malady (an individual devotes his or her professional career to learning and applying decision analysis or artificial intelligence or some other family, and is understandably reluctant to admit that other tools may be equally or more appropriate to the problem at hand) and the fact that methodology experts tend to live in methodological ghettos (operations research types do not talk to computer scientists, etc.). Regardless of the etiology of the syndrome, it can have genuine costs and poses a significant danger to decision aiding. Rather than a method searching for an application (the familiar solution-in-search-of-a-problem syndrome), the process should be requirements-driven.

The lure of technology push (over the more mundane but more valid strategy of requirements pull) is a pervasive danger; given the faddish nature of preferences for given methodologies and technologies, the law of the hammer is a constant against which system design teams must be vigilant.

Today's favorite, of course, is artificial intelligence, offered as a panacea for almost every imaginable kind of command and control decision problem. A recurring C^2 problem, for example, is target assessment and weapon-to-target allocation and/or assignment, a domain which has attracted the interest of a number of decision aid projects. Two of the recent aids are (Tactical Air Target Recommender) (TATR), an intelligent assistant expert system for tactical air targeting under development for the Defense Advanced Research Projects Agency by Rand (Callero et al., 1982, 1984) and Battle, also a rule based expert consultant system for weapon allocation (Slagle, 1984; Slagle et al., 1983) a Naval Research Laboratory effort.

TATR matches targets and weapons via an analytical integration of information about friendly and enemy force posture, capability, order of battle and objectives to determine effective and efficient courses of action; TATR's rules are used to develop attack plans, update the data files and control the sequence of program events. In weaponeering, effective combinations (aircraft types, munition loads, delivery tactics) are determined by rules; the system calculates probability of arrival and probability of damage. Battle generates weapon allocation plans for (Marine Integrated Fire and Air Support System (MIFAAS) by using artificial intelligence (AI) tools to determine the effectiveness of each individual weapon against each prospective target and then construct a weapon allocation tree to devise good allocation plans for the set of weapons. The system's computation network features rules specified by a military expert.

In analytical methodology, a very basic distinction can be made between *algorithmic* and *heuristic* methodological "meta-families," a bifurcation which constitutes the most fundamental branch on the methods tree (algorithmic and heuristic correspond closely to qualitative and quantitative techniques, respectively). The emphasis on rule based expert systems for the target selection/weapon-target allocation aids may exemplify methodological overkill. Cost-benefit techniques, other operations research (OR) methodologies and selected decision analytic (DA) tools would suffice. This is essentially a bounded, analytically tractable, algorithmic domain, the very opposite of domains where AI is expected to contribute. Alternatively, it may make sense to blend AI in the form of embedded rule based systems and knowledge bases with OR/DA methodologies, exploiting different families of methodologies for their unique assets and implementing a broad gauged strategy of methodological synergism. The increasing emphasis on incarnating integrated suites of methodologies within decision aids is applied to the Air Force tactical problem domain in Courand, et al. (1984) and discussed more generally in Hopple (1986c) and Madni (1985). Against this backdrop, the concept of a decision aid methodology selector may be introduced.

Those familiar with the literature on forecasting methodology will readily recognize the idea of a matrix designed to facilitate the selection and application of analytical methods tailored to substantive problems. Especially in management science, works such as Wheelwright and Makridakis (1980) and Armstrong (1978) provide elaborate, structured matrices which match families of methodologies with clusters of forecasting situations (amount of time available to generate a forecast, the time horizon of the forecast, problem complexity, etc.). A simple but illustrative forecasting methodology selector appears here as Table 1, where seven families of forecasting methodologies (ranging in complexity from intuition to econometric model based projections) are cross-classified with five problem assessment criteria (talent required, time and support available, cost and data requirements). The problem context characterized as "low" on all of the criteria—which corresponds to unaided crisis decision making in command and control—is associated with intuition; in contrast, multiple regression and econometric modeling make high talent, time, support, cost and data demands. Note that the extrapolative (time series) family dovetails very well with its more complex methodological "cousins," *except that* this family costs a lot less in a relative sense.

In order to select a methodology, it often is necessary to descend to a lower analytical level and determine the most appropriate methodology *within* a family. For example, suppose that the forecaster had decided that a Bayesian methodology were the preferred tool for his or her problem and needs. Figure 2 depicts a Bayesian forecasting method selector, which advises the user to select "basic Bayes" (Bayes' Theorem) if the problem is simple, the data requirements are limited and the time available to generate a conclusion is circumscribed. A morc involved Bayesian based technique, hierarchical inference structuring, is recommended for situations characterized by high complexity, extensive data requirements and more time. Influence diagramming falls between the two.

The concepts of selecting *across* and then *within* families of methodologies furnish the context for the development of a generic analytical methods inventory for command and control decision aiding. The generic methodology "tree" is displayed in Figure 3.[4] The matrix is organized initially around four very broad gauged meta-families:

- Qualitative;
- Hybrid/bootstrapping;
- Quantitative; and
- Information (computer) science.

METHODS \ CRITERIA	TALENT		TIME		SUPPORT		COST		DATA	
	HIGH	LOW	HIGH	LOW	HIGH	LOW	HIGH	LOW	HIGH	LOW
INTUITION		●		●		●		●		●
DELPHI		●	●			●		●		●
CROSS IMPACT		●	●			●		●		●
BAYESIAN	●	●	●	●	●	●	●	●	●	●
EXTRAPOLATIVE	●		●		●			●	●	
MULTIPLE REGRESSION	●		●		●		●		●	
ECONOMETRIC MODELING	●		●		●		●		●	

Table 1. Exemplar Forecasting Method Selector.

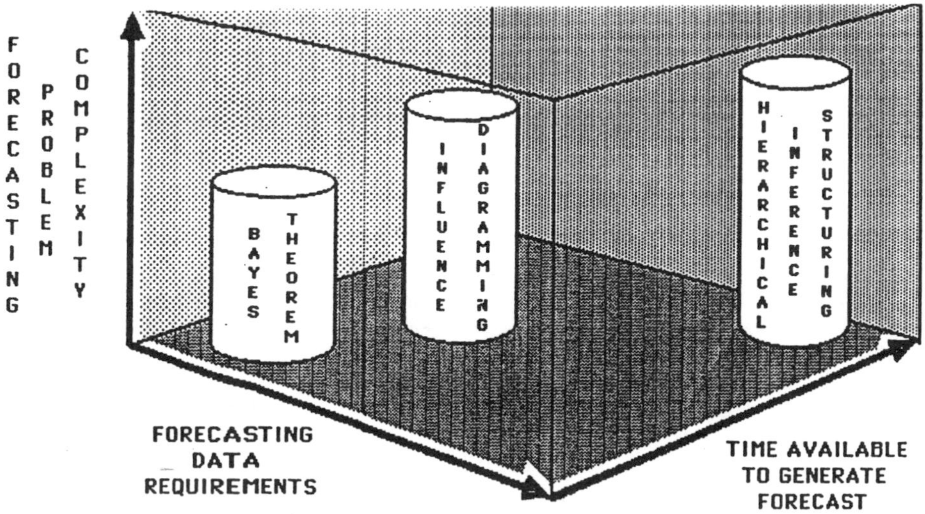

Figure 2. Bayesian Forecasting Method(s) Selector.

The four meta-families, in turn, are broken down into specific families of related tools and techniques. The qualitative family is comprised of two families: subjective assessment and structured qualitative. The hybrid/bootstrapping family is a generalized analytical strategy which blends the qualitative and quantitative; it does not yield any discernible families, and will be treated as both a meta-family and a family. The quantitative branch features four distinct offspring: time series/extrapolation; stochastic/probabilistic; statistical/operations research; and causal models. Finally, the information science block spawns two computer based but otherwise very dissimilar families: information management (conventional computer science) and artificial intelligence. Overall, then, there are four meta-families and nine specific families.

In discussing the methods inventory, the following terminology will be used consistently. A meta-family is equivalent to a meta-method, the most basic kind of methodology decision confronting the designer of a command and control aiding system. A family (within a meta-family) constitutes a method. Tools and techniques within families are referred to as submethods. Thus:

Meta-family = meta-method;
Family = method; and
Offspring = submethod.

By the end of the requirements analysis phase, the design team should have settled on the appropriate meta-family(ies); more generally, selectors of computer based problem solving systems (off-the-shelf or built in-house) should have a clear idea about the methodology base of their domain *before* they begin to review candidate software or decision aid systems.

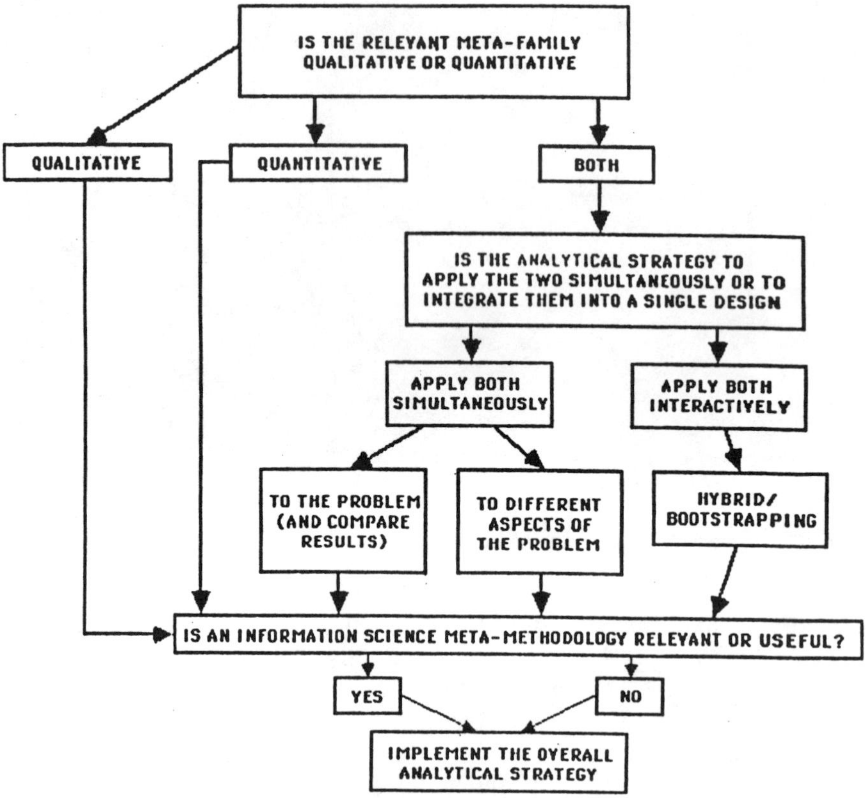

Figure 4. Simple Methodology Selection Flowchart.

The first and most fundamental methodology decision a system design team must make is: Will the aid be embedded in a qualitative (subjective, soft data based) or quantitative (objective, empirical, hard data based) meta-family? Or, will both be implemented simultaneously (by applying the two meta-families to the same problem *or* to different aspects of the problem) or interactively (by adopting the hybrid/bootstrapping route to analysis)? Note that an information science approach can be combined with either a qualitative or quantitative meta-family strategy; this, then, is the second general methodology decision. (This also implies, of course, that decision aiding may or may not be computer based.) The simple flowchart displayed in Figure 4 is a useful roadmap for the early methodology decisions.

It also is possible to offer some well established recommendations about the relative potential utility of the meta-families and families to different epistemological classes of command and control problems and decision contexts. As an example, Figure 5 ranks the nine families from high to low on the criterion of applicability to the class of open decision situations, which is by nature deductive in its epistemological foundations and tends to involve probabilistic (rather than deterministic) judgments and subjective (rather than

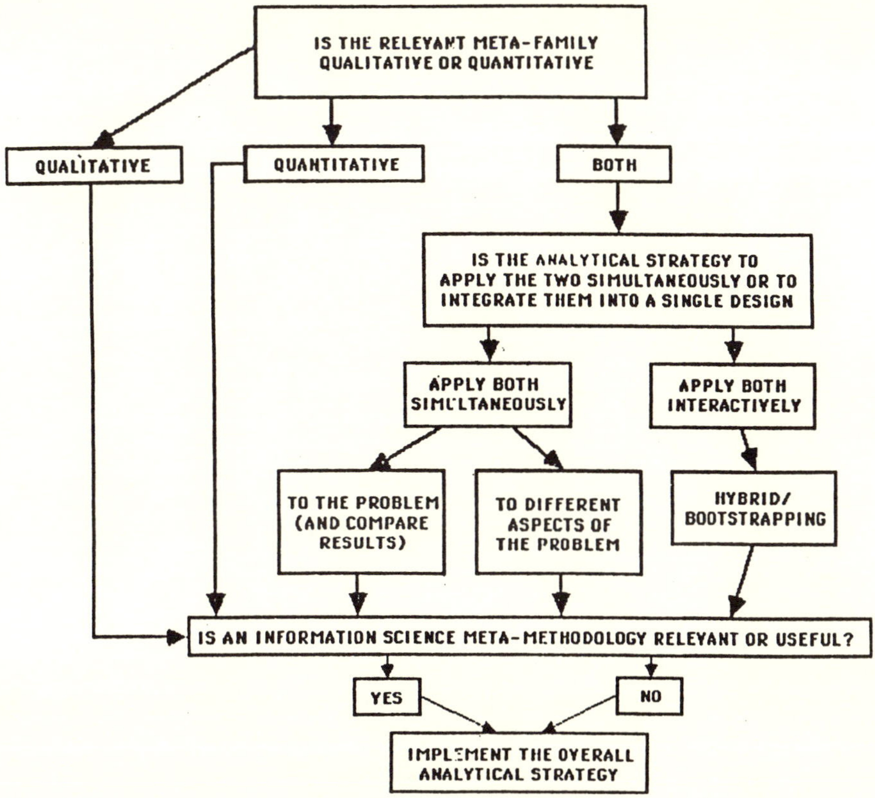

Figure 4. Simple Methodology Selection Flowchart.

The first and most fundamental methodology decision a system design team must make is: Will the aid be embedded in a qualitative (subjective, soft data based) or quantitative (objective, empirical, hard data based) meta-family? Or, will both be implemented simultaneously (by applying the two meta-families to the same problem *or* to different aspects of the problem) or interactively (by adopting the hybrid/bootstrapping route to analysis)? Note that an information science approach can be combined with either a qualitative or quantitative meta-family strategy; this, then, is the second general methodology decision. (This also implies, of course, that decision aiding may or may not be computer based.) The simple flowchart displayed in Figure 4 is a useful roadmap for the early methodology decisions.

It also is possible to offer some well established recommendations about the relative potential utility of the meta-families and families to different epistemological classes of command and control problems and decision contexts. As an example, Figure 5 ranks the nine families from high to low on the criterion of applicability to the class of open decision situations, which is by nature deductive in its epistemological foundations and tends to involve probabilistic (rather than deterministic) judgments and subjective (rather than

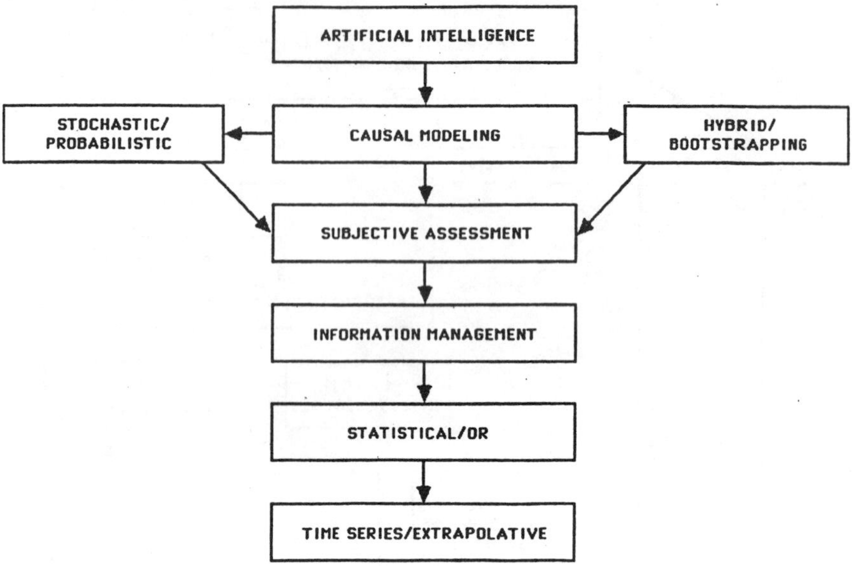

Figure 5. Methodological Family Rankings for "Open" Decision Problems.

empirical-quantitative or "objective") decision procedures. An open decision problem is one in which the potential solution set is by definition unbounded; in contrast, a closed decision problem may be pervaded by uncertainty, but the options can be delineated in advance.

The family ranked first is the structured qualitative, which is primarily deductive, probabilistic and subjective. The least relevant family, in contrast, is the time series/extrapolation cluster, which is inductive, deterministic and objective (a configuration which is the polar opposite of the open decision situation pattern). Not surprisingly, time series techniques, which operate on the assumption that some pattern recurs over time and that it is possible to identify the pattern exclusively on the basis of historical data, do not perform credibly with respect to open or unstructured decision situations where the solution set is, by definition, "unboundable."

The other families fall between these two. The AI family is ranked second, just below the structured qualitative family (albeit because of AI's promise for this class of C^2 decision situations, not as a result of its current performance). Three families cluster at the third tier in the relevance hierarchy: stochastic/probabilistic; causal modeling; and the hybrid or bootstrapping family. Open decisions are intrinsically stochastic, and some of the specific submethods under this rubric may be worth exploring for decision aids in this area. Causal models may capture the underlying dynamics of such situations if they are carefully developed, and may offer viable mechanisms at least for structuring such problems. Hybrid models which combine the strengths (and compensate for the respective liabilities) of humans and computer based systems may also make a contribution (although, as with AI, the bootstrapping family ranks high on the basis of promise for the future rather than current performance). Below this tier are two families which have some limited utility for open decision situation aiding: subjective assessment and information management.

Finally, two families are ranked sixth and seventh, respectively: the statistical/OR and time series families. Both are inductive epistemologically, objective and deterministic. This, of course, is completely incompatible with the epistemological character of the domain, suggesting that decision aids which are based on these families would be useless, or even dangerous, to the C^2 decision maker facing an open situation.

The full matrix is shown in Figure 6, which duplicates the meta-families and families of Figure 3 and simultaneously displays the more specific aiding tools or *submethods* for each of the nine families. There are 55 identifiable offspring or submethods available for command and control decision aiding.[5]

Qualitative methods rely primarily on the wisdom, intuition and experience of experts. The qualitative meta-family is ideal for illuminating the rich dynamics of a C^2 situation or process and for making judgments and inferences based on a detailed understanding of concrete aspects of particular cases. At the same time, qualitative tools and techniques are less precise than quantitative methods; furthermore, although the qualitative methods consist of discrete steps, the quantitative methods are more precisely specified.

However, most of the structured qualitative methods combine expertise with an explicit methodological template and a detailed embedded analytical model. For example, cross-impact analysis, Bayesian analysis and decision analysis all take expert judgments and assessments as inputs and filter and process them through a predefined methodological "blackbox" (e.g., the Bayesian formula for updating probability estimates and automatically generating revised probabilities on the basis of expert estimates).

Often, it is advisable to combine the two qualitative families in a single decision aid methods design. They are natural complements since the structured qualitative family is more deductive epistemologically and the subjective assessment offspring tend to have a more inductive basis; both families are stochastic or probabilistic and rely primarily on subjective rather than objective data.

Compared to the other families, qualitative methods are *on the average* fairly inexpensive in terms of almost invariably scarce time and cost resources. Subjective assessment submethods are generally faster and less expensive than their structured qualitative cousins, reinforcing the recommendation that, whenever feasible, both should be used to tackle a C^2 problem and the results should then be compared explicitly.

Of the nine families, *subjective assessment* methods most closely resemble "traditional" or non-methodological C^2 decision making; this family in fact provides a very useful bridge between purely intuitive decision processes and the more formal families on the other branches of the C^2 methodology tree. Some of the subjective assessment submethods rely on groups of experts.

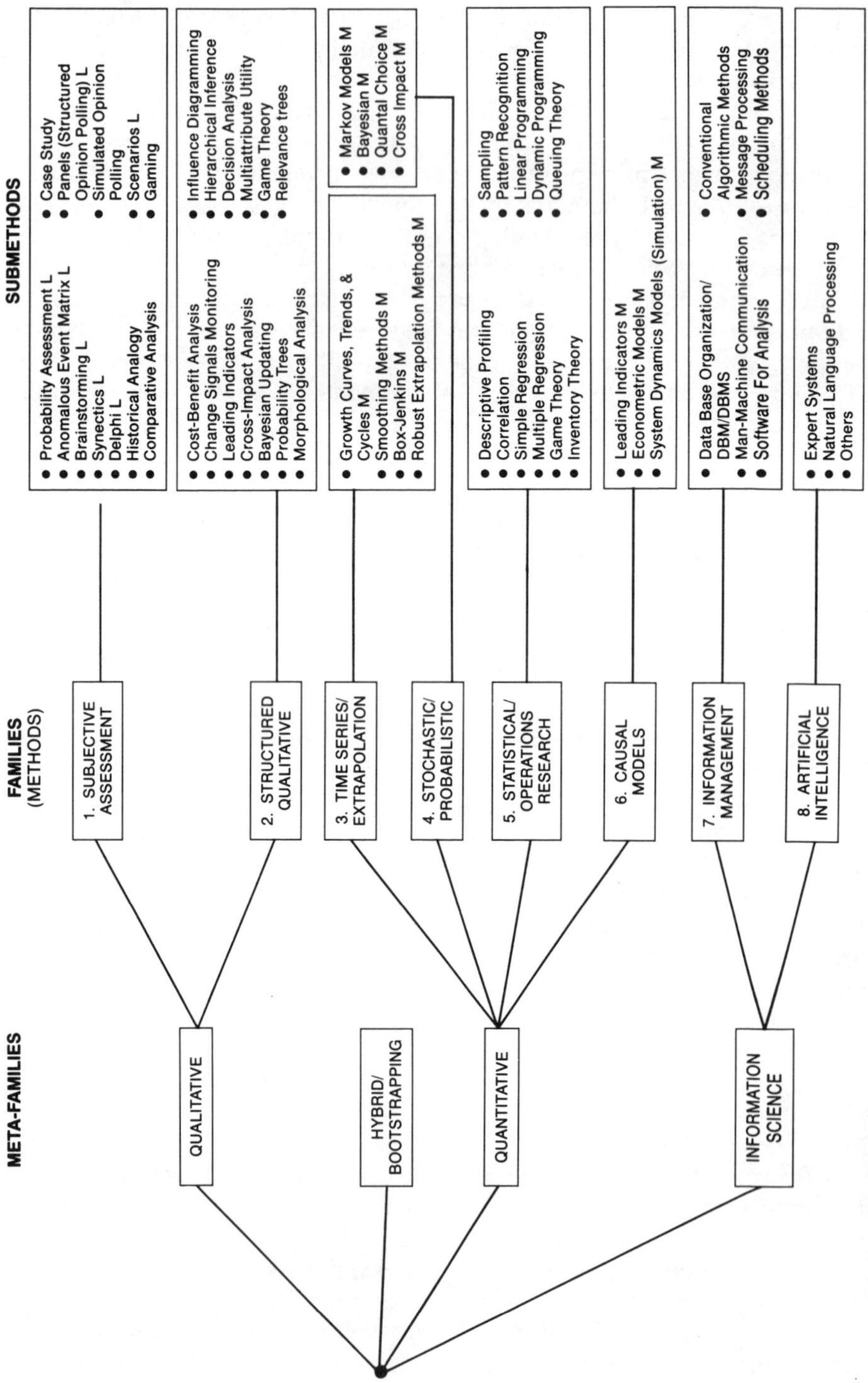

Figure 6. Methodological Meta-Families, Families (Methods) and Submethods.

Groups are particularly valuable for solving complex, interdisciplinary problems (e.g., combined logistics-operations decisions) and for identifying and structuring problems.

One important distinction within this family should be underlined: consensus oriented submethods such as panels and Delphi, are designed to produce convergence (and perhaps false or premature consensus); and scenario approaches are designed to intentionally emphasize uncertainty and highlight varying but plausible forecasts. Whenever possible, two (or more) subjective assessment submethods should be applied simultaneously.

A key, but not so obvious, difference separates the *structured qualitative* from the subjective assessment family; many of the former submethods combine expertise (subjective data) with a quantitative routine or set of algorithms. Just as the subjective assessment family combines traditional expertise with explicit methodological tools and thereby serves as a bridge between intuition and self-conscious methodology grounded C^2 problem solving, the structured qualitative family blends the qualitative and the quantitative—playing an analogous role as a bridge across the two primary branches of the decision aiding methodology tree.

The *qualitative* meta-family subsumes a variety of disparate techniques which all share a common characteristic: reliance on data and the utilization of a statistical or mathematical procedure to produce results. Essentially, all of the quantitative methods and offspring invoke the same sequence of analytical operations:

> Past data values → Set of embedded rules (algorithms, formulas, invariant statistical procedures employed to search for patterns) → Output (forecasts of future values, past or current data configurations, some kind of measurement based profile).

For all quantitative methods, there is a set of data believed to be relevant to the command and control analytical task (description, explanation, forecasting, prescription or option selection, evaluation or some combination thereof) and a theory of some kind which links the data to the future. Note that both characteristics assume the use of judgment to assess the nature and relevance of the data and to develop and verify the underlying theory.

The *time series/extrapolation* family forecasts the future values of a variable on the basis of its past values. A law or relationship is assumed to hold between the past of a time series and its future. Submethods housed here take a purely deterministic stance. Further, no causal linkage is assumed; if such a mechanism exists, it operates "behind the scenes." All of the offspring of this family assume that some pattern (or combination of patterns) recurs over time, and that it is possible to identify the pattern solely on the basis of historical data.

A time series is a sequence of numerical data in which each item is identified with a specific instant in time. Examples include daily temperatures at a weather station, the monthly price of some commodity or quarterly counts of Soviet missile tests. The data points in a time series do not need to be equally spaced; and, in some cases, the variables are not continuous, but instead change in connection with discrete events (a not uncommon pattern in the command and control domain). Quite different submethods are used for equally spaced and other data.

Stochastic/probabilistic methods emphasize a probability distribution over possible outcomes (e.g., the probability of an attack at time **t** is .50). This family is relevant when there is no one-to-one correspondence between the historical data or time series and the outcome; instead, an uncertain (but not wholly random) or probabilistic linkage exists. Each possible future state is assigned a probability on the basis of the historical data. In contrast to the other families in the quantitative meta-family, stochastic methods assume explicitly that there is no unique relationship linking past and future; the present state of affairs may be followed by any of several future states, and each future state may, in turn, arise from several possible prior states.

There is a kinship between this family and many of the submethods delineated under the subjective assessment and structured qualitative rubrics of the first meta-family. For example, simple probability assessment and all of the stochastic/probabilistic submethods involve the estimation of probabilities of occurrence of some event, situation or condition. The criticial difference is that subjective assessment probabilities are elicited from one or more experts directly, while the probabilities here are all statistical–objective in nature.

The qualitative stochastic affinities are underlined by the presence of two submethods under both the structured qualitative and stochastic family headings: cross-impact analysis and Bayesian analysis. Both can be used with either subjective or objective data inputs (or some combination of the two). Otherwise, the theoretical foundations are identical.

Two other points should be made about the offspring of this family. First, the other quantitative methods often are used along with stochastic methods. Second, this family almost always can be used when others in the meta-family cannot be employed at all because of the lack of appropriate data or because other prerequisites for their use are violated.

Statistical/operations research methods use explicit, rule based quantitative techniques to identify patterns in variables. Typically, a statistical method can supply information about the confidence that can be placed in analytical results and the exact likelihood of their being in error. Several comments should be underlined about the statistical via-à-vis other command and control decision aiding methodology families. First, the time series, probabilistic and causal model families all use statistical techniques. Within the quantitative realm, the aid design team has three basic model categories. In order from least to most expensive and complex, they are:

- The time series family, where discernible patterns are identified solely on the basis of historical data;
- The statistical family; and
- The causal model family, where the estimated values of one factor are a function of one or more others.

The probabilitic methods often are used as complements to these three.

In addition, to a greater extent than is commonly realized, the distinction between the qualitative and the quantitative is a nebulous one. Qualitative approaches can be rigorous; quantitative information is not invariably or even necessarily superior and more valid and the dangers of "false precision" are very real. In a more basic sense, nonstatistical and statitical approaches are merely logical extensions of each other; that which is measured is always a *quality,* and quantification is a method of expressing a quality with more precision. By the same logic, there is *implied measurement* in many "qualitative" concepts. For example, it is certainly possible to cast many (if not all) command and control issues in dichotomous or either-or (and therefore quantitative) terms (e.g., deception versus nondeception).

Causal modeling constitutes a quantitative family used to assess cause and effect relationships. Often used for forecasting purposes, a causal model forecasts the value of the variable of concern on the basis of other variables to which it is linked in a causal relationship (or network of causal interrelationships). A law governs the variable(s) to be forecast in that its values are determined by other (causal) variables.

Causal modeling methods are very decidedly statistical in nature, thereby underlying this family's affinities for statistical methods; econometric modeling, to cite just one example, builds explicitly on multiple regression analysis. Time often plays a central role in causal modeling applications, where long term trends and patterns are profiled and projected, suggesting a kinship as well between this and the time series family. Although causal models show a preference for determinism, they generally contain "error terms" and stochastic elements.

All of the submethods in this family share common characteristics. First, a causal model of any kind must be developed before it is tested. Model building or construction is always the first phase. Technically, this is called "model specification" in the econometric tradition. Judgment, training, experience, and even intuition are indispensable to this process, which involves the identification of the dependent and causal variables and the pinpointing of their interrelationships. This model construction process often is of inestimable value as a heuristic device for clarifying ambiguities, illuminating previously unsuspected relationships, and otherwise structuring the command and control problem terrain.

A second shared feature is complexity. Causal models tend to be elaborate, featuring multiple variables, many relationships (direct and indirect, positive and negative in direction), and large volumes of supporting empirical data. Some assume a nonlinear system; most concern the dynamic behavior of a large network of variables.

Compared to the other quantitative families, which tend to be inductive, causal modeling is deductive in nature. As a family, the submethods share three core assumptions:

- The model at the center of analysis represents a system of variables, interrelationships and a relevant environment—all linked to a real world system;
- Such systems can be expressed in terms of a formal statistical or mathematical model that can be validated using historical data; and
- The model can in turn be written as a computer program that can provide descriptions, explanations and forecasts.

The *hybrid/bootstrapping* family blends the qualitative and quantitative approaches. The bootstrapping method has been applied in a number of case studies in psychology, business and other fields (Armstrong, 1978), but has not been implemented in the C^2 domain.

However, several considerations suggest that bootstrapping represents a potentially viable methodology candidate. First, it combines the strengths of expertise (qualitative methods) and explicitness and precision (typical of quantitative methods), while simultaneously compensating for the respective weaknesses of the two families. Second, empirical results across a number of disparate domains demonstrate that bootstrapping performs impressively; the method always does as well as the unaided experts, and generally outperforms them—often dramatically.

Bootstrapping is based on the assumption that the decision or analysis process in a problem area can be made explicit. The methodology develops a "model of the expert" which typically is superior to the individual specialist at forecasting. For this reason, it is called bootstrapping—the analyst can lift himself up by his own bootstraps.

As Armstrong (1978) notes, bootstrapping follows the formal guidelines on how one should combine methodologies in problem solving. Subjective methods are used before quantitative or objective methods; the former *provides* the rules and the latter *applies* the rules. The objective model does the repetitive work by applying the expert's rules reliably and automatically (without being prone to such normal sources of human error as fatigue).

Deductive bootstrapping involves the direct translation of the expert's rules into a model, thereby making the subjective process explicit and operational. It is often sufficient to ask the experts what rules they are using (for example, decision rules used by doctors to prescribe drugs for patients). If the experts are not aware of how they make judgments or decisions (as is often the case with respect to command and control decisions), the methodologist asks the expert to "think aloud" as he or she engages in decision making and translates the verbalized process into specific rules (for example, for programming the rules used by an investment trust officer to select stocks).

Inductive bootstrapping takes the expert's conclusions or forecasts as the starting point and works backward to infer the rules used. The basic approach to inductive bootstrapping involves the development of a model where the expert's judgments are predicted by a quantitative model; the model uses the expert's judgments as the dependent variable and the factors that the expert uses are the causal variables. Typically, the procedure employed to do this is multiple regression analysis.

The qualitative, quantitative and hybrid/bootstrapping meta-families feature a variety of tools and techniques designed to take a body of input information and apply "off-the-shelf" analytical techniques to produce a judgment, estimate or recommended course of action. The *information science* meta-family focuses on the input to this process—information.

Information can be defined as "collections of symbols" (symbols in turn are "patterns that carry meaning"). The C^2 decision maker at any level deals with information in its many forms. Information is the raw material of information science. Operations on information are the procedures which can be performed on these patterned collections of symbols. Information can be created, transmitted, stored, retrieved, received, copied (in the same or another form), processed and destroyed. Information storage and retrieval operations refer to procedures for recording information for subsequent use (manually in paper files or on a computer) and for retrieving (accessing) stored information.

Finally, an information system can be defined as a computer based system which provides information to users and typically includes one or more large data bases. The system may be used for one or more analytical purposes (problem identification or structuring, description, explanation, prediction or forecasting, prescription or option selection, evaluation or documentation). Computer based problem solving systems are most likely to be successful if they result from an explicit, sequential design and development process (such as, the framework in Andriole [1983]).

Computer based problem solving systems may be broken down into two distinct families: conventional computer science information management and artificial intelligence (AI). In the most generic sense, *information management* refers to the management of diverse data bases (and their interrelationships), including the organization, storage, retrieval, display and processing of such data bases. The submethods in the family, which consist of data base organization/management (and DBMSs), conventional (algorithmic) analysis techniques, man-machine communication (and MMI, or man-machine interfaces), message processing, analytical software and task scheduling methods are really strategies for structuring information systems, interacting with such systems (via message processing activities, the application of algorithms and software, etc.) rather than "methodologies" in the classical sense of specific analytical techniques like Bayesian analysis or time series analysis. In fact, most of the submethods classified under the first three meta-families can be applied to computerized systems via off-the-shelf software packages or specially programmed software.

The AI family houses an array of tools and techniques designed to produce "smart" computer based information and problem solving systems—smart in the sense of representing at least a certain level of human intelligence. While information management can be equated with "conventional computer science" and is therefore essentially quantitative in how it handles and processes information, *artificial intelligence* deals with aspects of human mental activity that are approximate or heuristic and abstract and symbolic in nature. For example, information management involves data bases (with predetermined structures), while AI works with knowledge bases (comprised of unstructured, almost isolated facts—with paths formed to relate facts as needed to solve a particular problem).

This definition of AI suggests that it should have undeniable relevance to command and control, where decision makers often encounter issues cast in approximate, heuristic, abstract terms. This has been argued forcefully elsewhere, and a number of prototype AI based C^2 problem solving systems can be catalogued across the application areas of expert systems, natural language systems, intelligent information management, planning and general problem solving.[6] Although the potential relevance of AI is high, its immediate utility to the designer of command and control decision aids is generally quite limited because of the time and effort required to develop, test and transfer to the user AI based problem solving systems.

Methodologies and the Military Decision Process

Methodologies are never applied in a vacuum, and the relevant context here is the domain of C^2 specifically and military decision making more generally. Wohl's previously delineated SHOR taxonomy breaks the decision making process down into four components or stages; the S or stimulus and H or hypothesis generation tasks correspond to inference making and the O or options and R or response phases can be aggregated into the larger function of option generation, evaluation and selection.

The inference aspect of C^2 is essentially a task of problem structuring, an area of decision aiding and analytical methodology which has received relatively little attention—but a crucial concern if decisions are to be effective.

The CIA's Office of Research and Development (ORD) has been sponsoring work on a computer based problem structuring aid for several years. The aid is designed to facilitate the unconfounding of hypotheses, evidence items and assumptions—the three building blocks of inference and structures of argument. Central to the design concept is the idea of featuring sophisticated graphics capabilities, a desiderata that is attractive from the perspective of portraying uncertainty and, in addition, has been found in the military planning arena (such as, research reported in Andriole [1984]) to capture the way that planners think, reason and make inferences and decisions. The visual depiction of the problem structuring process—and its stages and elements—is a core system design guideline. Data management functions (such as, sorting, filing and categorizing data) are relegated to the aid's executive function and downgraded (so that the need to organize data does not cloud or drive the entire inference process). The user will be capable of incorporating an audit trail of his or her interaction sequences with the aid.

All of these system design principles coalesce in the generic idea of a problem structuring aid rather than an expert system, a perspective which is fully compatible with a requirements-driven aiding design process for tactical military C^2. While such an aid would be expected to include knowledge bases, expert system inference engines and other tools and techniques of AI, the system also would feature quite a few decidedly non-AI methodologies to support problem structuring and inference structure mapping. In a more philosophical sense, such an aid (or ensemble of aids) would be designed to augment rather than replace the human inference maker or automate the inference analytical process. What is envisioned, then, is a man-machine partnership, a machine-aids-man (or team) paradigm rather than a machine-driven computer based problem solving system.

Typically, as Pitz (1983) points out in a very useful review, decision aids make few assumptions about the nature of the user's problem(s). The prototypical decision aid is anchored in the technology of what Pitz calls "orthodox decision analysis," a blend of expected utility assessment and its extension to the class of multiattribute utility (MAU) models, that has been incarnated in so many decision aids and support systems. Such aids generally assume an effortless transition from the decision maker's problem to the formal representation of the latter in the actual aid. Such an assumption, like so many of its cousins in tactical inference, is both risky and frequently unwarranted.

Some effort has been made to build problem structuring routines into the aiding process in order to facilitate the key objective of illuminating the interrelationships among the elements of the problem. In problem structuring, the user must specify the range of options available (and the possible outcomes to be expected from each) and identify (and clarify) his or her goals or objectives. Notice the isomorphism to military planning or plan generation of any kind, where—prior to deriving an attribute structure for evaluating options and outcomes—we must initially specify a theme (in military terminology, higher level guidance), a situation (a mission) and one or more deduced goals, all of which set the stage for a plan. The theme/situation → goal → plan → plan evaluation/replanning hierarchy pervades military doctrine and practice and is reflected in the designs for the many tactical planning and other C^2 aids (See Andriole and Hopple [1987]).

The simplest and most straightforward way in which a decision aid can facilitate problem structuring is to ask the user to provide a list of options and a set of attributes; an early aiding system, QVAL, takes this lack of implementing a very direct approach (Weiss, 1980). MAUD, discussed in Humphreys and McFadden (1980), goes beyond this by offering a more elaborate set of procedures for eliciting attributes; MAUD makes available numerous techniques to help the user develop a multiattribute representation of the problem.

More contemporary decision aids are less frequently grounded in MAU and other decision analytic technologies and have become less dependent on decision trees and analogous DA tools for structuring problems. A particularly relevant example in this context is GODDESS or the Goal Directed Decision Structuring System (described in Pearl et al. [1982]). GODDESS, which borrows techniques from both DA and AI, is a goal directed (and therefore planning oriented) approach to decision structuring.

GODDESS encourages the use of real time Delphi methods for aggregating the opinions and estimates of different experts, and is therefore appealing from the vantage point of an aid targeted at all stages of tactical inference and option selection, where G2, G3 and other inputs must be incorporated. More importantly in this context, it is a computerized, domain independent decision support system with a structure for representing problems that allows the user(s) to state relations among aspects, effects, conditions and goals (in addition to actions and states, the elements of the traditional decision tree approach). The constraint of representing knowledge in tree form is, it should be noted, a serious limitation on structuring; as planners, analysts and commanders often perceive tasks not in the form of a time sequence of decision alternatives and event outcomes, but as a static network of influences surrounding issues and factors. GODDESS is designed to "explicate the underlying causal network rather than to hypothesize and evaluate various action/event scenarios," according to Pearl et al. (1982: 251), an orientation fully compatible with cognitive science views of the planning and decision making processes.

A problem structuring aiding system must confront at least three distinct issue clusters. First, it is necessary to facilitate the retrieval from the user's memory of information that can provide templates for generating options and describing objectives and attributes. Second, the aid should help the planner to go beyond the discrete information available (for example, by providing abstract typologies of options, outcomes, objectives and planning factors). Third, the aid must convert the planner's inference structure of assumptions, hypotheses and evidence items into a form that is compatible with the analytical model(s)—AI based, decision analytic or some other combination of tools—which exercise the problem solving system's internal (and usually transparent) processing routines for plan generation and decision making.

As is the case with uncertainty portrayal and propagation, no validated suite of methodologies exists for problem structuring (or even identification). However, there are some relatively useful typologies of decision problems, classification schemes that have at least some utility for the vital initial inference task of translating a poorly defined but perceived problem into one that is amenable to analysis and inference. Von Winterfeldt (1980) provides a review of this literature and expands on the idea of prototypical decision analytic structures.

The problem structuring process features three distinct phases. The first is problem identification (the substantive features of the problem, the generic classes of alternatives, objectives and events, the salient characteristics of the problem environment). The second step entails the development of an overall analytic structure, yielding potentially viable analytic problem representations. Tree structures are perhaps the most popular, but many others exist (simulation models, causal models, linear programming representations, and so forth). From a methodological perspective, the most useful analytic problem representation vehicles for tactical inference and option selection military problems include influence diagrams, hierarchical inference models and MAU models (along with a variety of other candidates from the generic problem solving methodology tree). The final phase involves the formalizing of the parts of the overall analytic structure; the problem elements and relations identified in the first step are refined. If necessary, detailed decision trees, event trees and goal trees are constructed. This step coincides with the more conventional image of restructuring.

Problem identification typologies and taxonomies typically group decision problems by analytic categories (such as, whether a problem is single or multiattributed). Such typing schemes identify an unknown element (such as, a future oriented defensive tactical decision problem) with a *class* of problems (that is, planning). Critical at this point is the distinction between ill-structured and well-structured problems. Uncertain, complex and conflicting problems typify the former category and resemble open decision contexts; decision aids and methodologies are especially rare for this genre of problems.

To an extent, problem structuring aids either rely on abstract structuring concepts like decision trees, inference trees, influence diagrams or MAU templates, where the user must force-fit his or her specific problem into the precanned problem structure parameters (See Kelly's chapter in Hopple and Kulhman [1981] for an overview of some prominent decision analytic based aids.) or are narrowly problem-specific (such as, aids to calculate the value of artillery targets).

Von Winterfeldt has attempted to compromise between the generalized empty structures and the overly substantive aiding structures by identifying generic classes of problems. Four typical problem classes are facility siting, contingency planning, budget allocation and regulation; these four—and many others—would serve as the fulcrum for developing prototypical decision analytic structures. A valuable byproduct would be the ability to move beyond abstract structuring contexts like simple decisions under uncertainty or multiattribute issues.

Problem identification/structuring are related to and flow into hypothesis generation. The former initiates the process of amassing assumptions; the latter phase of inference making moves into evidence node identification and hypothesis development (although tentative hypotheses are of course entertained from the very outset of the process, it makes sense to distinguish between structuring and hypothesis generation).

Hypothesis generation is a deductive and, under optimum circumstances, a creative process. The extensive literature on creativity and hypothesis sets is summarized in Tong et al. (1983). In the course of specifying options, outcomes and attributes (and other elements of the planning task), the decision maker inevitably explores new ideas (Pitz, 1983: 212). At this point, memory and inference processes are intertwined; current theories of memory acknowledge this fusion by arguing that human memory is organized on the basis of knowledge structures such as schemata and scripts, a theme that is explored from a cognitive psychology perspective in Nisbett and Ross (1980) and from the vantage point of military planning in Hopple (1985). Schemata facilitate the retrieval (and understanding) of information; often, knowledge structures are productive mechanisms for producing plans and decisions, but they also can impose excessive structure and lead to mental fixation—sometimes with deleterious or disastrous consequences (failing to foresee massive deception or a successful surprise attack, not anticipating enemy behavior on the battlefield, and the like).

The idea of embedding preestablished hypothesis sets in aids as part of the problem structuring/inference making aspect of C^2 aiding is explored in Andriole and Hopple (1987), where there is a more comprehensive discussion of empirical research on hypothesis generation.

By now, it is clear that the heart of decision making—in theory and in practice—is option selection (preceded by the generation of a set of alternative options). Tong et al. identify six discrete elements of phases of the decision making process: problem recognition (identification) and problem structuring set the problem up for analysis, option generation follows, the actual analysis phase is next, and implementation of the choice and review (evaluation of the quality of the process and the actual outcome via feedback) complete the process. For each phase, there is a corresponding task:

(1) Problem recognition	→	Information gathering
(2) Problem structuring	→	Model building
(3) Option generation	→	Problem solving
(4) Analysis	→	Outcome evaluation
(5) Implementation	→	Plan generation
(6) Review	→	Status monitoring

Very simple (automatic, repetitive or stimulus-response) decision contexts may move directly from problem recognition to implementation. Complex decision occasions—typical of many C^2 decision contexts—warrant attention to all six of the stages (usually in considerable detail). For the "typical" military decision problem, options are not self-evident; plans are complex and contingent rather than straightforward (and involve cascaded or hierarchical inferencing, potential multiple goal conflicts and outcomes contingent on the behavior of an intelligent adversary as well as "nature" or the planning environment); even the recognition of a problem (and its concomitant structuring) can be problematic.

Methodologies, which—as noted—fall into various families, can be applied at all six stages. Problem recognition or information acquisition (the intelligence phase of decision making), for example, often involves the use of computer based information systems or MISs (management information systems), an increasingly popular topic in tactical intelligence. Model building features an array of model "templates" ranging from simple checklists and problem solving templates to a variety of qualitative and quantitative modeling strategies (influence diagramming, MAU, and other decision analytic models, the AHP or analytic hierarchy process technique, and so forth). Problem solving tools for generating option sets include a wide range of methodologies designed to encourage the user to go beyond his or her prevailing mental set. Examples include analogies, brainstorming, and synectics. Analysis or outcome evaluation makes use of such well tested techniques as simulation and decision analysis.

Generally, however, there is a paucity of work on aiding systems and methodological foundations to facilitate option generation (as opposed to option selection, which is treated rather comprehensively by a variety of decision analytic and cognate tools). In addition to problem structuring, this emerges as a realm of command and control in need of more basic and applied research and development.

Conclusion

Quite a few aiding systems and a host of methodologies are available to the C^2 decision maker at all levels. Increasingly, emphasis is moving toward hybrid methodologies and the incorporation of AI tools into larger computer based system architectures; analytical methodology is much more than just AI. Andriole and Hopple (1987) and Hopple (1986a) survey the recent trends in the arcane milieu of methodologies for C^2 problem solving.

This cursory overview has provided some feel for the dynamic world of advanced analytical methodologies and command and control decision support. Many items remain on the research agenda for the future, including the need for aids, support systems and new methodology applications to problem structuring and option identification.

I would like to thank Dr. Stephen J. Andriole of IIS and George Mason University for his support and encouragement. [Any errors of fact or interpretation are the responsibility of the author.]

Footnotes

1. On aiding systems generally, see Andriole (1986); on command and control aiding systems, see Andriole and Hopple (1987) and Rockmore et al. (1982).
2. The literature on command and control is voluminous. The following sources are especially useful: overviews are presented in Harris (1979) and Wohl (1981); the decision making perspective or human dimension of C^2 is discussed in Andriole and Hopple (1982), Hopple (1986b, 1986c), and Teates (1982); C^2 decision aiding methodologies are the subject of Hopple (1986a, 1986c).
3. C^2 decision aiding taxonomies are developed for the Air Force in Wohl (1981), for the Navy in Miller et al. (1979), and for the Marines in Crolotte and Saleh (1979).
4. This section draws on Hopple (1986c). The general methodology tree is based on several dozen sources on analytical methodology and the national security problem solving process. The most relevant works include: Andriole (1986); Armstrong (1978); Central Intelligence Agency (1982); U.S. Army (1983); and Martino et al. (1981). Also very useful are: Andriole and Hopple (1984); Clarkson (1981); Hopple and Kuhlman (1981); Hunter (1983): and Wheelwright and Makridakis (1980).

5. The 55 command and control decision aiding submethods are profiled in the sources listed above in note 4 and elsewhere; it should be pointed out that this is a reasonably comprehensive, but by no means exhaustive list. Furthermore, creative decision aid design teams frequently "invent" methods and combine two or more submethods in a unique or application specific way.
6. On the AI/C[2] interface, see Andriole (1985) and Tullington (1984).

References

Andriole, S. J. (1986) *Handbook for the Design, Development and Application of Interactive Decision Support Systems*. Princeton, New Jersey: Petrocelli.

Andriole, S. J. (1985), ed., *Applications in Artificial Intelligence*. Princeton, New Jersey: Petrocelli.

Andriole, S. J. (1984) "The Design and Development of an Intelligent Planning Aid." Perceptronics and International Information Systems, Woodland Hills, California and Marshall, Virginia.

Andriole, S. J. (1983) *Interactive Computer Based Systems Design and Development*. Princeton, New Jersey: Petrocelli.

Andriole, S. J. and G. W. Hopple (1987) "Decision Theoretic Modeling for Tactical Planning." Marshall, Virginia: International Information Systems.

Andriole, S. J. and G. W. Hopple (1984) *Revolution and Political Instability: Applied Research Methods*. London and New York: Frances Pinter and St. Martin's.

Andriole, S. J. and G. W. Hopple (1982 "They're Only Human: Decision Makers in Command and Control," *SIGNAL*, 36 (March): 45–49.

Armstrong, J. S. (1978) *Long-Range Forecasting: From Crystal Ball to Computer*. New York: Wiley-Interscience.

Callero, M., D. A. Waterman and J. R. Kipps (1984) "TATR (Tactical Air Target Recommender): A Prototype Expert System for Tactical Air Targeting." Santa Monica, California: Rand.

Callero, M., L. Jamison and D. A. Waterman (1982) "TATR: An Expert Aid for Tactical Air Targeting." Santa Monica, California: Rand.

Central Intelligence Agency (1982) "Handbook of Problem Solving Techniques for Intelligence Analysts." Office of Training and Education.

Clarkson, A. (1981) *Toward Effective Strategic Analysis: New Applications of Information Technology*. Boulder, Colorado: Westview.

Courand, G., C. O'Reilly and J. R. Payne (1984) "OCA (Offensive Counter Air) Mission Planning." Advanced Information and Decision Systems, Mountain View, California.

Crolotte, A. and J. Saleh (1979) "Development of a Decision Taxonomy for the Marine Command and Control Environment." Woodland Hills, California: Perceptronics.

Harris, D. R. (1979) "Final Report: OSD Sponsored Colloquies on Command and Control." Arlington, Virginia: ESL.

Hopple, G. W. (1986a) "Assessment of the State-of-the-Art of Commercial and Military Decision Aids and Support Systems: Steps Toward a Coherent Acquisition and Investment Strategy." Marshall, Virginia: International Information Systems."

Hopple, G. W. (1986b) "Command and Control." Washington, D.C.: Synergy.

Hopple, G. W. (1986c) "The Law of the Hammer, and Other Maxims." IEEE *Transactions on Systems, Man, and Cybernetics*.

Hopple, G. W. (1985) "Intelligent Support Systems for Command, Control, Communications and Intelligence (C[3]I): A Prototype Causal Schema Knowledge Based System." Presented at the Human Factors in Manned Systems Integration Conference, Kansas City, Missouri.

Hopple, G. W. and J. A. Kuhlman (1981) *Expert Generated Data: Applications in International Affairs*. Boulder, Colorado: Westview.

Humphreys, P. C. and W. McFadden (1980) "Experiences with MAUD: Aiding Decision Structuring Versus Bootstrapping the Decision Maker." *Acta Psychologica*, 45: 51–69.

Hunter, D. (1983) *Political/Military Applications of Bayesian Analysis: Some Methodological Issues*. Boulder, Colorado: Westview.

Kaplan, A. (1964) *The Conduct of Inquiry*. Scranton, Pennsylvania: Chandler, 1964.

Madni, A. (1985) "Methods and Models for Representing Decision Processes," in the Third Annual Workshop on Command Decision Aiding Technologies for Military Command and Control, jointly sponsored by the Space and Naval Warfare Systems Command and the Office of Naval Research in cooperation with the Joint Services Working Group on Decision Augmentation, November, 1985.

Martino, J. P., et al. (1981) "A Survey of Forecasting Methods." Dayton, Ohio: University of Dayton Research Institute.

Miller, A. C., III, T. R. Rice and M. R. Metcalf (1979) "An Analytical Characterization of Navy Command and Control Decisions." Menlo Park, California: Applied Decision Analysis.

Nisbett, R. E. and L. Ross (1980) *Human Inference: Strategies and Shortcomings of Social Judgment*. Englewood Cliffs, New Jersey: Prentice-Hall.

Pearl, J. A. Leal, and J. Saleh (1982) "GODDESS: A Goal-Directed Decision Structuring System." *IEEE Transactions on Pattern Analysis and Machine Intelligence,* PAMI-4 (May): 250–261.

Pitz, G. F. (1983) "Human Engineering of Decision Aids," pp. 205–221 in P. Humphreys, et al. (eds.), *Analysing and Aiding Decision Processes*. Amsterdam: North-Holland.

Rockmore, A. J., et al. (1982) "Decision Aids for Target Aggregation: Technology Review and Decision Aid Selection." Palo Alto, California: Systems Control Technology.

Rouse, W. B. (1985) "Design and Evaluation of Computer-Based Decision Support Systems," in S. J. Andriole (ed.), *Microcomputer Decision Support Systems*. Wellesley, Massachusetts: QED Information Sciences.

Slagle, J. (1984) "BATTLE" Resource Allocation Project: Expert Advisor for Weapons Allocation." Presented at the Artificial Intelligence/Intelligence Community Annual Symposium, Central Intelligence Agency.

Slagle, J., M. W. Gaynor and H. Hamburger (1983) "A Decision Support System for Fire Support Command and Control." *SIGNAL* (August): 52–60.

Teates, H. B. (1982) "The Role of Decision Support Systems in Command and Control." *SIGNAL* (September): 45–49.

Tong, R. M., A. Arbel, S. O. Cioffi, C. M. Kelley, J. R. Payne and E. Tse (1983) "Option Generation Techniques for Command and Control." Mountain View, California: Advanced Information and Decision Systems.

Tullington, B. J. (1984), ed., "Proceedings of the U.S. Army Conference on Application of Artificial Intelligence to Battlefield Information Management, April 20, 21 and 22, 1983." Battelle Columbus Laboratories, Battelle Washington Operations, Washington, D.C.

U.S. Army (1983) "Handbook on Strategic Intelligence Analysis." Intelligence and Threat Analysis Center.

von Winterfeldt, D. (1980) "Structuring Decision Problems for Decision Analysis." *Acta Psychologica,* 45: 71–93.

Weiss, J. J. (1980) "QVAL and GENTREE: Two Approaches to Problem Structuring in Decision Aids." McLean, Virginia: Decisions and Designs, Inc.

Wheelright, S. C. and S. Makridakis (1980) *Forecasting Methods for Management*. New York: John Wiley and Sons.

Wohl, J. G. (1981) "Force Management Decision Requirements for Air Force Tactical Command and Control." *IEEE Transactions on Systems, Man and Cybernetics,* vol. SMC-11 (September).

Processing Environments for Dispersed Command and Control

Ronald W. Foss

The past 200 years of military history have demonstrated a continuous evolution toward the employment of larger forces, deployed over wider areas, moving with greater speed and capable of inflicting greater destruction over greater distances. During the same period, however, the ability to control fielded forces has not kept pace with the combat power of those forces.

Attempts have been made in recent years to apply the information processing power of modern technology to stem this historical trend. Many of these attempts have fallen far short of their objectives; many have been abandoned without delivering anything to the troops in the field. Explanations for these disasters are myriad, yet one thread runs through many of the postmortems for tactical command and control (C^2) system developments: The system architecture could not keep up with the evolving requirements of the user.

The concept of evolutionary development for theater C^2 systems has been developed as a means of better accommodating continuously changing user requirements. The tools with which such a system would be implemented have not received as much attention as has the evolutionary development methodology itself.

**This article originally appeared in the April 1986 issue of SIGNAL.*

The objective of this article is to present a set of information processing concepts that could serve as the conceptual framework for the definition of machine assisted theater C^2 systems that avoid the pitfalls of the recent past. This article does not define decision aids, correlation tools, etc.; rather, it seeks to identify a processing environment in which mechanized assistance effectively can support the continuously evolving needs of tactical commanders and their staffs. It seeks to identify a suitable processing environment for the development and operation of an evolving, machine assisted C^2 system.

The characteristics of such a processing environment stem from the powerful modularity concepts that will be used in developing future C^2 systems; those modularity concepts stem, in turn, from C^2 system requirements that inadequately are satisfied today.

Requirements

Operational constraints on the information processing needs of tactical C^2 systems tend to fall into three categories: those imposed by force structure, those imposed by threat and those imposed by support structure. These three areas are unique for each situation.

> The command and control system of each commander is . . . a unique, situation specific, living web of components in its own continuing process of evolution. Because no two forces are exactly alike, no two [C^2] webs will be exactly alike. [The] commander is deeply concerned with the full web. His mission performance depends on it.[1]

In the past, developers have tended to ignore this observation, to the peril of C^2 system developments. One of a kind systems, for a single user (Commander in Chief, North American Air Defense Command; Commander in Chief, Strategic Air Command; etc.), successfully have been implemented. Low-level systems for air defense, field artillery direction, etc., successfully have been implemented. Attempts to construct generic solutions to theater level C^2 needs generally have failed. For example, the theater level Tactical Air Control Center (TACC) may be occupied by a battle staff of between six and more than 100 persons. In past exercises, the functions of a TACC have been performed at other facilities, such as the Air Support Operations Center (ASOC), in support of low-level contingency or casualty backup scenarios.

Existing transportable theater C^2 facilities are targeted much too easily by hostile forces. These facilities are too large, centralized and cumbersome to be concealed and, once located, they are too difficult to move to escape destruction. Transportable C^2 facilities deployed in support of contingency forces are constrained severely in terms of the airlift assets allocated for their support. All theater level C^2 systems must accommodate the C^2 interfaces of allied forces, and they must be able to exploit the civil infrastructure (for example, local commercial telephones) to the degree necessary to maintain continuity of operations.

It is too easy to define C^2 requirements in terms of peacetime needs. As a nation, the United States has little experience in performing theater level C^2 in a hostile environment.

> Vietnam, with its environment benign to C^2 systems and to logistics and other [support] systems as well, misled us badly. The C^2 system environment of, say, a NATO war will not be benign. In part because of Vietnam, but also for other reasons, the chaos and confusion of battle does not figure in many analysts' calculations.[2]

Due to both technical and financial constraints, advanced theater C^2 systems must be developed and fielded on an evolutionary basis. Not only will initial elements of a new system be operated in combination with a varying mix of existing facilities, they also must operate within the organizational structure and operating principles of existing fielded systems until such time as system users gain sufficient "live" experience with the capabilities of the new system to determine optimal tactics and procedures to exploit their potential. Furthermore, the life cycles of C^2 systems are extremely long. Existing facilities of the Tactical Air Control System procured under the 407L program will have served with active duty forces for over 20 years by the time they are replaced. When this service life is combined with the 10 to 15 year development cycle, and the potential for continued service with reserve units, the total system life cycle becomes impressive.

Providing C^2 systems that are capable of meeting evolving operational needs over extended life cycles will require either astounding prescience or the enforcement of an architecture that can support insertion of advanced operational and technical innovations. This will require more than a preplanned product improvement (P^3I) program; how does one define "preplanned" improvements based on technology insertions that may be available in 15 years or on doctrinal changes brought on by the availability of radically new weapons systems or altered political relationships? A C^2 architecture for the future must provide for true evolutionary growth in unexpected directions.

In today's C^2 systems, concepts of modularity frequently are employed to allow facilities to vary in capabilities to meet mission requirements. In today's data processing systems, concepts of extensibility frequently are employed to allow machines or programming languages to vary in capabilities to meet user needs. The following architecture is based on the fusion of those two concepts to meet the C^2 needs of the future.

Modular Dispersed C^2 Facilities

Modularity is a term familiar to C^2 users and developers; modular systems successfully have been developed and fielded in the past. However, until recently it has been necessary to partition modular C^2 systems on the basis of hardware resources (Figure 1). For example, in the AN/TSQ-91(v) Control Reporting Center/Post, facilities were partitioned into data processing modules, display modules and ancillary modules. All the data processing capability of the facility was resident in one data processing module. In the event of failure of that module, all computer supported functions of the facility were lost. Furthermore, the facility had no leap-frog or echeloning capability. In addition, a facility could contain only one data processing module. Facility modularity was defined only in the number of console positions available, four per display module.

Until recently, module partitioning based on classes of hardware resources was necessary due to the size and weight of the equipment. It was not practical to integrate processing, display and communications equipment into a single shelter.

In the late 1970s, the availability of large-scale integrated circuits with acceptable reliability and thermal characteristics allowed the size of processor based equipment to be reduced significantly. Accordingly, it was possible to partition C^2 systems on the basis of function rather than on the basis of hardware resource types. The TAOC-85/Modular Control Equipment is an excellent example of function based partitioning.

In the multiresource modules shown in Figure 1, each module contains an identical suite of processing, display and communications equipment. A C^2 facility can be assembled by replicating modules until the desired set of resources is obtained. The proportion of resources within a module is predefined, based on the results of functional analyses to determine the proportion of computational, communications, information storage and information display resources required to control tactical air defense missions.

As functionally partitioned modules are applied to C^2 activities that differ from the original function set, however, the proportions of required resources will vary from those provided in the module. For example, battle management functions will require greater data storage and ground-to-ground communications resources and proportionately less ground-to-air radio communications resources than will air defense functions. In replicating modules to provide all required resources, some resources will be superfluous. Not only is the cost of the facility thereby increased, but the size of the facility is increased as well.

The previously discussed modularity is in terms of a standard tactical shelter. The shelter is by no means the only feasible basis for modularity. In particular, C^2 systems may be required to operate in physical facilities other than a standard tactical shelter, such as a permanent facility (a bunker), a temporary structure of convenience (a house or a hangar) or a vehicle.

In order to provide a C^2 module with greater physical flexibility, work stations sometimes are discussed as the fundamental unit of modularity for systems. In this approach, all voice communications, data processing, display and storage resources are contained in each work station. Each work station may, therefore, operate independently or as part of a larger local network.

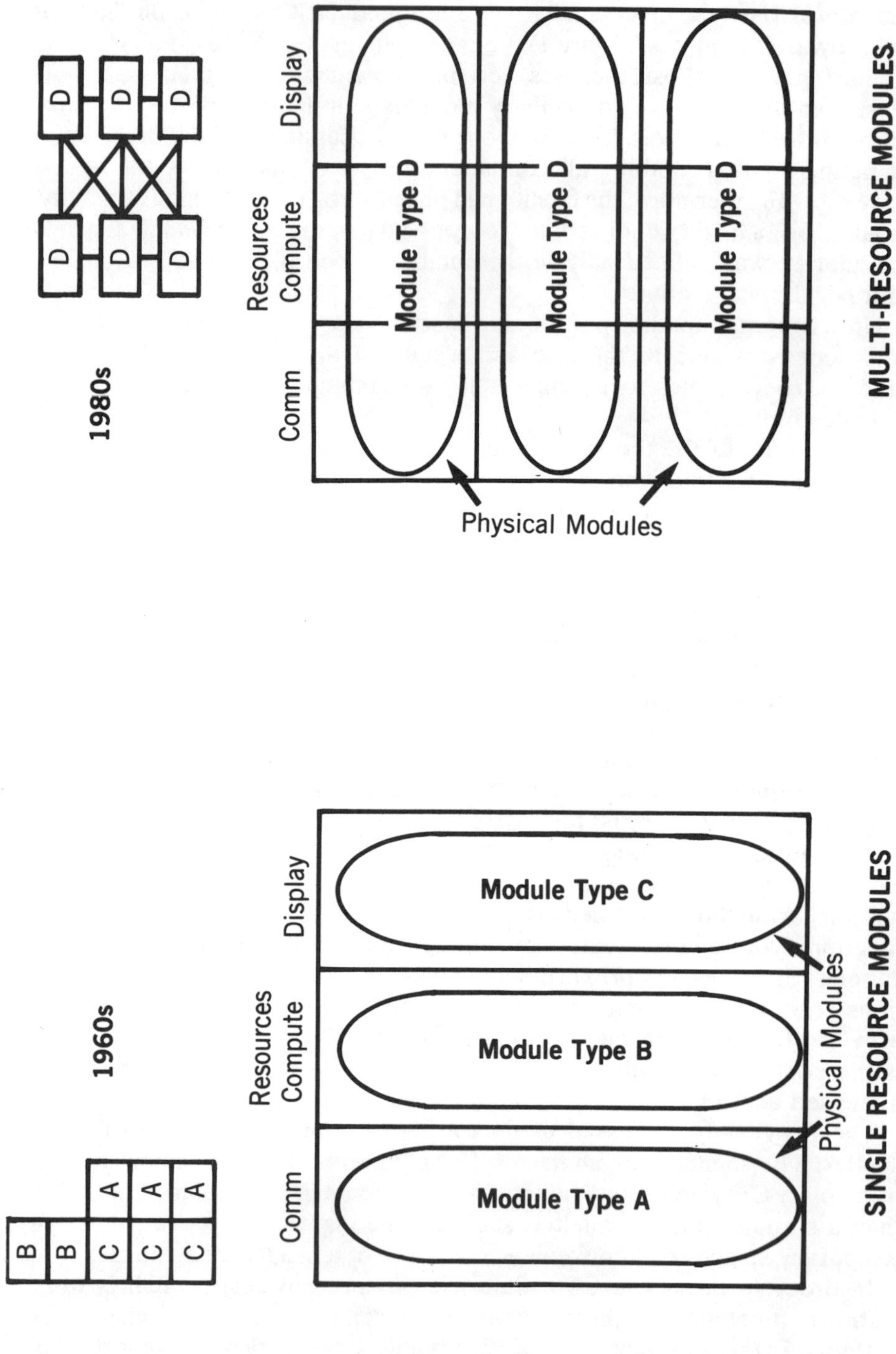

Figure 1. The evolution of modular C² concepts.

Even in a work station, the proportions of processing, display and storage resources largely are predefined. The potential for inefficient resource utilization remains; only the individual pieces are smaller and less expensive. Furthermore, since the individual resources are coupled tightly in a single physical unit, it may be more difficult to upgrade them independently without redeveloping the entire module.

A hybrid approach to modularity therefore may be necessary to achieve an evolutionary C² facility module. In the "21st Century Tactical Command and Control Study (TC2-21)," recently published by the U.S. Air Force Systems Command, just such an approach was recommended.[3] A set of basic resources, processing, display, storage and communications were defined to be organic to each work station. Shared resources were added to individual work stations or to work station clusters as necessary to support any set of assigned C² functions. This approach provided a more flexible way of configuring modular resources to meet the needs of C² systems (Figure 2). A set of standards, necessary to allow the interconnection of work station and resource modules, also can be used as the basis for evolutionary growth of the architecture over time.

Modular C² Processing

An architecture incorporating a series of processor modules will require a carefully constructed set of interface standards to be supportable. Resources may be configured to support a given set of functions in a given environment to implement a given segment of a C² system. The fact that those resources (work stations, embedded support processors, shared processing/storage resources, etc.) have been optimized for a given set of functions does not mean that they are dedicated to those functions. Since modules of the same types may be combined optimally to support any set of C² functions, it follows that any arbitrary collection of modules can perform any set of C² functions, albeit inefficiently or with reduced capacity.

This capability of the processing system has important implications, both operationally and technically. Most importantly, from an operational standpoint, any facility, or segment of a facility, could be reassigned to perform any other set of C² functions. Thus, any facility could support the functions of any other facility. Casualty recovery, deployment planning and facility echeloning will be far more manageable under such an architecture than with present systems. From a processing technology perspective, the ability of any element to assume, however inefficiently, the responsibility of any system function means that every element must support all classes of processing services provided by the system. These classes may include communications processing, correlation processing, display processing, rule processing, data base processing, maintenance processing, system and security control processing, among others in addition to the general-purpose computing used in present systems.

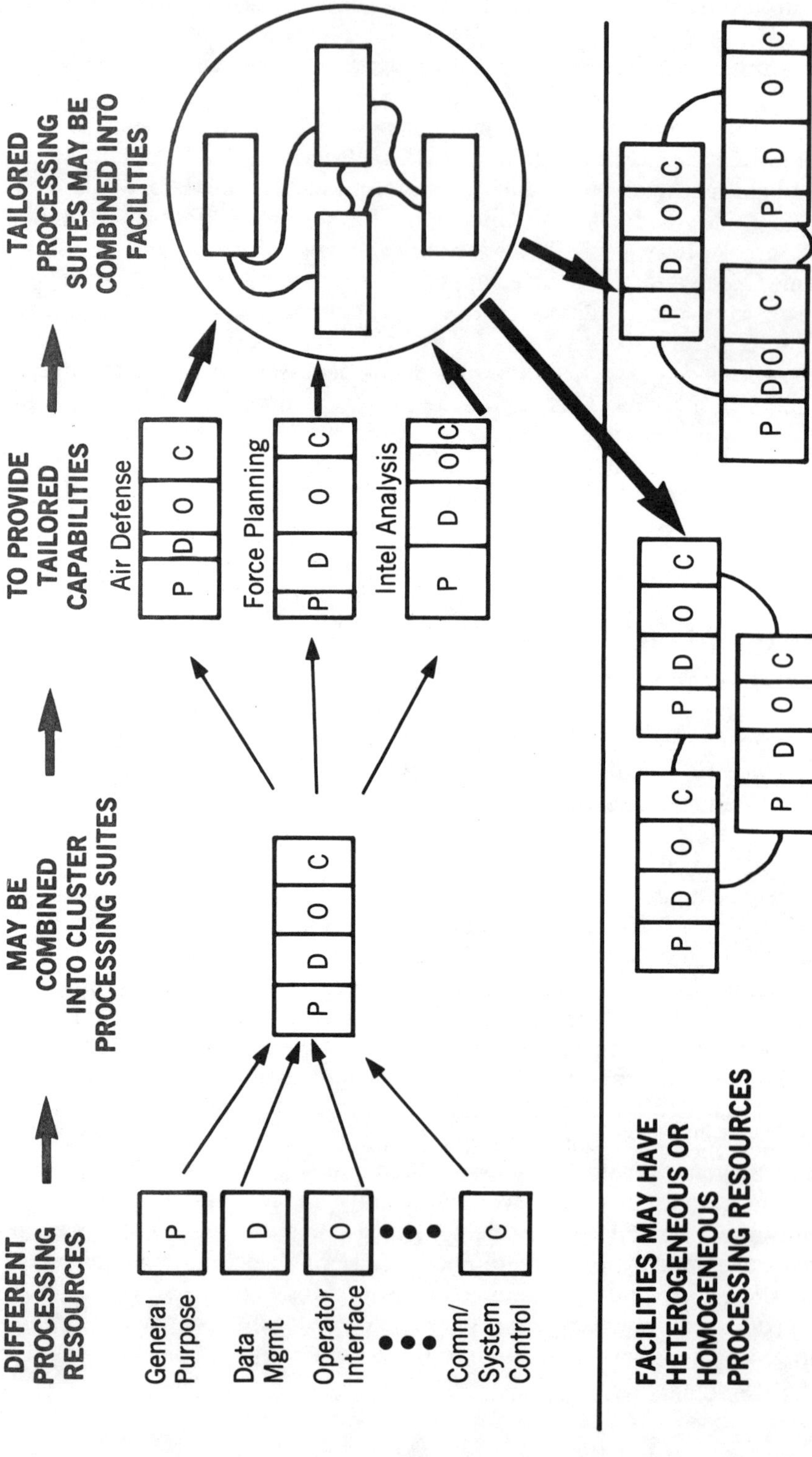

Figure 2. Architecture based on assembling problem specific sets of modular resources.

Depending on the processing loads demanded by the set of functions being performed, any given C^2 facility may require extensive, limited or no support from a class of processing service. Since support of all functions is highly desirable by all elements of the C^2 system, these processing classes may be supported in one of three ways at any element. First, services may be provided by a physical processor within the element if needed to support high volume demand. Second, services may be provided by accessing a physical processor attached to the element as a cluster server or shared processing resource. This may be preferred where several functions within a facility generate modest demands for the resource class. Finally, services may be provided by a software or a firmware emulator located within a general-purpose processing resource of the element.

To preserve the intra-system transportability of application software packages, the existence of these distinct mechanisms for providing a processing service must be transparent. The TC2-21 study described this concept in terms of Virtual Processing Resources.[4] Each class of processing resource in the system appears to each resident application software as a specific machine attached to the local host as a co-processor. In fact, each resource may be implemented by a real local physical processor, by a remote real physical processor, or by an emulation actually executing on the host. The decision to implement each processing resource class with a real or emulation processor is made based on the throughput required on a case by case basis. Since application software is presented with an identical virtual resource, independent of the manner in which that resource is implemented physically, applications thereby are isolated effectively from the physical implementation of the system. This means that hardware components can be replaced without affecting the software of the system. It also means that the physical size of the processing facilities can be varied without affecting the functionality of the system, only the throughput.

Thus, the processing resources of the system may be regarded as an extensible run-time support environment. In traditional systems, the run-time environment (RTE) provides other software elements of the system with a uniform means of accessing the input/output hardware of the system, management functions (for example, task scheduling, inter process communications) and common primitive functions (for example, mathematical transforms). Today, these functions strongly are dependent on the configuration of the host hardware.

In advanced, modular C^2 systems, the run-time software environment must provide a broader range of functions than those provided by a traditional RTE. It must:

- be capable of supporting high-level applications in a highly heterogeneous processing environment;
- provide transparent access to the full range of processing classes;
- provide integral query processing and data management functions to exploit a data base engine;

- provide high-level direct inference engine support to exploit a rule processor;
- provide search and match functions to exploit a signal processing engine; and
- provide for a range of sophistication in the access it provides to the hardware resources of the system.

The notion of a processing environment composed from sets of dissimilar processing engines is no longer radical. In the commercial world, attached processors to perform communications management or data management are extremely common. There is an increasing trend toward including attached processors under the umbrella of the host operating system, as with add-on array processors attached to general-purpose minicomputers and floating point processors attached to general-purpose microprocessors. A recent National Bureau of Standards study on future information processing technology concluded that:

> [future computing systems] will contain multiple computers dedicated to specific functions of the system. The specific function of each processor . . . will usually be determined by alterable microcode. . . . Users will be able to choose degrees of increased cost to obtain increased levels of fault tolerance. There will no longer be a single, shared storage serving all the processors. Instead, storage will be dedicated to individual processors or clusters of processors. . . . such a modular computer system will be able to employ widely varying sets of functional processors. . . . Within limits the processor orientations can be changed via the supervisory processor to meet changing workloads.[5]

The report postulates specialized processors in classes similar to those previously described—supervisory engines, data base engines, associated processors for searching, pattern recognition engines and processors to support knowledge based decision support systems.

A physical element of a C^2 facility (Figure 3) would be assembled by configuring first a set of work stations with appropriate embedded special function engines. The work stations then would be linked via a high bandwidth local interconnect. Zero or more specialized resources then could be attached directly to the interconnect as shared resources.

The notion of a controlled system-wide virtual processing environment mandates the establishment of rigorous controls over the Instruction Set Architectures (ISAs) of the underlying physical resource providers. Given the long service life of tactical C^2 systems, the government should establish and control ISAs in order to maintain the integrity of the architecture over an extended period of time and to maintain a viable competitive procurement base for replication or enhancement of system modules.

Implicit in the definition of Virtual Processing Resources is the notion of multiple layers of resource access. Relatively low-level emulators, implemented in macro or micro instructions, form the lowest layer of system

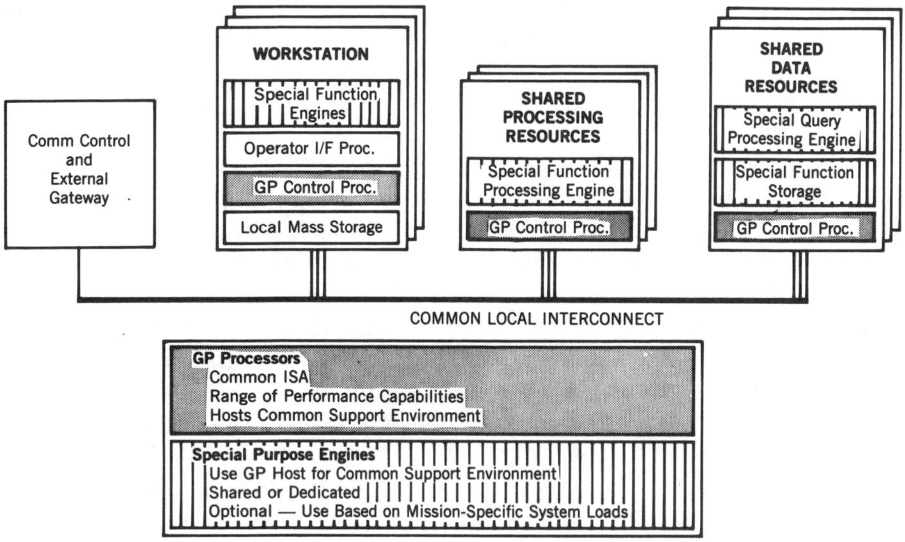

Figure 3. Hardware processing architecture.

software in the absence of the specialized hardware engines they emulate. Successive layers of Virtual Processing Resources provide successively higher levels of access to the underlying physical resources. Data management services, for example, may be provided at many levels. A process responsible for maintaining a correlated air track file might access directly air track reports in the data base, bypassing the higher level query processing and data integrity services of higher level data management functions to achieve minimum time delays. At the same time, the same information may be accessed by a higher level support process that is interested only in trend analyses. Such a relatively time insensitive but highly user tailored process most likely would make use of the highest level support processes to minimize the cost and time associated with development and user adaptation of the decision aid.

The multiple levels of processing services provided by the Virtual Processing Resource concept are analogous to the multiple levels of communications services provided by the Open Systems Interconnect (OSI) model of the International Standards Organization (ISO). Just as OSI protocols are being developed to improve the interoperability of netted data communications systems, so will development of Virtual Processing Resource standards improve the interoperability of the processing elements of future systems.

Implications for the Development Process

In past and present C^2 systems, the mission software components that directly support the human users ultimately are dependent on the supporting hardware structure. The requirements of present hardware structures are such that it sometimes appears that the objective of applications software is to adapt the human user to satisfy the interface requirements of the hardware and support environment (for example, rigid report generators, Procrustean query processors and arcane system control commands). The tremendous cost of supporting and updating existing computer based command systems are due in large measure to the close coupling of applications with the machine dependent support structure. The objective of the architecture previously presented is to provide a means of decoupling mission software from its underlying support components (Figure 4). The Virtual Processing Resource concept was defined to provide the needed interface between user driven mission software and technology driven support components. It is this interface that can provide the means by which user needs and implementation technologies can evolve independently to yield tactical C^2 systems that provide effective support to field forces over extended system lifetimes.

The decoupling of C^2 functions from the components with which they are supported also means that evolutionary system development will be made easier. Expansion of the Virtual Processing Resource concept into the development environment, as well as the operational environment, would ease greatly the transition of test bed developments into fully supportable field application. In. providing a streamlined process for migrating application developments from one host processing environment to another, and between environments of significantly different power and cost, these concepts for the future C^2 support environment complement the trends toward rapid prototyping and incremental development of user responsive tactical C^2 systems.

The systems needed to solve tomorrow's mobile, tactical C^2 problems will be highly modular. The challenge for the development community will be not only to define a set of building blocks, but also to define an integrated support structure capable of both meeting evolving user needs and accommodating evolving technological innovations. Unless this challenge is met, we may only continue the historical trend toward unsupportable and unusable theater command systems.

References

1. Cushman, John H., *Command and Control of Theater Forces: Adequacy* (Washington, D.C.: AFCEA International Press, 1985), p. 7.
2. Ibid., pp. 17–18.
3. "21st Century Tactical Command and Control Architecture" (Headquarters U.S. Air Force Systems Command/XRK, 1985), p. III-21.
4. Ibid., p. F-33.
5. *Future Information Processing Technology—1983,* NBS Special Publication 500-103 (Washington, D.C.: Government Printing Office, 1983), pp. 97–100.

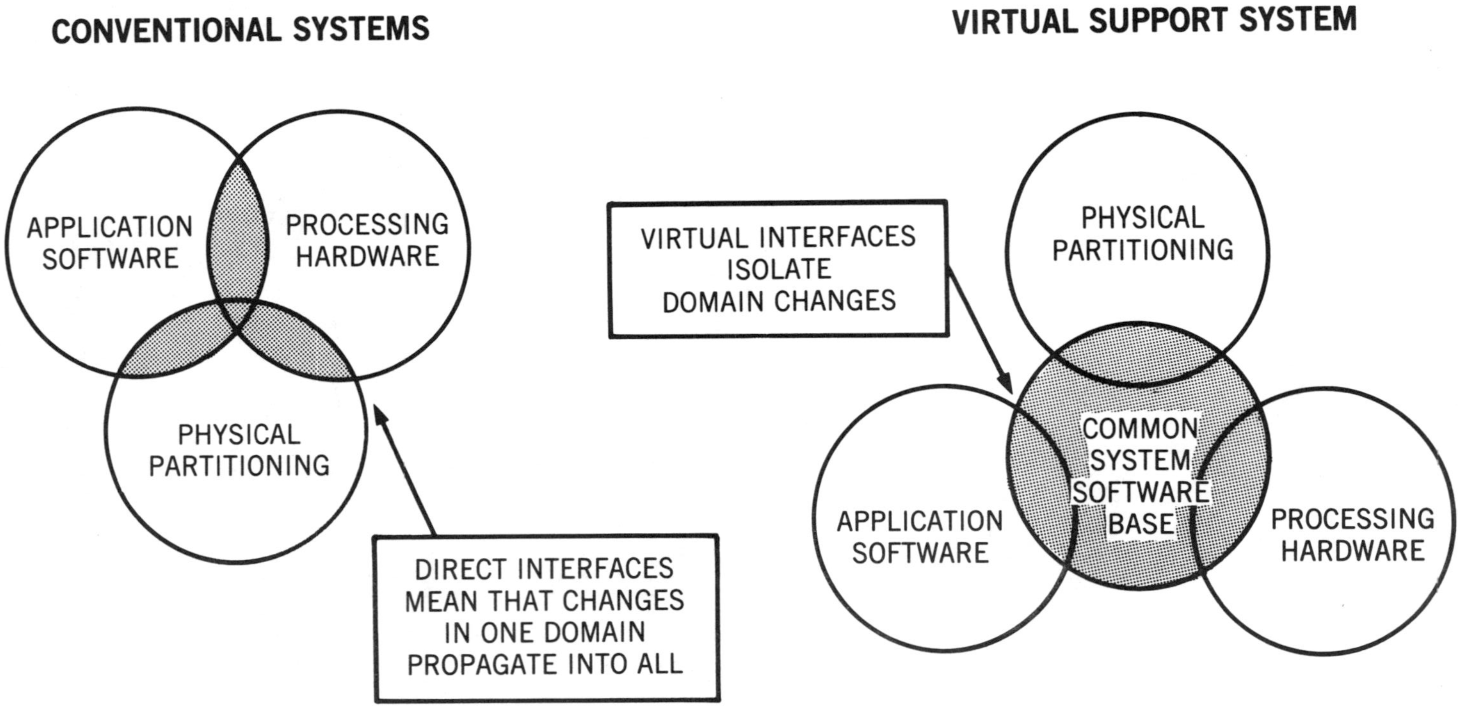

Figure 4. Layered run-time support environment.

Solving the Interoperability Problem

C. E. McKnight, Jr.

In today's world, U.S. military forces must be capable of performing single service functions, joint operations and combined operations with other nations' tactical and strategic forces. Most would agree that it is very unlikely that many operations will be conducted by only one service. The overall strategy of the United States includes the use of coalition warfare. Even within the U.S. force structure, different doctrines exist for the use of resources in air-to-ground missions. The problem of ensuring interoperability with the U.S. allies becomes even more complex. Solving an interoperability problem requires examining multiple doctrine, planning, procedures, hardware and operational techniques.

The need for the U.S. forces to be able to operate with each other and with our allies has never been greater. The Joint Chiefs of Staff (JCS) and the services are working hard to develop and to procure command, control and communications (C^3) systems that will enhance interoperability throughout the total spectrum of conflict.

But interoperability under all conflict conditions and at any worldwide location is difficult to achieve completely. It is difficult for many reasons, but most importantly because interoperability does not stand alone. It reflects the dynamic integration of planning, training, doctrine, procedures, force structure, experience and hardware.

As long as we continue to change our communications equipment to use different propagation modes, different modulation techniques, make it less "jammable" and more secure, we are unlikely to achieve full interoperability of tactical radios. But, we can achieve interoperability—and must—in any

**This article originally appeared in the November 1985 issue of SIGNAL.*

given operation. This requires advance planning for combat nets, radio frequencies, cryptographic devices—in short, it takes planning and logistics support in addition to compatible equipment sets to achieve interoperability.

Interoperability is not a new challenge. It has been around since before the Tower of Babel. War is the realm of uncertainty and operations seldom proceed completely as planned. Our services have been challenged throughout history to operate effectively with each other and with our allies. The lessons the United States has learned in previous conflicts are important today, even though some conditions may have changed.

Two definitions of interoperability are used by the Department of Defense (DOD):

> *Interoperability*—1. The ability of systems, units or forces to provide services to and accept services from other systems, units or forces and to use the services so exchanged to enable them to operate effectively together. 2. The condition achieved among communications-electronics systems or items of communication-electronics equipment when information or services can be exchanged directly and satisfactorily between them and/or their users. The degree of interoperability should be defined when referring to specific cases.

Communications experts have been quite successful in solving communications equipment interoperability problems when adequate resources were provided. The United States has done a good job of suboptimizing the use of equipment to solve specific communications problems. The concern is that we may have suboptimized a number of systems that could eventually become integrated into very inefficient macro systems.

Dynamic Environment

There are many users and manufacturers in the C^3 interoperability business. The degree of interoperability that can be achieved is impacted by the availability of hardware, well trained personnel, the available funding and the forces that must conduct joint or combined operations.

It is possible to state the categories of communication equipment that must be interoperable in order to ensure effective and efficient joint and combined operations in three ways: in terms of the tactical operations to be conducted; the function that the equipment serves; and the technical characteristics of the equipment. These three ways are interdependent and once one of the three ways is chosen, its relationships to the other two must be defined.

This procedure has been used within the joint community and it is documented in appropriate publications, directives and regulations. The U.S. problem has not been the overall procedure which in theory would ensure interoperability of communications systems; it is in the management of the total system. The United States has very few people who are capable of managing the total system in a macro sense. The United States does not have enough young people who understand the total system well enough to challenge the subsystem with which they are involved. The subsystem they directly are involved with must fit into the total set of systems that provide support for the command and control (C^2) function. This set of systems has evolved over the years and it is changing constantly based on the dynamic environment affected by budget decisions, new technology, changes in doctrine and tactics, national objectives and the actions of potential enemies. Due to budget constraints, technology changes and changes in requirements, the individuals managing the programs for a subsystem are faced with modifications to suboptimize their subsystem to stay within resource guidance, initial fielding target dates, etc., very early in the life of their program. When these decisions are made without an adequate macro understanding of the ultimate effect of the total C^2 system, problems later occur in integrating the subsystem with both the evolving C^2 system and other subsystems entering the system.

The United States needs to do a better job of developing individuals who can understand the macro environment so that effective decisions can be made when developing the subsystems. The total set of systems for C^2 are so complex that the United States needs to provide a mechanism for describing the total system. We have to develop models that show the total systems that are available; what we are doing with them; and what it is costing the nation to keep them alive.

The United States has spent a great deal of effort on master plans in an attempt to gain control of the problem. Long-term master planning should be used. The major problem has been that the master plans have not adequately addressed the financial and manpower constraints of the real world. Many of the master plans are unrealistic. A commander in Euorpe may have signed up for a new system. The manpower for the system may be available in the total force structure and now the commander in Europe faced with a congressional restraint on the number of personnel in Europe is faced with making a new decision on whether or not he can afford to deploy the new system based on what would have to be sent out of theater to meet the European Troop Strength constraint. An overall model that allowed analysis of total system impact would assist in making smart decisions early in the development cycle.

Key commanders have been indicating support for new communications equipment, but they have stressed the importance of affordability so the equipment can be introduced in a minimum time frame. They are concerned about the manpower requirements not only to operate the equipment, but also to provide the necessary maintenance support and related activities. Some one liners that have been mentioned by more than one commander include:

- Need to discipline the users;
- Lower development times for equipment;
- Do not size the pipes—size the needs;

- Suppression of user appetite for communications required. (Do not cave in to communications requirement—question commanders needs); and
- Challenge: Make the user understand that C^2 on the battlefield will never satisfy user desires.

Interoperability issues being worked on by the JCS, the Office of the Secretary of Defense (OSD) and the Services include actions ranging from significant revision of the tactical interoperability policy, DOD Directive 4630.5, all the way to the development of architectures in the joint arena. JCS recently provided OSD with a proposed version of DOD Directive 4630.5, Compatibility and Interoperability of Tactical C^3I Systems. The regulation was out of date and needed to be revised. Discussions have been held with the Federal Republic of Germany (FRG) to work toward achieving the maximum interoperability between the U.S. SINCGARS radio and FRG radios. NATO issues related to the Mark XV Identification Friend or Foe (IFF) program are being worked out in NATO panels. U.S. delegates are attempting to pull NATO messages and rules into synchronization with the Joint Intelligence Tactical Command and Control System (JINTACCS) message formats. A JCS action has been initiated to validate Pacific Command (PACOM) requirements for improved interoperability of C^2 systems in the Pacific theater. This step should form the basis for future actions. Efforts are planned for validating requirements for the interoperability of the Joint Special Operations Forces. These actions are a sample of the many things that are being worked on. Joint efforts between the individual services and efforts in combination with our allies constantly are involved in the over 300 C^3 programs monitored by the JCS C^3 Systems Directorate.

Many individuals want to solve a problem once and for all and then address an entirely new challenge. Since interoperability of both national and international forces are constantly changing relationships, they cannot be resolved once and for all. The problems must be worked out by identifying potential problems and minimizing the effort on the operational forces.

It is important to keep the following items in mind when working on interoperability problems:

> *Management Structure*. OSD has reorganized and now places functional responsibility across the board for policy and acquisition for C^3 systems in the Office of the Assistant Secretary of Defense (C^3I). On the Joint Staff, the C^3 Systems Directorate is charged with the functions related to C^3 systems. An OSD/JCS C^3 Executive Committee and a C^3 Review Council have been created to solve C^3 related problems. The Review Council has representation from OSD, JCS, the services and the defense agencies. A series of panels provide support to the C^3 Review Council. Each of these panels has the inherent responsibility to examine interoperability problems. A little over a year ago, DOD created the Joint Tactical C^3 Agency (JTC^3A). Its primary mission is to ensure the interoperability of tactical C^3 systems for joint or combined operations through the development

and maintenance of joint architecture, interface standards and interface definitions. The management structure is there. We do not need to expend effort creating a new organizational structure to solve interoperability problems. We need to make effective use of the existing structure.

Common Equipment. One method that will reduce interoperability problems is to have people use common equipment. The word reduce is used because even when people have common equipment they may not be able to communicate with each other if they do not have a common language, common procedures and a common understanding of how the other operates. The degree to which exactly common equipment can be used is limited by operational economic, geographic and political factors. There may be operational reasons why special equipment is needed. Due to different priorities, a Service's ability to purchase a piece of equipment in sufficient quantities may preclude the Service from scheduling the installation or deployment of the equipment at the same time. Both economic and political factors can hamper using a common item of equipment. Although NATO has agreed in principle that common equipment is desirable, each of the countries faces political pressure to ensure that employment, balance of payments, etc., are considered prior to agreeing on the "common item." Even if the common item was acceptable to each country, some may not be able to afford the most recent technology, and even if they could afford it, based on individual deployment schedules, commonality may not be achieved. This is an area where we must continue to work to reduce interoperability problems.

Common Standards. With the increased complexity of C^2 systems, common standards are required to ensure interoperability. Without common standards, problems will occur when systems developed by different Services or other nations must be integrated to accomplish joint or combined operations. Major efforts have been initiated, but emphasis must continue to obtain common standards.

Doctrine, Tactics, Techniques and Procedures. When the operational forces have a common doctrine and tactics, the ability to interoperate is improved. Getting the United States to achieve a common doctrine and common tactical procedures is unlikely given the many countries with which the United States operates. Each country has valid reasons for their methods. However, an investment in educating both U.S. and allied leaders in each others doctrine and tactical procedures can provide a big payoff in the interoperability of the combined forces. Within the U.S. force structure, a major effort is underway to develop joint doctrine. These efforts need continued support.

Unified and Specified Commands. The U.S. operational forces function under the Unified and Specified Commands. Many critics of the existing system for development of C^3 systems have argued that the real users on the C^3 systems do not have an adequate impact on the development, acquisition and fielding of the systems. A major effort has been taken to increase the influence of the commands throughout the life cycles of systems. The commands are not involved in the review of requirements, and they have had increased participation in the planning, programming, budgeting and execution process. Within the JCS C^3 Systems Directorate, there are action officers who work closely with the commands; they are charged with ensuring that the input from the commanders in chief are considered when decisions are made on C^3 systems. Each year, representatives from the commands, the Services and JCS meet in Washington, D.C. and review the Program Objective Memorandums submitted to OSD to identify potential problem areas. These problem areas are worked on by the service staffs and OSD to ensure maximum input from the users of the C^3 systems—the commands. Since the budget process involves the allocation of limited resources, everything that a commander would like to have cannot always be provided, but maximum effort is made to ensure that the commander's inputs have an impact on the decisions that are made.

Future Actions

The long-term solution to improving tactical C^3 interoperability is to develop and enforce a joint tactical C^3 architecture and associated standards. The development and maintenance of a joint tactical C^3 architecture (a mission that has been established for the JTC^{3}A) would provide the basis for measuring new requirements and examining changes to ongoing programs to identify the effect on required interoperability. JTC^{3}A is in a transition period. It will be working on the overall architecture problem while it also is working on individual problem areas that ultimately will contribute to the long-term solution.

The need for macro managers in the C^3 area was mentioned previously. We are sending between 25 to 30 students a year to an 18 month C^3 concentrated masters degree program at the Naval Postgraduate School in Monterey, California. In addition, we train approximately 150 action officers a year at a six week C^3 oriented course at the Armed Forces Staff College. These efforts are excellent, but we need to ensure the continued development of these individuals by providing career opportunities that will make them experienced macro managers.

The Army Signal School at Fort Gordon, Georgia, has been involved in assisting a number of states in developing secondary school electronics programs. It is in our interest to support these efforts to ensure that adequate numbers of young people develop the necessary skills in the electronics area. Industry and the Services will be competing for these key skills as the available pool for recruitment decreases.

The DOD share of the demand for electronics skills has decreased over the years. As our society has increased its use of electronics, the major market is on the civilian side. With the rapid obsolescence of new electronic devices, some would say that new generations of equipment are being introduced every three to five years. The current development and acquisition process extends well beyond the generation change in technology. Although we have been emphasizing the need to examine off-the-shelf procurement, we will probably need to go further than that to achieve affordable C^3 systems that contain the latest technology. There is a need to examine the use of subcomponents, which are being mass produced for the civilian markets, and ensure that anything that is required as a unique military component cannot be satisfied by an existing item manufactured for the larger civilian segment of the market. This procedure should result in lowering the initial cost. Support for the component could be minimized since components and expertise would be available in the civilian sector. If the development time and the cost of new systems can be decreased, systems will be fielded and can be procured in larger quantities and thereby decrease interoperability problems.

With the continued increase in cost for modern sophisticated military equipment to include C^3 systems and the limited resources available, efforts to limit expenditures to systems that provide only the minimum essential interoperability requirements will continue. If the "nice-to-have" systems and features are not scrubbed out, we will not be able to purchase sufficient quantities to equip all of the forces that require the system. The net result would be an increase in interoperability for the forces equipped with the new system, but a decrease in interoperability when all the forces must work together. This is particularly true when examining the introduction rate of new technology in U.S. units versus allied units. In many cases, we may move from a condition of limited combined interoperability to a condition of no interoperability between the old systems used by one nation and new technology used in another nation's units.

The bottom line in interoperability decisions is to determine how much interoperability is required and what resources should be used to obtain that degree of interoperability. The value of the interoperability to be achieved versus its cost will then determine whether one insists on solving the interoperability problem. If the decision is that we must purchase the interoperability, then we must select the method to obtain it. Since maintaining interoperability is a dynamic situation depending on the integration of planning, training, doctrine, procedures, force structures, experience and hardware, and since each of these areas are impacted by politics, affordability and personalities, it will require concentrated efforts to ensure that the interoperability provided via C^3 systems is the best that can be achieved to increase combat effectiveness for the forces of the United States and U.S. coalition partners.

Why Not Fuzzy Measures of Effectiveness?

John T. Dockery

Those in the command, control and communications (C^3) business know what measures of effectiveness (MOEs) are. Contracts call for them, consultants supply them and meetings are held to discuss them. MOEs are important because, most of the time, people are using something that lacks a mathematical definition. A highly subjective entity must be treated as though it were something rather concrete. A mode of helpfulness pervades the subject matter. It is captured in the often repeated conundrum that "you can't compare apples and oranges," yet the requirement to do so is encountered daily. For example, what is the force effectiveness tradeoff between an improved communications satellite system and a new aircraft design? This article directly confronts the aforementioned conundrum and attempts the comparison.

Unless one is talking about something fairly obvious such as the performance of a piece of equipment, argument ensues. Even cost as the final arbiter has its pitfalls, especially when one is dealing with social or military programs. When the entity in question is part of something called infrastructure, the problem gets worse. Effectiveness of the military can be evaluated only with respect to mission accomplishments. These sorts of mission outcomes must be measured in terms such as prosecution of combat or the avoidance of combat. What is even more difficult to quantify is keeping the peace. Quantifying anything is particularly difficult for C^3 since it does none of the foregoing directly. It is always a second order phenomenon that serves to facilitate other actions.

**This article originally appeared in the May 1986 issue of SIGNAL.*

Analysts usually end up substituting something related to the quality of the communications support for the mission outcomes, reasoning, one presumes, that if people talk and send messages, it must help. What never seems to be analyzed is that thousands of messages sent do not lead to thousands of actions. In fact, they lead only to a relative handful of actions which, upon examination, do not seem always causally related to any particular action. If money were plentiful and if decision makers did not have to make painful choices across not only C^3 programs but also C^3 programs in relation to combat arms, perhaps no problem would arise.

How do decision makers handle the problem? They, and/or their staffs, often end up making subjective judgments about the value of a particular C^3 system to war fighting capability. What really seems to pervade discussions of MOEs is the necessity of knowing the context. Context, in these cases, can be viewed as a convenient shorthand for "subjective."

What if imprecision itself is sometimes a state of nature? Could we then acknowledge MOEs for what they often are: very imprecise statements of relative ranking? More importantly, could such heresy be incorporated within a mathematical framework? The answer lies with a mathematical discipline, which itself is a kind of heresy. This is the mathematics known as the fuzzy set theory.

Background

The choice of the name fuzzy sets, although eloquently descriptive, produces a number of well cataloged reactions. Among the milder are disbelief or disdain. The more serious reactions occur among those to whom fuzzy is a term of opprobrium, e.g., fuzzy thinker. Fuzzy set theory is none of these. The term was introduced first in 1965 by Lotfi Zadeh of the University of California at Berkeley to deal with a problem arising in the pattern recognition of printed letters of the alphabet. When magnified, the letters or words in a newspaper take on a fuzzy quality around the edge. Hence, the name fuzzy. While the physics community deals straightforwardly with names like charm and beauty for the point-like particles called quarks, the operations research community, which sometimes lives too close to the world of public relations, was not receptive to either the name fuzzy or the concept. Although still surrounded by controversy within the mathematics community, fuzzy set theory has matured in the 20 years since its introduction with the widespread community of scholars taken together with those who have experimented with applications of the theory, the total community numbers in the thousands. It is much stronger outside the United States than within. There is also a formal journal entitled *Fuzzy Sets and Systems* by North-Holland and two irregular publications. One tracks the People's Republic of China's increasingly important contributions to the field. The other is French based and called *Busefal,* a dual language acronym.

Regardless of the popular image it might have had, in the late 1970s this author determined to try to apply the theory of fuzzy sets to a fundamental investigation of MOEs. The theory spoke to the problem as nothing else to be found. If offered a way finally to compare apples and oranges. The result was a published report entitled "A Theory of Effectiveness Measures" (STC TM-729). The report sets forth the premise that MOEs are fundamentally fuzzy entities whatever else they may be. This article is based on that work.

Fuzzy Sets

Upon first encounter, fuzzy sets seem to fly in the face of logic. An object either is contained in a conventional set or is not contained. There is no ambiguity. In terms of logic this corresponds to a "yes" or "no" answer. One can even introduce a bit of probability at this point by betting whether the objection in question is in or out; but they still are only two choices. Real life decision situations, however, are mostly grey. Rarely are things squarely in or out of contention. The decision maker most often says "maybe." In other words, he/she sees features that warrant membership in the set and things that militate against it. It is not a question of probability. A simple example couched in prosaic decision making terms is a judgment on the "goodness" of membership in a club. The spectrum runs from phrases like "he's a full time member" to "he's a member in name only." If "goodness" of membership were an MOE, then "full time" rates higher than "in name only."

The mathematics of fuzzy sets capture the foregoing distinctions between varying degrees of "goodness of membership" by assigning a numerical value, called the "grade of membership," in the fuzzy set to the "goodness" MOE. Expressed simply, this is a number between zero and one. If this numerical assignment were all that occurred, fuzzy sets might well merit some of the scoffing that they have engendered. Fortunately, much more is involved. First of all, by definition, the numerical values assigned are tied to the local impression of what constitutes "goodness," for example, the number of activities in which the member is engaged. Different clubs may have different ideas of high and low activity levels. Fuzzy sets provide a formal mathematical framework for making such comparisons. Moreover, the theory permits the comparison of different concepts of "goodness of membership," such as one tied to a member's cash contributions versus one tied to his service activities.

The most elementary sort of fuzzy set operations use maximum and minimum operations for combining membership values. These are fuzzy union and intersection respectively. They correspond to fuzzy ~OR and ~AND operations that permit combinations of similar or dissimilar MOEs where the tilde (~) under the options indicates they are fuzzy. (In this article the tilde will precede the option.) Generalizations of these two operations admit the required apples and oranges comparison.

The sequencing of several fuzzy operations leads to a situation of indeterminate understanding. For conventional MOEs, there is no limit to the length of a string of operations. In real decision making, there are limits to concatenation before the result becomes mush. Fuzzy MOEs capture this experience.

Much more complicated operations exist. In fact, a considerable amount of conventional mathematics has been generalized to incorporate fuzzy set concepts. Fuzzy sets also have been a source of new concepts. One concept of particular interest to MOEs is something akin to multiplication called composition. It is interpreted to mean the intensity of the relationship between two different kinds of fuzzy sets. In fact, fuzzy set theory is about relations rather than functions. Relations are more general than functions. What one encounters in the foregoing statement is a kind of paradox. In mathematics, more primitive is more powerful. Fuzzy sets fit this latter mold.

Another concept is the extention of fuzzy set ideas to purely linguistic forms. Fuzzy sets also have led to the postulation of a theory of possibility as a complement to probability. It is an attractive method of representations since it appears that decision makers deal with possibilities as often as they deal with probabilities.

Aspects of Fuzzy Theory of Effectiveness Measures

One of the first requirements for a fuzzy theory of effectiveness measures is a formal definition of such measures. One was constructed as follows: A measure of effectiveness is any mutually agreeable parameter of the problem which induces a rank ordering on the perceived set of goals.

The effect of this definition is to embed at once MOEs directly into the modeling process, while at the same time defining them only with references to the subjective context of the problem at hand. This definition then was set into the very formal context of something called category theory, which may be thought of as a kind of super set theory. Some consequences of the definition were as follows:

- A formal link with fuzzy sets was established through a theory of modeling based on category theory.
- It incorporated the common occurrence of a ranking process that may not be repeatable as can happen to committee deliberations, which often can yield MOEs used only by the committee.
- It accommodated rank ordering, and hence MOEs, that may occur on such surrogates as degree of goal attainment or means to goal attainment, the latter being very common in practice.
- It showed that MOEs in both a formal and practical sense serve to encode the essential structure of a model (or decision process) and to pass it along to higher levels of modeling. (Functionally, they do so by

aggregating results of models. Limits on this aggregation are set by the theory of fuzzy sets.)

The identification of MOEs with fuzzy sets permits certain top-level checks, which are based in fuzzy set algebra, to be made regardless of which MOE is involved. A list of some of the more interesting characteristics follows. The term in parentheses represents which of the various algebraic properties is being invoked.

- Once an MOE is first or last, additional operations do not change this fact. Thus, various rescaling and/or intensification operations often used to generate new MOEs leaves the set unchanged. (Idempotency)
- No particular measure should appear more than once, otherwise, information carried by other MOEs may be lost. (Absorption)
- The sequence in which measures are used at a fixed point in time are not important. (Commutativity)
- In comparing multiple measures, they may initially be grouped in an arbitrary way to facilitate comparisons. (Associativity)
- MOEs either may be ranked among themselves and then against a standard or may be compared separately with the standard. (Distributivity)

A rather interesting branch of fuzzy set theory is fuzzy arithmetic. A simple example of this kind of arithmetic follows. If you ask someone "how much is 'about two' and 'about two'," he/she will invariably answer "about four." A conventional fuzzy set notation would be: $\sim 2 + \sim 2 \approx 4$. Now ask the same individual another question: How much is 'about two' and 'about 100'." Unless he/she is overly precise, chances are excellent he/she will answer that it is "about 100." A notation would be $\sim 2 + \sim 100 \approx 100$.

With the proper definitions of fuzzy numbers, such operations may be made routinely. (For mathematicians, this article notes fuzzy numbers can be defined on a ring rather than on a field. Ergo, more limited mathematical operations are allowed.) In fact, one could add such numbers as "very close to 50" and "something like 50" and get consistent results. However, certain common operations like division contain surprises or may be undefined. The latter caution is extremely important to note because the operations research community usually treats the numerical value of fuzzy values as if they were conventional numbers. And, numerical MOEs sometimes are such real numbers (e.g., miles per gallon); trouble comes when they are not. Fuzzy arithmetic handles conventional numbers as a special case of fuzzy numbers. Mathematicians inside the area of fuzzy sets work as hard to challenge their own assumptions as those outside their camp. In fact, they work harder, for they work to tie the subject firmly into the larger context of other mathematics.

Application Examples of Fuzzy MOEs

Very little has been said to distinguish command and control (C^2) and C^3 MOEs from the general run. Some special features of these MOEs, which can be accommodated by fuzzy sets, now will be introduced. Perhaps most important is the degree to which an evaluation of the total C^3 posture includes such intangibles as the effect of training, the influences of "better" procedures or the skill of the operator. The former coexist with semi-quantitative items like degree of connectivity. All the foregoing coexist with hard numbers like things transmitted or received. Only the context can tell the degree in which each contributes to the total. Fuzzy set theory thus permits operations between things characterized by hard numbers and vague, but equally important estimates of processes, all in the proper context. For instance, fuzzy set based MOEs permit the replacement of the often meaningless number of messages disrupted with the degree to which operations are disrupted. The latter has the advantage of incorporating military judgment and scenario information as well.

Another interesting possibility explored in the theory is weapon evaluation in context. Thus, in the area of combat arms, one first defines the context free tank in terms of MOEs like survivability, mobility, etc. One then adds environmental context such as various terrain possibilities (e.g., swamp, desert, urban) to get the true value of the MOE. By an extension of the procedure, one then may add the communications and C^2 environment to generate extended fuzzy values for survivability and mobility. In fact, an interesting feature of the fuzzy MOE is that additional descriptive MOEs can be added to the set without sacrificing estimates already in place. Mathematically, this is possible because, as descriptors are added, their membership function values may be anything between zero and one. Compare this with the conventional approach in which you are forced to divide unity by an ever larger value of *n,* where *n* is the number of attribute descriptors.

By way of illustration, consider the case of a tank. Without communications and C^2, it might be characterized by three MOEs called mobility, invulnerability and firepower. A tank scores high on all three, and a common effectiveness scheme might assign values of 0.33 to each feature. If C^3 is added, or at least a concept of it equal in importance to the other three, each element is then rated at 0.25. Various weighting schemes for features only make matters worse. Carried to its illogical conclusion, the conclusion generated is similar to one famous from economics (and now outdistanced by inflation) that "everything in the universe costs a dollar a pound." The picture is quite different using fuzzy effectiveness measures. For instance, each of the original four features can retain their original values of 0.8, 0.8 and 0.8, while adding C^3 at say 0.7. Moreover, various combined MOEs can be constructed using fuzzy operations like AND and OR. For instance, the result of the first operation (~AND) is a lumped MOE of 0.7 while the second (~OR) yields 0.8. In fact, the decision maker doing the combining can be modeled as there exists a whole hierarchy of combining operations each representing different styles of decision making. In effect, the concept of fuzziness is extended to operations between the fuzzy sets.

Fuzzy set MOEs combined with a technique borrowed from mathematical sociology, called q-analysis, has permitted evaluation of such difficult C^3

concepts as the following: comparison of the degree of degradation of the connectivity of mixed communications and processing architectures; comparison of reduced command structure functioning; and mixed electronic warfare and physical damage effects on a system. One final example is included to show that fuzzy set MOEs can produce counter intuitive results. By investigating the ability of a C^2 system to generate and to exploit shared training concepts versus the generation of multiple alternatives, one can argue that architectures exist for which jamming before an attack is critical, but for a certain period of time after the attack has started, it may be more of a nuisance than anything else. This is quite the contrary to U.S. presumptions under which jamming before an attack is troublesome, but jamming during an operation is very serious. Consider a military culture that is very centrally organized versus one that sets great store by local initiative. The first requires precise information, which jamming makes fuzzy, before an engagement. The second requires precise communications, which jamming renders fuzzy, during the course of the engagement.

Conclusion

Fuzzy set theory is a rapidly maturing discipline that has attracted an international collection of excellent scholars. Fuzzy sets speak to decision making as it is found in practice not theory. It is an excellent framework for improving our understanding of MOEs. Fuzzy set theory provides valuable insight into, and what is more a coherent framework for dealing with, the subject of MOEs. The theory provides unique computational possibilities for C^3 tied to the real world decision maker.

Bibliography

1. Dockery, J. T. "Fuzzy Design of Military Information Systems," *International Journal of Man-Machine Studies*. 16: 1–38 (1982). Has a worked military example.
2. Dockery, J. T. *Theory of Effectiveness Measures*. STC TM-729, SHAPE Technical Centre, The Hague, Netherlands (January 1985). Technical Memorandum from which this article is derived.
3. DuBois, D., and Prade, H. *Fuzzy Sets and Systems: Theory and Applications*. New York, New York: Academic Press (1980) Best overall treatment, but a hard place to start.
4. Gupta, M. M., Sardis, G. N., and Gaines, B. R., eds. *Fuzzy Automata and Decision Processes*. New York and Amsterdam: North-Holland (1977). The first of several such collections, and still current.
5. Gupta, M. M., Ragade, R. K. and Yager, R. R., eds. *Advances in Fuzzy Set Theory and Applications*. New York and Amsterdam: North-Holland (1979). One of several such collections, and still current.
6. Kandell, A., and Lee, S. C. *Fuzzy Switching and Automata*. New York, New York: Crane Russak (1979). An interesting book especially since there are now experimental chips based on fuzzy logic.
7. Kaufmann, A. *Introduction to the Theory of Fuzzy Subsets, Vol. 1*. New York, New York: Academic Press (1975). Still the gentlest introduction and replete with numerical examples.
8. Kickert, W. J. M. *Fuzzy Theories on Decision Making*. M. Nijhof (Leiden) thru Kluwer Boston, Inc. (1978). A nice readable introduction if you can obtain a copy.

9. Mamdani, E. H., and Gaines, B. R., eds. *Fuzzy Reasoning and Its Applications*. New York, New York: Academic Press (1981). Concentrates more on the logic of fuzzy reasoning.
10. Negoita, C. V., and Ralescu, D. A. *Applications of Fuzzy Sets to Systems Analysis*. New York, New York: Wiley and Sons (1975). One of the best, but also the most intensely mathematical.
11. Wang, P. R., and Chang, S. K., eds. *Fuzzy Sets: Theory and Applications to Policy Analysis and Information Systems*. New York, New York: Plenum (1980). Mostly description and discussion.
12. Zadeh, L. A., et al., eds. *Fuzzy Sets and Their Applicaiton to Cognitive and Decision Processes*. New York, New York: Academic Press (1975). A watershed collection.
13. Zimmerman, H. J., Zadeh, L. A., and Gaines, B. R., eds. *Fuzzy Sets and Decision Analysis*. New York and Amsterdam: North-Holland (1984). A nice assortment in a managerial idiom.
14. Proceedings of the First International Fuzzy Sets Association (IFSA) Congress. Held in Palma de Mallorca, Spain, July 1985. The most recent collection to be found.

Force and Force Control—In Pursuit of Balance

Alan D. Campen

The Department of Defense (DOD) is under heavy and often vitriolic attack for shortcomings in its efforts to improve military capabilities. Much of the criticism focuses upon allegations that command and control (C^2) capabilities cannot support the military strategies and tactics needed to underwrite national security policy.[1]

It is proper that C^2 be assessed critically; absolutely credible C^2 is a keystone of credible deterrence. Some of the concerns being expressed are legitimate: C^2 capabilities, while much improved these past few years, are nowhere near the level of performance where most agree they should be and some think they could be. The need to bring force and force control capabilities into better balance is a tenet in Defense Guidance, but the means to that end are not as clear as some critics assert. This article presents one view of this problem: why progress has been so tortuous; what has been learned that might guide continuing efforts to improve C^2 during the more difficult times that lie ahead; and a framework for constructive analysis and criticism.

Keeping Up is No Small Task

Urgent effort began in the waning days of the Carter Administration to improve the operational capability and effectiveness of combat forces, mainly through weapons modernization and the development of new tactics to exploit emerging technology. However, the expected gain in combat force effective-

**This article originally appeared in the March 1986 issue of SIGNAL.*

ness usually is dependent upon a concomitant improvement in the performance of command, control, communications and intelligence (C^3I) systems. This dependence, this vital linkage between force and the ability to focus combat power, is acknowledged more readily in rhetoric than during apportionment of defense resources. The result is a paradox that has confounded successive administrations: significant and steady progress in improving C^2 support to selected, high priority missions, yet an overall C^2 posture that continues to lag force modernization in several respects—notably interoperability, security and survivability. This condition persists despite affording C^2 the top priority in President Reagan's strategic force modernization program; substantial increases in resource allocation to C^3I programs; reorganization in the Office of the Secretary of Defense (OSD), the Organization of the Joint Chiefs of Staff (OJCS) and the unified commands to provide better focus and higher priority to joint C^2 matters; and the fortuitous emergence of cheap commercial solutions to many of the seemingly intractable problems in collection, storage, processing and transmission of information—the essence of the C^2 problem. What then lies in store for C^2 with a plunging DOD budget; intensified competition for fewer personnel resources; and escalating costs for C^3I systems? What must be done to maintain the high level of support to C^2?

It is not clear that national security interests are well served by the ongoing public debate over the health of U.S. military control mechanisms; the essential elements of this esoteric subject are, must and will remain hidden from all who lack the genuine need to know such things. It is clear that those interests are ill-served if public perceptions of this critical national security issue are distorted by shallow research, wildly inaccurate conclusions, simplistic solutions, and, in one pernicious instance, the inference that a weak strategic C^2 posture results from a military conspiracy to frustrate presidential nuclear policy. The peril of uninformed and unchallenged public debate is that it rivets attention on dramatic but specious issues at the very moment that consensus and added resources are needed badly to solve a few well understood, prosaic, but nonetheless persistent problems.

C^3I system programs suffer not only from those ills that plague all force modernization efforts, but from some peculiar ailments that have impeded the best efforts of talented and dedicated officials for years. Responsible debate concerning C^2 should address these unique burdens; assess how well they are being managed; and provide constructive and practical recommendations. Only through such insight can we hope to garner the additional resources needed to close the gap between weapons and control capabilities.

Wholesale overhaul of the defense structure, as proposed by the staff of the Senate Armed Services Committee (SASC), may be warranted and might happen, but it would do little to ameliorate the special impediments to C^3I systems improvements.[2] Indeed, the situation could backslide if the peculiar needs for C^2 management are lost in the turmoil that will attend a reorganization of OSD and OJCS. It has happened before.

Five Root Causes

There are five root causes of C^2 deficiencies:

• *Requirements Definition and Non-responsive Acquisition.* These are shibboleths that the C^3I community finally is beginning to master. Having long agreed that precise articulation of C^2 requirements is difficult, if not impossible, many now believe it to be unnecessary as well. There are no "requirements" waiting to be graven. There are only poorly defined, unbounded expectations and ephemeral aspirations that are biased by culture, doctrine, antiquated procedures and faulty assumptions about technology. Requirements must be discovered and honed in a continuum of interaction between user and provider within an environment that proximates combat, is fault-tolerant and encourages experimentation with doctrine, procedures and prototypes that display the best in emerging information systems technology. Further, the acquisition process must be tailored to facilitate this iterative interplay between user and provider. The heretical notion that systems that support the function of command need tailored acquisition is not so threatening now as when it was introduced in 1978 by the Defense Science Board (DSB) and subsequently was validated by an AFCEA study. But evolutionary acquisition (EA) still is discretionary and its adoption is sporadic and circumstantial. Efforts to reinforce EA through elaboration in DOD Instruction 5000.2 were defeated, but some authorities believe current OSD policy is adequate to effect and to guide implementation (see "Evolutionary Acquisition of Command and Control Systems," *SIGNAL,* September 1985, page 39). Indeed, this ongoing debate over EA is a mystery to many field commands in the United States and overseas who perceive EA as alive, well and capable of producing immediate solutions to many of their most urgent C^2 problems. Those with the courage and the conviction to break with tradition are finding that, while there still is much to be learned about effective coupling of the requirements definition and acquisition processes, it is not the conundrum once feared and it is certainly no longer an acceptable excuse for the failure to produce affordable increments of improved C^2 capability. While EA may not require further policy elaboration, it is time to mandate its application to C^3 systems.

• *Technology Transfer and Affordability.* Much of the electronics technology needed to enable combat units to fight "smarter," finds its way more readily and cheaply into consumer products than into military equipment. Defense requirements at one time dominated the electronics industry and influenced the technical directions and pace of the industry. They no longer can do either. Instead, the commercial marketplace has prevailed and the role of defense has been relegated to searching for ways to exploit this cornucopia of innovative solutions to management problems, which are not all that different from the problems of the military commander: the rapid, secure and accurate handling of information being central to both civil management and C^2.

It is true that civil technology does not fully meet all military requirements—a weakness shared by many military systems as well—and it is a disquieting fact to some that widespread tactical use of commercial products is going to force the rethinking of such basics as the traditional military approach to sustainment. Those are the facts. They also are largely irrelevant. The appetites of the combat forces have been whetted by the radically lower costs and incredibly fast responses to their immediate needs by the hardware and software adapted from commercial products. While this new approach has many uncertainties, there is not likely to be a turning back. The low costs and ready availability of commercial equipment give the military users the means to acquire directly improvements to their own C^2 capabilities. Troops will forage, if they must.

The commercial market is able to produce information system products that will satisfy all but the most stringent military requirements. By coupling DOD and civil needs (as the National Security Agency is doing with the next generation of secure voice equipment), we can experience economies of scale denied in the limited DOD market and as a direct result, can meet immediate military operational needs while releasing DOD resources to pursue those peculiar requirements that practicably cannot be imbedded in commercial product lines.

- *Interoperability*. Nearly three decades after President Eisenhower's dictum on the improbability of separate ground, sea and air warfare, we still encounter enormous difficulty in conducting joint operations with combat units able to intercommunicate in benign, let alone stressed, electromagnetic environments. To be fair, the present lack of operational compatibility is not solely a technical matter. There are no simple electronic remedies for incompetent planning for the brigading of combat units known to possess incompatible signaling systems. On the other hand, it is no longer necessary that all military organizations be stamped from the same mold, be commonly equipped or be instructed in common dogma to enable them to function jointly with other services or allies. Emerging information systems can end babel in ways far different from the procedural regimentation of the joint interoperability of tactical command and control systems (JINTACCS).

Commonality may be the only solution to some joint and combined interoperability requirements, such as with munitions and aircraft fuel, but modern digital technology renders the utopian policy goal of common information systems anachronistic. The bias against buffering and translative devices made some sense in the 1967 edition of DOD Directive 4630.5 on interoperability, but it seems oddly out of step with reality when repeated in the 1985 update of that directive. Open systems architectures and commercial buses can provide logical and manageable partitioning, and microprocessors can buffer and translate among tactical systems that should remain functionally unique.

The lack of joint C^2 interoperability came very close to being a systemic problem, with everybody—thus no one—in control. There was no one to define joint requirements, to select joint programs, to adjudicate trade-offs, to enforce decisions and, above all, to apply sanctions in the face of noncompliance. DOD has taken actions to correct the structural weaknesses that have indulged the fielding of incompatible systems. The obsolete and ineffective 1967 DOD directive on tactical C^3I systems interoperability has been overhauled; JCS has formed a Joint Requirements and Management Board to elevate these matters to the level where they now can be adjudicated and monitored; and the Joint Tactical C^3 Agency (JTC^3A) has been formed and chartered to identify and help solve interoperability problems through the development of standards. The added discipline enforced by reduced DOD budgets certainly will help curtail fanciful technical excursions, and in the background lies the threat of radical organizational surgery to end service bickering as proposed in the SASC report.

The causes of joint/combined interoperability problems are well understood within DOD, and the adverse impact on military capability is sensed, but is not understood all that well by the public and their representatives in Congress. Basic changes already have been made in an effort to deal with serious structural deficiencies. Critics who would further the cause of improved C^2 capabilities would do well to assess patiently and studiously these changes and to eschew dramatic reforms that fail to deal with root cause problems.

• *Priorities*. It has been said that DOD has only one problem—affordability. The refurbishment of the defense establishment, begun around 1979, involved an effort to fix everything at once: readiness, sustainability, force structure expansion and modernization. It was a difficult task under the most favorable financial conditions. Today, questions of affordability are resolved through priorities that emerge from spirited advocacy. Initiatives for combat force expansion and modernization are shepherded by skilled, broadly based and tenacious constituencies, which the C^2 community never can hope to rival but with whom it must, nonetheless, contest for resources.

Who speaks for joint C^3I systems, this sine qua non of combat force effectiveness? The military services? No, and one can argue that joint C^2 matters are outside their charters. The OJCS? That body evidenced an extraordinary lack of interest in C^2 in the mid- to late 1970s, until it was prodded to form what has become an increasingly effective C^3 Systems Directorate. The Chairman of OJCS? As the most vocal and effective military spokesman for joint C^2, the Chairman only recently has been given the staff support to compete as an equal in Defense Resource Board (DRB) debates. The unified commands? Obviously, but they are not staffed adequately for this role and until FY 85, the program/budget process granted them but a miniscule voice in resource allocation. The OSD staff? Certainly. Successive administrations have differed over the form, but never the need, for strong OSD technical and program direction over C^2 in general and joint C^2 in particular. But what of operational guidance in these matters so basic to military operations? Must this too reside—by default—with the civilian staff of the Secretary of Defense? The most eloquent technical explanation of a C^3I program cannot help but be buttressed through the support of the individual accountable for its employment in combat. Significant progress is being made to increase the voice of OJCS and the unified/specified commanders in the resource allocation process. Congress may accelerate the rate of this change.

• *Realistic Operational Evaluations*. It is astonishing that military commanders who would never consider the commitment of untested forces, weapons or tactics to combat, seem apathetic about the viability of the C^2 capability that they concede is essential to mission effectiveness. Have we become so hostage to rhetoric that we accept, on faith, an asset so manifestly fragile before the effects of war? Military campaigns since World War II have not tested the mettle of control capabilities, and field or command post exercises have failed to fill the void. Indeed, it is not easy to emulate faithfully the rigors of war in an exercise and still provide sufficient C^2 to meet training objectives. But the frailties of men, machines, plans and doctrine are accepted in military exercises. When failures occur, they provide the justification for additional resources and corrective actions. By our own declarations, C^2 is a key factor in combat force effectiveness, yet it is subjected to test criteria that does not represent the combat environment. This must change. The testing of C^3I systems is not a matter for off-line observation by the "wizards" against static, technical criteria. Viable C^2 is a command problem and operational staffs must set the evaluation criteria, conduct the test and the critique, establish the priorities for any repairs and be lead advocates for the resources to correct the problems. The Commander in Chief, Strategic Air Command (CINCSAC), for one, did this and the results support the thesis that C^2 gets fixed when the commander is confronted with the problem in realistic exercises; when he is convinced of the need and becomes a spokesman for corrective actions.

The above list of root causes is not exhaustive—there are as many views as there are viewers. However, it is more than exemplary because these are basics that reach to the heart of the C^2 problem: identifying real operational needs and finding timely and affordable solutions. What the public debates now lack and this article seeks to provide is a focus upon a manageable set of specific problems and a general theorem to guide progress assessments and recommendations for future actions.

The User Oriented Approach

The Reagan Administration took up the reins some five years ago with a commitment to C^2; a full appreciation of past attempts to deal with these root problems; and some ideas about engaging the support of operational commanders who must live with the consequences of any imbalance between force and force control capabilities.

C^2 is a command problem. Strong, centralized technical and program direction from OSD is demonstrably essential in C^2 matters, but it is not sufficient and what it lacks can be provided only by those accountable for mission accomplishment—the operational commanders. Unfortunately, they have had precious little voice in C^3I system matters. In the zeal to centralize oversight of C^3I programs, the voice of the commander and his operational perspective of requirements was lost. As C^3I programs took on their own identity in the planning, programming and budgeting system (PPBS) process, there was no convenient means to remain linked with the forces and joint missions to be supported. We gained visibility over the cost of C^3I systems and lost it over their worth. Programs increasingly were examined with an accounting mentality and decisions pivoted on issues extraneous to the central question—what is the impact upon force effectiveness? C^3I program advocates were obliged to grope alone through the resource allocation maze, usually perceived as competitors to force modernization. They were far better equipped to give exquisite technical descriptions of their programs than provide credible assessments of their critical impact upon combat force effectiveness.

For example, communications and connectivity are two nouns describing a basic C^2 requirement, but only the latter strikes at the heart of the problem; that is, the need to link certain force elements, at designated times, to achieve specific operational objectives. Programs to improve strategic C^2 took on new life when commanders of strategic naval and air forces joined the C^3I community to field teams able to explain both the "what" and the "why" of strategic connectivity programs.

The resource allocation process has been opened over the past few years to permit operational commanders to become involved in the development and review of service Program Objective Memoranda (POM) and to present their views on resource matters directly to DRB. However, this exposed another weakness in PPBS, in that resources are packaged and labeled in ways that bear little relationship to the military missions to be performed. C^2 is penalized additionally by this program/budget structure because it obliges constituent elements to be considered separately from each other and in isolation from the weapons they are meant to control. A dichotomy must be resolved. Enormous effort is expended in mission area analysis during the planning phase of PPBS, but is abandoned—stripped away by irrelevant categorization—at the precise moment it would have the greatest impact upon resource decisions. The result is stand alone C^3I programs, devoid of essential linkage to strategy, doctrine, weapons and plans for force application. The program/budget process makes it virtually impossible to discover critical dependencies and interrelationships among things that must mesh on the battlefield. The question of balance between force and force control and the dependency of military operations upon control capability will remain difficult to assess until means are found to couple the respective resources during the allocation process.

Mission Area Perspective of C^3I Resources

The aforementioned SASC report concludes that organizational change must come to OSD and OJCS to force a change from functional to mission orientation and from fixation on inputs (budget) to outputs (mission). Such an organizational change may be warranted, but it is not clear that the unique problems of C² automatically would be solved thereby. Indeed, there is substantial latitude for mission area assessment within the present structure. The Under Secretary of Defense for Research and Engineering (USDRE) sponsored several efforts in 1982–83 review cycles to introduce mission area packages to the DRB deliberations, and in 1984, the OJCS and OSD staffs collaborated to see if special packaging of resource issues might provide needed operational insights into the particularly troublesome question of the deep interdiction (Follow-on Forces Attack [FOFA] in NATO terms) mission area.

Any mission area package will cut a wide swath across many categories and service budgets; will involve sensors, processing and decision aids, communications and command centers and alternative air, sea and ground launched weapons; and all will overlay upon an approved operational concept of employment. The deep interdiction package was sought to do this and to quantify expected combat capabilities in terms that could be related to POM data and the characteristics of enemy targets. The desired output was a crude approximation of increments of improved operational capability from each level of additional investment. Additionally, it was hoped this approach would expose any gross inconsistencies and mismatches among system capabilities; highlight doctrinal implications; and show incipient demands for interoperability. The contribution of this experiment to the resultant resource decisions will be left to the judgment of those it sought to assist, but it does suggest an alternative to standing the Pentagon on its end to achieve a mission perspective of programs and budgets.

Responsibility for Joint C²

Despair over the inability of the military to solve the interoperability problem led successive administrations increasingly to centralize such matters in OSD. The Reagan Administration opted for one more effort to get those most effected to take the lead in solving what only can be described as a major operational problem. Management principles were set out in the Carlucci Memoranda of March and June 1981. Critics immediately zeroed in on three elements that called for emphasis upon improved strategic planning; decentralized program execution; and participative management. The critics contended that this restored C² to the inept service hands from which it progressively had been wrenched.

Quite the opposite was intended and the results are fair game for continuing assessment. These management principles clearly recognized the unique nature of cross-service, cross-command and cross-mission programs. They specifically cited C³ and intelligence as major exceptions to the general policy of decentralized program execution. OJCS and the Defense Intelligence Agency (acting as J-2) were tasked to take the lead in planning for joint and combined

C^3 and intelligence systems to include requirements and priorities. The impact of this new challenge upon OJCS was appreciated fully then and it should not be forgotten now in any judgments as to the results, for it thrust OJCS into resource issues that traditionally have been the uncontested domain of the services and OSD and into matters where joint procedures and skills were not adequate to the task. There were and still are doubters who hold that fundamental, systemic flaws in the OJCS structure render it incapable of taking positions and enforcing decisions contrary to service desires.

Necessary organizational and procedural steps were taken to cope with those new responsibilities for crosscutting the issues. These actions, along with other changes to strengthen the role of the operational commands in developing joint doctrine and in formulating priorities for DRB, could do much to restore the operational perspective sorely lacking in joint C^3I affairs. The central thrust behind the initiatives should be remembered in any critical assessment of the results and certainly in any proposed tinkering with the organization of OSD, OJCS and the operational commands. It has taken decades of trial and error to discover that C^2 is indeed different and that classic management structures and processes have proven to be manifestly inadequate. The questions to be pondered are whether the problem assessment has been accurate and have the solutions adopted been appropriate.

Summary

This article has sought to add substance and dimension to a vacuous and counterproductive debate over the state of C^2 and to suggest structure and boundaries for assessment, criticism and recommendations. It also has attempted to make these points:

- The problems that have impeded more rapid improvements to our C^2 capabilities are neither mysterious, insidious nor intractable; but solutions are not simple and any solution will require an infusion of far more financial resources than are provided thus far.
- C^2 is, perhaps uniquely so, a military operational problem that will not be solved without a deep and persistent involvement in the requirements, resource allocation, acquisition and testing processes by those ultimately accountable for mission accomplishment—the operational commanders. Command has no surrogate.
- Achieving better balance in force and force control requires a mission area approach across the full spectrum of PPBS activities; a task well within capabilities of the present OSD and OJCS organizational structure.
- Inherent risks in the increasing dependence of combat doctrine upon survivable C^2 must be demonstrated to commanders by simulations and exercises that place realistic stress upon C^3I systems.
- Evolutionary acquisition and the leveraging of commercial information systems technology are proven means to satisfy most military C^2 require-

ments and they are probably the only ways to lower costs and produce near term improvements substantially.

- Informed public opinion is a legitimate and useful input to decisions on matters this vital to national security, but the true state of our C^2 capability is, in any case, unknowable and deterrence is best served when it remains so.

Footnotes

[1]C^2, as used in this article, means the functions of commanding and controlling combat forces. C^3I means the technical systems and procedures that support those functions.

[2]"Defense Organization: The Need for Change," Staff Report to the Committee on Armed Services, U.S. Senate, October 16, 1985.

21st Century Battle Management: Penetrating the 'Fog of War'

Donald C. Latham

It is doubtful that any of ancient history's great military leaders ever separated command from control with so much as a hyphen let alone a conjunction. The size and composition of their forces involved in battle, the physical dimensions of the battlefield and the limitations imposed by the range and lethality of their weapons, made it possible to control events in a fairly direct manner.

Technological advances in the science of warfare over the last 130 years have expanded the scope of battle in every dimension. The field at Waterloo could accommodate the set piece, chessboard tactics mandated by early 19th century weaponry, while at Kursk, the battle raged over a field of incomprehensible dimensions, both spatial and temporal. Not only had technology spread the field of battle over hundreds of miles, but it filled the sky with incredibly lethal weapons capable of striking even farther afield. Perhaps of greatest importance, it denied the commander his most precious commodity of all, time. Not only could the commander not directly control events as they unfolded at the main point of the battle, but also the rapid pace of warfare that came with technology deprived the commander of the time necessary to "read" his intelligence and issue orders which could be executed rapidly enough to influence the battle. The commander's ability to control the battle was limited to the state of training of his subordinates and the planning he had accomplished prior to setting his forces in motion.

This article was written especially for **Principles of Command and Control.*

The term "command and control" (C^2) first appeared in the 1950s to describe the system being built to automate the air defenses of the United States. Throughout the next quarter century, "command and control systems" development was driven by and geared toward advancements in technology. Motivated by cybernetic theory, soldier and scientist tried to replicate the thought processes, mechanisms and procedures that had proved successful for heroes of earlier battles and campaigns. The objective was to shrink the battlefield in all dimensions, including time, by using computers to help manage, sort, catalog, recall, recombine and utilize the surfeit of information available to the commander; then, by using these same computers, to quickly disseminate decisions. These command and control systems were the means by which purposeful activity within a military force could be carried out. These are the procedural aspects of C^2.

Command and control systems also provide the structure for carrying out military actions. The traditional chains of command, staff functioning, institutionalized concepts and Service doctrine also are C^2 systems. These are the structural C^2 systems, and technological innovations in warfare have inspired as many changes to command and control structure as it has to procedures.

The most visible changes in command and control at the dawning of the 21st century will be procedural and will impact most directly on the techniques and tactics relative to maneuver, mass, speed, security and surprise. The net effect of the changes will be to shrink the size of the battlefield, giving the commander the ability to "see" the location of all his forces and those of his adversary to the maximum extent that the "fog of war," and the achilles heel of electrical communications, allow.

The ambitious and extensive command and control program improvements begun by the Department of Defense in the early 1980s will have been fielded by the turn of the century. However, structural command and control changes will be necessary before the full capabilities of these systems are realized. By the year 2000, joint doctrine, restructuring of tactical organizations, and the redefinition of roles and missions will be fairly complete. Anachronistic levels of command will have been eliminated allowing commanders to receive, process, and act upon vital information in a more rapid fashion. The increased reliability of communications and data networks will be used to issue orders directly to the forces capable of executing them without the delays injected by a series of intermediate headquarters.

Massing of forces and firepower of sufficient strength to overwhelm an adversary is always crucial to success. In order to commit his forces effectively and efficiently at the precise moment that will ensure the greatest opportunity for victory, a commander must know the precise location of every element he commands, its state of preparedness, and its ability to prosecute the attack order. Precious moments lost in updating charts and searching for the nearest unit capable of sealing a penetration or able to lend the extra combat power necessary to seize the offensive could mean the difference between victory and defeat in the highly mobile style of warfare that will characterize the year 2000.

On the fluid battlefield units will report their position automatically without thought or concious effort. The Enhanced Position Locating and Reporting System (EPLRS), and the Joint Tactical Digital Information System (JTIDS) will provide both radio communications and automatic battlefield location reporting. With the positions of their master stations accurately fixed by the Global Positioning System (GPS) constellation of satellites, EPLRS and JTIDS will provide unit locations accurate enough to use in the control of indirect fire weapons or to guide offensive support aircraft on to a target from hundreds of miles away.

In a related area, it will be far easier for the twenty-first century commander to recognize and distinguish hostile aircraft from allied aircraft in spite of the similarities in airframe design. Identification Friend or Foe Navigation (IFFN) systems will make it easier for air control agencies, air defense missile units, Forward Air Defense (FAAD) teams and intercept aircraft to isolate and engage attacking aircraft while minimizing fratricidal conflicts. On the ground, the NATO Identification System will include the ability to interrogate friendly armor using lasers and other techniques. Anti-tank hunter-killer teams will sort out friend from foe quicker and engage hostile targets with fire and forget weapons and withdraw before the enemy can react.

To employ his forces effectively, the commander must know the enemy situation, his strength and weaknesses, and likely courses of action. A variety of sophisticated sensor processing and reporting systems will provide the commander with a continuous flow of real-time intelligence tailored to his specific needs. Neither night nor foul weather will offer the enemy any respite from the relentless penetration of his airspace as manned and unmanned airborne vehicles, satellites and manned orbiting space vehicles will scrutinize enemy activity to a depth of several hundred or more miles from the Forward Line of Troops (FLOT). Passive and active sensors of remotely monitored battlefield sensor systems delivered manually, by indirect fire weapons, or by unmanned aircraft will augment and complement these airborne data collection systems. The steady flow of information provided by these various systems will be collected either via real-time data links or when called from sensor memory by interrogation units of overhead aircraft.

The enormous amount of raw intelligence information provided by these sensor systems will be received, processed, analyzed and reported on at tactical, theater and national levels in formats tailored to the needs of the user in near real–time.

Cataloging, storing, retrieving, sorting and combining data with other intelligence sources will be accomplished at automated analysis and reporting centers. A hierarchy of combat information and intelligence systems, each responsible for a specific sensor system or area of combat information will ensure thorough analysis of data while streamlining reporting and expediting dissemination of the product. Correlation of this information and synergism of effort will be accomplished at the nexus of Service combat information and intelligence systems such as the Air Force's Enemy Situation Correlation Element (ENSCE), the Marine Corps' Intelligence Analysis Center (IAC), and the Army's All Source Analysis Center (ASAS).

The IAC, at the heart of the Marine Air-Ground Intelligence System (MAGIS), will receive near real-time intelligence information from a number of subordinate systems. The All Source Imagery Processor (ASIP) will exploit data from reconnaissance aircraft equipped with Side Looking Radar (SLAR), infrared, and electro-optical sensors as well as national level imagery received via the Defense Dissemination System. The Integrated Signals Intelligence System (ISIS) and the Electronic Intelligence Support System (ESS) will provide information on the enemy order of battle through intercept, analysis and reporting of communications and non-communications emitters.

Accuracy and effectiveness of weapons at the turn of the century will ensure not only greater lethality, but greater economy of effort. For the first time in history, artillerymen may be able to repeatedly produce first round hits on all manner of targets from area troop concentrations to moving armored vehicles. Sensors, both manually and remotely operated, will report target locations to the nearest meter directly to fire direction computers. Fire support will be more responsive and more flexible. The Battery Computer System (BCS) and the Army Field Artillery Tactical Data System (AFATDS) not only will produce a quicker solution to the gunnery problem, but also will provide different solutions for each howitzer if the situation demands it. Data links between fire direction centers and weapons platforms will adjust automatically the launch tube to the proper firing setting. Once launched, "smart" projectiles will adjust their trajectory to pinpoint accuracy, guided at the final stages by ground based and airborne target designators like the Modular Universal Laser Equipment (MULE). Direct fire weapons and air-delivered ordnance will be self-guiding using image, infrared and electromagnetic characteristics of intended targets to guide them.

Command and control systems will also enable the commander to convey orders and directions more rapidly to his widely separated forces. Synergism of intelligence systems, fire support coordination systems, logistic and equipment status reporting systems and air defense systems will provide information in a format and sequence that will lend itself to rapid conversion to air tasking orders through a combination of EIFEL, the German Command and Control System, and the USAF Wing Command and Control System (WCCS). Once the aircraft are airborne, JTIDS will provide enroute and terminal control agencies with information on fuel and weapon status, aircraft location, as well as the enemy and friendly situation as seen by the aircraft's radar. Orders will be relayed from the air commander to the pilot by means of data exchanges between computers of the various allied air command and control systems, all of which are fully interoperable through common link protocols and bit streams.

In the year 2000, the Western Alliance, retaining a technological edge, still will be able to provide a shield for the security of its tactical forces from aggressive Soviet manned and unmanned reconnaisance assets. Deception will be elevated to a new level of sophistication and effectiveness. Active and passive deception techniques and systems will become as valuable on the battlefields of the twenty-first Century as the equipment and systems they mimic. Sophisticated emitters will "spoof" enemy sensors, creating illusions of troop and vehicle concentrations, radiating patterns from communications nodes that don't exist. Sensors and "smart" ordnance will be dumbfounded by the requirement to choose from among multiple targets as decoys copying the electronic or physical facade of the genuine article proliferate on the battlefield. In short, much greater emphasis will be placed on combat deception operations which require a robust and capable C^3 system.

Deception alone will not be enough to stop a determined adversary. Active countermeasure technology and techniques will be employed to seek out and destroy enemy command and control systems. Antiradiation missiles will home in on enemy radars and communications emitters. Especially disruptive of enemy capabilities will be advances in U.S. electronic combat systems with heavy emphasis on the disruption and denial of enemy C^3I capabilities.

Communications, just as it is today, will be the critical link in the effective operation of the command and control systems at the turn of the century. Increased capacity, greater resistance to electronics countermeasures, simpler methods of operation, more secure but less complex crypto systems, and greater ranges will be present. Digitalization of all communications media, to include voice, will have been accomplished. ECCM resistance will be included in all single and multichannel radio systems. Mobile Subscriber Equipment (MSE) will route tactical radio messages automatically as units move rapidly across the battlefield.

The replacement of separate, dedicated, static, low-capacity analog communications links for command and control systems will be almost complete. Shared, switched, high capacity digital systems using Integrated Services Digital Network (ISDN) and Open Systems Interconnection (OSI) protocols and interfaces will be interconnected via commercial, tactical and strategic communications networks without regard for location of physical transmission systems.

Encryption will be employed throughout to enforce the security principle of "least privilege"; that is, to allow each subscriber access to the minimum amount of information necessary to perform its functions and mission. Electronic transfer of keying variables and the ability to create privacy nets within security systems will eliminate procedural interoperability that plagued the 1980s.

The integration of tactical and strategic systems will be almost complete. The facility with which tactical commanders can access other tactical subscribers via cellular-like telephone systems will be duplicated in and extended to strategic networks. Routing agents in both networks will check periodically the correctness of current subscriber directory listings by corresponding with other elements of the system.

As it has been since the beginning of warfare, the success or failure of an army rests, in large measure, on the ability and prescience of its commander.

The commander in the year 2000 will enjoy an ability to command and control on a battlefield shrunk to spatial and temporal dimensions unprecedented in modern times. Born of a generation that has an almost innate familiarity with computers and information management systems, it will be his knowledge of C^3 systems and his ability to use them to penetrate the "fog of war," that will be one of the primary keys to success on the battlefield of the 21st Century.

Equipment, Personnel and Procedures—Foundations for Future C² Architecture

Robert T. Herres

Command and Control

Since the earliest times of recorded history, organized armed conflict has been a major feature of man's activity in this world. The details of such activity probably have been documented or preserved in some accepted fashion as thoroughly as any other human endeavor. Undoubtedly, the deadly and decisive nature of warfare has been one of the driving reasons for mankind's insatiable fascination with the whole business of war. Robert E. Lee was understood to have said upon riding through a civil war battlefield, "It is well that war is so terrible, else we should grow too fond of it!" Yet today, threats of armed conflict seem to persist as an apparently inevitable dimension of man's quest for a more secure life on earth—or perhaps a manifestation of his zeal or greed. Thus we are compelled to understand its ingredients and the keys to success in battle.

This article was written especially for* ***Principles of Command and Control.*

Students of warfare have devoted much scholarly work to analyzing records of military engagements as far back in history as records will permit. Central to all these analyses is a critical review of the behavior and performance of commanders. While training, logistics, mobility, weapons and firepower technology are important elements of any comprehensive historic work, few aspects of warfare receive the kind of scrutiny and microscopic review as the style and manner in which commanders—particularly those with more than their share of success—direct and control their forces. This, of course, is what command and control is all about.

The purpose of this chapter is to consider C^2 in the year 2000 within the context of the principles which govern the command and control process. To do this, we shall first examine the process with an eye toward fully understanding the "basics" which transcend the influences of contemporary technology, and then consider how the relevant principles have been manifested in operative equipment and activity as the command and control discipline has evolved. These considerations then lead to speculative comments about how we might anticipate trends in C^2 systems to point the way toward the year 2000 and beyond.

The single most important characteristic of command and control is the fact that it is a *process* rather than an "entity" or a "thing." This is the most significant point made in this paper, and it is a fundamental principle of command and control, i.e., that C^2 is a dynamic *closed loop process*. Implicit in the concept associated with a process is the idea that activity must take place; in fact, several categories of activity take place as a commander directs and controls or regulates the activities of his forces. While the characteristics of the activities which take place in this process will vary at different levels of a tiered structure and within the dimensions of different missions, the fact that a process does take place is reality. These activities are interactive—so much so that one must recognize the dynamics of the process with its constant interaction and frequently unpredictable events. The purpose of it all, of course, is to prevail in combat by assuring that the right amount of firepower can be delivered at the proper place and time. When accepting his tasking, a commander assumes that he has the capacity to prevail—even against a superior force—if he applies the right amount of force at the correct location with proper timing, and he further assumes, consciously or not, that a responsive command and control capability is available for him to make sure that happens. Without the imposition of an effective controlling process to overlay the combat environment, the forces themselves can provide little more than an aggregate of disconnected and uncoordinated effort with chance the only determinant of success: an armed mob in quest of variously defined objectives!

What are these activities that form the basis of the process we call command and control? The preceding discussion in regard to the purpose of the process itself provides a basis for describing these activities. The commander must bring together knowledge about his forces and their capabilities, combine that with a review and understanding of his adversary and select the timing, locations and force structure for engagements. Implicit in all of this is the idea that the knowledge and understanding referred to has both breadth and depth in its context. To give a simplistic example: It is one thing to know that the enemy has 100 tanks at position A and that friendly forces have artillery and anti-tank weaponry at position B, but it is from a broader base of knowledge that one understands whether or not the tanks involved have significant vulnerability to the artillery and anti-tank weaponry on hand. Thus, the activities of force status reporting and monitoring as well as intelligence collection and analysis are quite critical to basic C^2 activities, but they are useful only in the broad sense rather than a simple "bean counting" mode. Thus, force capabilities are an essential, implicit aspect of this status monitoring process.

If the commander's force employment problem is sufficiently complex either because of the size, variety and sophistication of the force structure, the geographic distribution of the engagement activity, or for a number of other plausible reasons, then a conscious advance planning phase of force employment is prudent. We call the product from this activity, "operations plans" or "ops plans." Such "preplanned-plans" are developed under circumstances which suggest that certain scenarios or situations are sufficiently likely to develop so as to justify the advance preparations involved in detailed *a priori* planning. Obviously those situations or eventualities that we must be certain we can deal with are selected for detailed advanced preparation because of our expectation that plans prepared and rehearsed in advance are more likely to be successful. While generally not considered an integral part of the C^2 process, using information developed during this activity dictates the existence of a tight connection with the planning activity. Thus, force employment planning, like intelligence collection, analysis and production is an essential feeder to the command and control process. Even the most basic engagement activities are based upon some employment concept and plan; some are planned, coordinated and documented well in advance, while others emerge as developing situations unfold.

Altogether too often taken for granted is the crucial activity of issuing directives and guidance to the engaging forces. Disseminating force execution directives is often the weak link in the C^2 process, and the criticality of this phase is frequently overlooked or lightly regarded. Field forces must have clear, unambiguous guidance and direction, ranging from general guidance on rules of engagement to the specificity of a precise execution order which sets in motion carefully timed employment activities. This is the fourth and last of the major activities which underpin the command and control process. With these brief observations in mind, let us now relate the entire closed loop process to the mission and tasking aspect of force employment.

The C^2 process takes place in a hierarchical structure at each level of command. While this may not seem on the surface to be an especially profound observation, there is a common tendency to regard command and control activity as a single, vertically continuous process when it should be viewed as a multi-tiered series of closed loop (but interdependent) processes

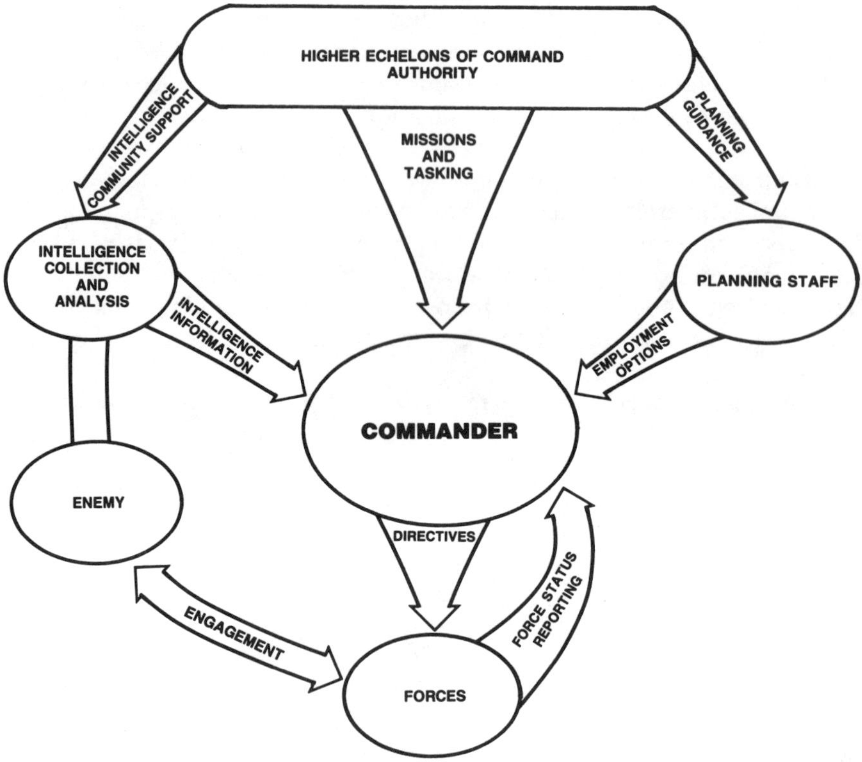

Figure 1. The command and control cycle

at each echelon of command. The output of the process at each level provides the input or tasking to the next level. A thread of continuity must run from the theater commander through each echelon. Mission and tasking statements begin at the top of the hierarchy and are the connecting functions that stimulate or set into motion an identically cyclical process at the various tiers in the structure and thereby induce activity in a sequential fashion.

The simplified diagram shown in Figure 1 should be helpful to summarize the relationships between the activities just described and a representative tier of the process itself. The commander is the nexus of activity—the focal point of the decision making process. He responds to a variety of directives that range from his static mission statement to the dynamic tasking that provides direction for specific operations on an *ad hoc* basis. Planning guidance and intelligence support are inputs (but not forcing functions!) to the commander's staff to facilitate their respective roles in supporting the decision making role of the commander.

During the process of developing and selecting courses of action to carry out his mission or respond to specific taskings, the commander forms his judgments by considering the relationships among the status, disposition and capabilities of his own force and similar knowledge about his adversary.

Intelligence information is basically non-cooperative force status reporting on the disposition and condition of the enemy. Implicit in the knowledge about one's own force status and that of the enemy is a comprehension of the various forces and system capabilities. A second and similar example to the one used earlier should help drive the point home: knowing that an enemy has 20 Model "X" fighters available for use at Base "Y" is not particularly helpful without some knowledge of the performance characteristics of these fighters. Much of this kind of knowledge we tend to take for granted as "trade knowledge," but it must be accounted for and understood as an inherent part of what it takes to support the commander's decision making process.

When the commander has chosen a course of action to be executed by his forces, he needs to communicate his decision to them. His directives set in motion the engagement activity—or alter the course of that activity, should engagement already be underway. Moreover, *changes develop* in both the status of the forces and the posture and condition of the enemy *as a consequence of engagement*. The cyclical dynamism of the process begins to show clearly at this point. There is obvious anxiety to know the outcome of a strike or directed action—both in terms of its effect on the enemy and the losses suffered by friendly forces. Natural curiosity aside, this data forms the crucial basis for requirements and options for future courses of action in the engagement. The dynamism is most pronounced when real time feedback of force status and intelligence information is available during the course of an action, making possible "mid-course" corrections to what has previously been directed, intelligent commitment of resources held in reserve, quick follow-up or exploitation of an unexpected vulnerability, and so on through an array of other obvious possibilities. *Timing* and *clarity* become increasingly important during the various stages of the process, so that the communications and associated procedures which support the dissemination of directives serve an absolutely vital role. New execution directives, changes in the rules of engagement or planning guidance and other direction from the commander to his forces will set off new relationships as fresh courses of action affect the postures of both friendly and enemy forces. Thus, the continuous nature of the process develops as a post-engagement phase sets in and activities shift in pace and emphasis.

The preceding explanation provides a basis for discussing the concept of process stratification from a broad perspective of viewing activity at a number of command echelons. As the command and control structure works its way downward from the senior echelons, the number of entities engaged separately (but hopefully not independently) in the process depicted in Figure 1 increases both vertically and laterally. Each entity consists of a commander directing his forces on the basis of his mission and tasking, basing this direction upon the condition of his own forces and the status of the enemy. Complexity of the individual activities within each cycle involved in the aforementioned process differs considerably where a battalion commander might be involved as opposed to that of the theater commander, and the interaction between entities can also vary considerably. Nonetheless, the *pattern* of activity remains the same.

At this point, another word is in order about how the relationships between the intelligence and the operations planning functions fit into the command and control process. Both functions have grown increasingly dependent upon sophisticated equipment and complex catalogues of information and detail. Concentrations of the resources that provide such material are not economical at all levels, or on a widespread basis, so that higher echelon intelligence staffs must share products and information by providing "intra-community" discipline, unique support outside the command and control cycle. Similarly, mobility planning, lateral command support, logistics management, etc., are planning considerations of increasing complexity in a modern military force structure, and the resources (people, data and equipment) that are necessary to develop viable operational planning options are too costly and elaborate to be widely available. Thus, the operational planning community has evolved a supporting structure of guidance and planning references and data that can facilitate option development at several organizational echelons simultaneously. Again, this operations planning activity is conducted in parallel with the C^2 process, yet provides important ingredients to the decision making component.

Having established the nature and character of the command and control process, let us now consider how C^2 systems fit into what we have discussed. Conveniently, the Department of Defense (DOD) provides us with a second useful definition:

Command and Control System: "The facilities, equipment, communications, procedures, and personnel essential to a commander for planning, directing, and controlling operations of assigned forces pursuant to the missions assigned." (JCS Pub 1).

As the tools with which warfare is conducted grow more complex, so do the means which facilitate the command and control process. When warfare was carried out in hand-to-hand combat, the majority of the C^2 activity was embedded in the commander himself. The battlefield was normally in his full view as he would seek to position himself on a rise in the terrain or achieve height advantage by other means for better viewing; both the status of his own forces and that of the enemy were readily apparent by direct visual inspection. Directives were disseminated by voice and later by combining elaborate arrangements with signaling devices such as flags (which also played an important role in the force status monitoring function). Wind instruments such as horns and bugles supplemented messengers who moved on foot or the back of an animal and became the stock in trade for communicating among the forces. Horsebacked messengers delivering written instructions and reports with hand drawn diagrams evolved to provide more elaborate communications as the size of the battlefield grew substantially. When gunpowder made the "stand off" weapons of the day a reality, commanders sought ways to see and view at the ranges this new technology permitted him to engage. Hence, telescopes and binoculars became standard tools for commanders as well as for their artillerymen. And so, for a good many generations, the standard equipment that facilitated command and control activities were horns, bugles, flags and telescopes. The human messenger on horseback endured for some time as a critical means for delivering directives as well as the planning and intelligence information needed to insure proper force direction and coordination. Interference with such communication was recognized in turn as a way to gain advantage over an otherwise well coordinated adversary. In fact, a common technique of the U.S. Civil War era was the use of cavalry patrols to interdict messengers to exploit knowledge of their message content or, at a minimum, to assure that the messages never arrived. Because of this vulnerability, it was learned that coding messages could prevent such exploitation.

Although useful for the "long haul" transportation of information, early telegraph communications were not flexible enough for field use; however, they did provide for message traffic between static headquarters, and could be used as communication "trunks" or "back bones" with manual message delivery for communications "tails" from telegraph stations to deployed field forces. While we have just briefly reviewed how C^2 methods have been transformed through the technologies of contemporary times, the illustrations used should serve the purpose of portraying the evolutionary nature of the growing inventory of instruments and vehicles which facilitate these activities.

The point of all of this is, of course, that the command and control process and systems which support that process are not creatures of modern technology—rather the systems are a natural manifestation and product of the need to direct and control forces so effectively that the most telling results can be achieved with true economy of force commitment. That this process did not take on the trappings of a well defined and discrete discipline in the past is understandable when we consider that much of the process took place in the mind and person of the commander himself with the direct support of his immediate staff; to a large degree, helping with command and control was what his staff was all about. The occurance of failure from time to time is understandable in light of the fact that not all commanders executed the process well. But those who had made provisions for a tight loop of well timed information flowing to and from their forces, *and* who exploited every opportunity to collect and utilize information about their enemy, are commanders whose place in history has generally been viewed with favor because they functioned effectively when they made the right decisions. However much a C^2 system may facilitate decision making, no system can substitute for wise choices by the decision maker himself; and conversely, no matter how wise the commander's choices may be, a poorly functioning command and control process can readily deny him success.

Today the dimensions of the battlefield and time sensitivity of the actions which take place are many orders of magnitude larger and more sensitive than the scope of operations in the era of the bugle, telescope, flags and humans on horseback. Commanders know from their own experience and study—particularly of recent history—that good decisions and sensible direction to one's forces must be based upon an array of near real–time information on friendly and enemy force status and actions. The dynamic aspects of modern warfare place extraordinary demands upon today's C^2 systems. The potential devastation and far reaching implications resulting from employing contemporary weapons has caused defense managers to take early cognizance of the C^2 requirements of new systems during the system definition process. These time and space factors are instrumental in generating pressure for more sophisticated command and control system components.

As we have seen, the earliest "hardware" items to evolve in direct support of the command and control process were communications devices. Our look at its evolution suggests that the most urgent demand has been for better ways to communicate in order to direct and control forces, to coordinate planning and to transport intelligence information from sources to users. However, the sophisticated requirements of modern weapons and breadth of potential military activity, as well as interaction of complex supporting forces and systems, caused automated data processing to become an indispensable element in supporting the C^2 process. Thus, today's modern inventory of command and control system hardware elements are made up principally of communications and computer systems. Considerable complexity and flexibility are brought about by the software that must be integral to both of these types of systems.

Some readers may be surprised by the exclusion of sensor systems from the preceding discussion. Sensor systems are devices or elements of a variety of military equipment which in many cases, directly support or are integral to a C^2 process at some level. However, they are functionally difficult to categorize; and more often than not, they are multi-functional systems. Sensor systems are used extensively, of course, in the business of intelligence collection—a function that is obviously a critical contributor to the command and control process, but one more often regarded as a functional entity unto itself. Sensor systems perform the vital operational function of surveillance. They are the telescopes and binoculars of the modern era and, as such, provide the critical data needed to get weapons on properly located targets. Whether or not a sensor system is a C^2 system depends upon its functional application at the time of its use and its role relative to the supported commander's mission.

A contemporary example may serve to illustrate the difficulty of categorizing the role of sensor systems. The U.S. Space Command operates a world wide inventory of terrestrial and space based missile surveillance sensor systems to detect promptly ballistic missile launch activity, as well as the supporting systems which process the aggregate of sensor inputs to the Cheyenne Mountain Complex in Colorado Springs. The task, of course, is to support NORAD's mission of ". . . warning and assessment of aerospace attack . . ." This mission statement drives the distinction between the role of these sensor systems as C^2 system elements or as basic elements of Space Command's force structure. From the viewpoint of Commander in Chief,

Space Command (CINCSPACE), these sensors are not command and control system elements, but rather the *basic components of his force structure*. On the other hand, systems by which he monitors the status and capabilities of these sensors and directs their configuration and alert posture are his command and control systems. However, to the National Command Authorities with a much broader mission into which these sensors fit, they are normally regarded as C^2 system elements within the context of the national system. This example serves simply to illustrate the fact that sensor systems are not easily categorized but, either tangentially or directly, provide essential support to the command and control process.

In looking toward the future of C^2 in the 21st century, the most relevant reality for us to keep in mind is that the year 2000 is not far away. Given the recent history of the total elapsed time required to develop and field new equipment—particularly the software intensive variety—most of the significant command and control systems that we will be fielding and using past the turn of the century already have reached Milestone I (concept phase) in the development cycle. Just as important, the personnel who will be our future field commanders in the year 2000 now are in active service. Only the procedures which will tie these two C^2 elements together remain to be conceived.

Since the purpose of this paper is to speculate about what will be the demands on these three elements in the future, a review of the projected operating capabilities and characteristics can provide useful insights. Intuitively, we can conclude that the basic principles and activity components associated with the overall command and control process as described earlier are not going to change. Hence, they can form the foundation from which to consider the operational needs and trends that are likely to evolve in the future.

Equipment is the first element to be considered. Communications systems, automatic data processing equipment and displays and interface devices compose this group. Communications systems will unquestionably continue to be regarded as the lifeblood of any C^2 process. The importance of communications is richly evident in all forms of combat systems support; logistics, mobility and intelligence, to name a critical few. But nowhere in the serious business of warfare is the art and science of communicating information as fundamentally vital as in the command and control process. The scope and nature of warfare have driven the state of the art of communications equipment more decisively than any other category of systems. This tendency for such heavy dependence will not be altered. Trends evidenced in the past to improve such characteristics as speed, accuracy, reliability and security will not change abruptly. As advanced weapons systems are deployed, dependence upon an effective force control process and enhancement of systematic methods to support it will continue to be emphasized. Forces on both sides of any future engagement are certain to be well aware of the force multiplier effect of this process. Adversaries will certainly exert well planned and executed efforts to interdict the communications which facilitate the commander's mission execution, and this expectation drives the need for protection.

Indeed, speculation about the nature of conflict in the year 2000 points toward several important attributes of communications which still must be improved and further developed. Communications, by their very fragile nature are inherently vulnerable, physically as well as electronically, and those favorable operational characteristics such as the flexibility of large footprints and equipment mobility are also features which can be readily exploited in a hostile situation. Thus, one of the most important attributes to be addressed will be vulnerability.

Developing new across the board anti-jamming techniques that will produce higher resistance to interdiction is a must. Eliminating single point failure nodes and automating current manual procedural limitations associated with distributing crypto keys are also needed and easily within grasp if given attention. Protecting communications against extended range weapons effects such as electromagnetic pulse (EMP) will be more difficult but certainly deserves continuing research. The need to operate for extended periods without revealing locations of user terminals is also pressing upon us. Said very simply, the foreseeable emphasis must continue to be on more jam-resistant, secure, EMP protected, low probability-of-intercept communications systems which can carry large volumes of data to users equipped with light, small, highly transportable terminals. Obviously, satellite communications systems will play an increasingly important role in meeting these demands, and satellite-to-satellite relay systems will be useful elements adding robustness to our capabilities.

The second equipment component is automated data processing equipment (ADPE). Automated data processing (ADP) systems are rapidly becoming the way of life for organizing and displaying vital information to commanders—even at the lowest echelons of command.

These systems replace portions of the staffs of the earlier days, and when used properly, will provide future commanders with extraordinary capacities to extend their impact upon the operational planning and force structure analysis aspects of command. Senior commanders at the beginning of the next century will be a computer literate breed of individuals who will have been operating and using software intensive systems for years. Exploitation of the attributes of automated data systems will come easier for this generation and the need for flexible, user responsive systems that are functionally efficient will be even more important than today. Said another way, the absence of or failure to improve such system attributes will be more noticeable than ever.

Computer languages will have to be user as well as developer friendly to the point that the user's manipulation of the system will be almost transparent to the software development process. That is, much real–time task—oriented software development will be accomplished through user manipulation. At the same time, software configuration management will have to be an automated process to insure that users have the flexibility and facility to exploit system capabilities without forfeiting system management and control discipline. Automated configuration control may be one of the most important opportunities leading to the sophisticated ADP developments of future C^2 systems.

Another major opportunity is the security from hostile exploitation of dependence upon ADP systems, i.e., computer security. While much work has been done in the field of automation schemes for securing ADP systems—particularly for command and control use—this work has not kept pace with the state of the art of parent systems. By the year 2000, there may be significant problems in the absence of creative solutions. The reason for such concern is that our dependence upon ADP systems to help us direct and control our forces efficiently is increasing so rapidly; this dependence spills over into the communications field as well, because modern communications have become quite software intensive. Communications processors, modems, crypto devices, switching systems, etc., are basically a variety of computers which have been programmed to function as communications system components. While this might lead to the conclusion that computer security and communications security problems can be solved using the same techniques, such is not entirely the case. A most significant difference is the growing interaction of the system user with the software in ADP which supports other C^2 related functions—particularly the operations planning and intelligence production activities. In effect, these users are very much a part of the systems they use for planning and analysis purposes—almost an extension of the system itself, even though the system is actually a tool of the planner or analyst. This phenomenon generates unique computer security problems that need to be recognized and dealt with.

We have witnessed a remarkable change in the down–sizing of computer equipment in just 15 years. Future needs will require this trend to continue. Forward deployed systems such as weapons, sensors, and C^2 battle management platforms will demand lightweight, high capacity, rapid processors with high fault tolerant reliability to accomplish the full spectrum of assigned missions. Space-based systems such as those envisioned for strategic defense will generate challenging requirements for positive command and control over some highly automated systems.

In order to achieve the real-time command and control system capability that the next century will demand, data communications will have to become a richer part of the voice/data mixture. This means that displays and terminals which present data–generated graphics must be capable of conveying higher order concepts routinely. It also means that considerable flexibility and responsiveness must be available to the users of such systems. The relative facility of data exchanges among decision makers will be measured by the degree of difficulty involved in weaning such users away from their secure voice telephone equipment. As new generations of leaders emerge from training, educational and work backgrounds in which computer knowledge and usage is explicit, the wide-spread use of person-to-person data (audio and visual) communications becomes an easier precept to implement.

A final word in this context needs to be offered regarding communications to data processing system interfaces. Already wide-spread in use, the direct computer-to-computer communications demand will drive the need for universal standards and protocols. Our expectation for more use of automated processing tools in future command and control systems simply highlights the fact that the problem will not be eased to any significant degree in the future. Our capacity to predict the need for interoperability in advance is never any better than our ability to anticipate specific scenarios, always a difficult problem. For this reason, we should expect interoperability problems to continue to be a way of life since they can almost never be solved without cost, and our anxiety to pay the price will be related directly to our concern and foresight relevant to the scenarios we expect to encounter.

If communications is the lifeblood of command and control, people are still the heart that pumps the blood. Personnel requirements in the years beyond 2000 will be even more critical than at present. The military service will have to select highly motivated and technically competent people to develop and operate the advanced systems. Not only extensive initial training and continued professional career education, but special career patterns will be an absolute requirement if career management and retention is to be successful.

At the very least, the services will have to begin to treat the broad discipline of developing and managing C^2 systems with a long-term view. They must create career specialties that serve the broad national interest and which permit military personnel to pursue and complete a satisfying career in this field. The idea that a two or three year career broadening tour in a command and control subsystem is an adequate approach to the problem belies the need for the development of C^2 professionals. Joint service interaction is an increasingly important area which defines a need for experienced personnel to reach beyond traditional disciplines. The complexities of procedural development remain an unknown as we prepare for the future operation of many weapon systems. Until these systems are delivered and operationally tested, employment procedures cannot normally be fully developed; and it follows that command and control procedures are to be correspondingly affected. Hence, the year 2000 will require great flexibility in procedure creation. Procedures will be embedded into software, built into the formats of operator displays, highly reliable in performance and simple to execute. To the maximum extent possible, error detection on procedural steps not taken in the proper sequence and other self checking patterns will be required. All of this points to the need for skilled people who understand the process and its complex systems well enough to manage these various interrelated tasks well.

Perhaps the greatest challenge of all future needs in command and control will be the requirement to develop a sophisticated understanding of the need for and use of the overall system capabilities in the absence of realistic exercises or major conflict. During wartime, this understanding develops quickly, but peacetime tends to dull our sensitivities to these realities. Users and operators of C^2 systems are frequently negligent in this regard, and the lessons of any failure to vigorously test and exercise full system capacity (people, procedures and equipment simultaneously) must not await the graphic consequences of failure or mistakes in combat. Failures in the future will be too decisive to arrange opportunities for recovery. While knowledge of the C^2 discipline is growing as a basic ingredient of broader military expertise among other bodies of knowledge, the rate of growth of this knowledge falls short of the rapid pace of the state of the art of our systems. Only by realistically exercising the hardware, people and procedures are flaws to be discovered and corrected.

The earlier discussion about sensors as either supporting or being force structure end items of a command and control system deserves a statement regarding their future role. As we move toward the next century, trends in sensor system evolution is likely to reveal dramatic technological progress and change. This assertion is based on two indicators: first, the rate at which progress has been made over the past twenty to thirty years has been astounding, so that an abrupt discontinuity would be necessary to see a reversal; and the operational demand for ever increasing levels of performance will be insatiable at the highest levels, generating pressures and priorities that will stimulate interest in the requirement to view and monitor non-cooperative targets and activities as routinely as is possible. Further, we will come to depend more and more on these multifunctional sensors to generate real-time inputs and feedback on force employment decisions at all levels.

As an example, arms control will be a fundamental supplement to overall national security policy. Verification techniques suggest the highest possible emphasis on exotic sensor technology development, deployment and use. Military C^2 systems will employ data from these same sensors to enable the commander to correlate other intelligence data and to adjust his forces over both the short and long term.

Obviously, sensors associated with potential advanced strategic defenses will be basic elements of network structures. Reliable and unambiguous data should be a hallmark of these sensors, as decision-makers will have to depend upon heavily automated systems to confirm data with independent sources prior to the commitment of forces.

The stated purpose of this paper was to discuss C^2 in the year 2000 within the context of the principles and parameters which govern the command and control process. To summarize, the text establishes the nature of the process and discusses its enduring principles and fundamental precepts as they have existed throughout the history of military activity. This treatment of the ways in which command and control activities have been conducted is offered as a means to suggest the existence of decisive factors that influence trends toward the future. The importance of communications in the command and control process is stressed, as is the recent application and continued growth of automated data processing, as a tool of the participants in the process. People and procedural components were addressed as essential in future command and control system trends. Sensor technology development is cited as an area in which we can expect remarkable new levels of achievement. Finally, it is important to note again that the technologies of command and control systems that will be in use in the year 2000 are at least in the laboratory now if not developed further. The challenge of the C^2 community is to relate these technologies to the basic need to facilitate this timeless process of critical military activity, and to intelligently develop and apply the most promising among them.

A View of Military Command, Control and Communications Systems of the Future

Walter R. Beam

Introduction

Prognostication about the far future is at once fascinating and frustrating. Individuals who consider themselves to be professional futurists often operate in what one might have termed until recently a "Star Wars" frame of mind. However unintended, that metaphor has been captured by particular U.S. defense programs, technological speculation about which might be seen as having political overtones. This notwithstanding, the professional futurist typically takes a flying leap into the future, the fewer concerns he has about how to get there, the better.

This approach may be contrasted to that characteristic of a long-range planner of government or industry, who is straitened by the need to build a credible path between the present and the postulated future. Often, the path must not only be *technologically* rational but also "affordable," a constraint

This article was written especially for **Principles of Command and Control.*

which often places insurmountable barriers in the way of what appears to be only modest steps of progress.

Though providing no real help to either school of future prediction, the real keys to future change most often are the unpredictables. At the time of World War II, there was no clue that the only recently developed electron tubes, which then were the keys to electronics, would within less than two decades be virtually swept into oblivion by transistors, and later by integrated circuits. The laser, which today fills an increasing number of roles in defense and industry, did not appear until a decade later. These certainly were *not* predictable technologies. Nor was it truly predictable during the prewar depression of the 1930s that the next decade would see the U.S. change from a second rate technological nation to an unquestioned world leader in scientific research and development.

Not only is there unpredictability in science and technology, but the future economic and geopolitical equations of the world must play central roles in determining the true future, as they have since societal organization advanced beyond local and tribal bounds. All skillful long-range planners begin their exercises with proposals for "environment" and/or "assumptions," which are tested for corporate or political acceptability. Thus armed, they can proceed to evolve detail which has near guaranteed acceptance. It appears that futurists take no such indemnifying precautions.

This writer's expertise lies primarily in technological, rather than societal areas. Though admittedly the latter will probably dominate future evolution of command, control and communications, technology will define its character and capabilities. Hence, the underlying assumptions for the sociopolitical environment must be "business as usual." On the technological side, the writer's experience as a long-range planner makes him uncomfortable suggesting, for example, the possibility that discovery of telepathic communication will revolutionize battlefield communications. However, there are important ongoing trends which give clues as to likely futures.

Technological Clues to the Future

The Unification of Data and Knowledge

By now, most of us have heard references to something called *artificial intelligence,* often promising computers which will program themselves or improved medical diagnostics by computer. Since their inception, digital computers have been described by "science" writers as "machines that think," or "giant brains." On closer examination of current realities in artificial intelligence (in years past known as *automata theory*), less meets the eye than might be expected; the heralded "expert systems" deal with narrow and specialized areas or do the sort of routine work in which computers are known to excel.

Since John von Neumann proposed the idea in the early 1940s, all computers have been able to store programs in the same memory and storage units in which they store data. When programs are being written they *are* data. But beyond that point, data and programs lead very separate lives.

This is now changing. Past and present programs, to a large degree, are monolithic and inclined to process all the data they encounter the same way. After all, this seemed to early computer developers to be a natural extension of industrial mass production to the new and powerful "tools" which computers appeared to represent. Perhaps not surprisingly, it has been *personal computers* which, to the greatest degree, have stimulated study and research into dealing with programs as collections of rather primitive data finders and manipulators. In the future, it will be possible to *combine* the work of many small and limited functional programs into a goal-driven hierarchy of computer capability which not only will deal differently with different kinds of arriving data, but also will evolve and adapt strategies for doing so.

Detailed modeling of human intelligence mechanisms is not a requirement and, in fact, could be a deterrent to progress in this area. But it *will* be necessary to understand human *problem solving techniques* in a more generic way, and to develop more open ways for computers and humans to interact.

There may even be a fallacy in the assumption that computers should attempt to solve problems using the same methods that are used by humans. Computers, after all, are far more *precise* than humans, and arithmetically they are incredibly *faster*. Over the past 40 years, we have learned how to put computers to work effectively on arithmetic problems, but we are only just learning how to apply them to the non-numeric problems (of which there are far more) with which we must deal. Nor do we know the best ways of using computers to deal with problems which are not, and often cannot be, stated in precise terms; yet humans deal successfully with these problems, mainly using experience to fill in missing information.

In the realm of non-arithmetic problems, in particular, the issue of what constitutes data and what constitutes a "program" is far less clear—all the more reason for a more even-handed treatment of these two forms of *knowledge*.

How will this evolving technology affect future command, control and comunication systems? Primarily, this technology will multiply the amount of information that humans will be able to deal with, and allow humans interacting with the systems to address information at different levels of detail or of abstraction than the lowest level, with which they must deal almost exclusively today. Understanding a military situation, for example, is compounded not only by visualizing enemy positions and movements from radar and other sensors, but also by relating these to observations, within a context of *experience*. Although the creation of computer programs directed toward doing this (albeit in some limited field of combat) has been attempted, the programs have lacked the flexibility and adaptability required for success. No program prepared wholly in advance of the real situation and operating with unchanging principles could be expected to deal satisfactorily with such problems. Hence, the evolution of computer environments that can make use of a wide range of ideas and historical trends and can assess their appropriateness based on current data may supply important future capabilities.

The Economics of Computing Power

In the past decade, computing power: (numbers of computer instructions executed per second) has plummeted in cost with the development of microprocessors. At retail prices of common microprocessor chips, one can purchase 300,000 instructions per second for an outlay of $3.00, or $10.00 per MIP (million instructions per second). This is not the entire story, however. If one is not satisfied with the small processor, 200 MIPs in a single package may cost $10,000 (memory, packaging and power supply included though by no means explaining the difference). Two decades ago, a 1-MIP computer with what would *now* amount to $25 worth of memory would cost about $300,000 per MIP.

Though at present (1987) the economical form of computer power—the microprocessor—is limited to about one MIP, continued technological evolution could bring 10-MIP or even 50-MIP computers into single chip form. Traditionally, at any stage in computer technological evolution, planners found it difficult to imagine how significantly more power or larger memory could be used. And it also has been traditional that the eventual use was in ways *not predicted* before the capability was actually available. Hence, rather than doing word processing much "faster" than currently possible, or dealing with even larger spreadsheets than are used, or completing payroll analysis in 3 seconds instead of one hour, it is more likely that new uses will evolve.

The command, control and communications arena has made its greatest use of computer power in processing sensor signals, maintaining records on available resources and supporting situation displays. Looking farther out in time, these usage patterns no doubt will change. It is very likely that large processors and their attached storage will be responsible for the reception and storage of situational reports and sensor information.

We can expect to find, as well, a large group of small but powerful "specialist" computer systems, whose expertise is derived not from computer hardware or software design but from continued attention to particular aspects of the overall command and control problem, use of certain historical data, concentration on certain subsets of available data and perhaps also a user-specified set of "biases," akin to those all humans bring to their work. Each of these in all likelihood will be composed of many microprocessors, each with access to both incoming and historical data. As a group, they will augment staff support to decisionmakers. Save for the possibility of one raising an alarm under certain conditions, they will continuously perform analysis. When asked for reports and/or "recommendations," they will be able to provide a response at any desired level of detail.

Although speech synthesis is now commonplace, speech *recognition* eventually will become a capability of computers in command and control systems, where its potential importance is greater than that of process control or weapon system applications. Associated with important command personnel in the future command and control center, there may be a *context processor,* whose function will be to monitor the context of activities in which the individual is involved, based on conversations, his use of display information, and the events taking place within the command. Its objective will be to make information available at the proper level of detail, with explanation given when needed for correct understanding. This *very* personal processor also would serve, along with a counterpart, as an "interpreter" in conversation with remote commanders (for example, when the individuals do not share common current context, or perhaps culture or a common natural language). In such communication, the "interpreters" for the individuals initially and briefly trade background information, so that after the direct (or indirect) human conversation, points of clarification could be exchanged. The objective in all of this will be amplification of the human resource represented by the commander and staff.

New ways will be found to make computers communicate with their human users through visual techniques, since for many objectives this is man's broadest band channel. Mere high resolution and highly accurate representation of battlefield (or air traffic, e.g.) detail quickly will reach a point of vanishing returns, since the human eye is often attracted by *detail* which may be irrelevant in the larger sense. Accordingly, visual information will be structured to convey situations in terms that emphasize situations in which meaningful action can or must be taken. Since bias is inescapable, the same somewhat redundant *situation analysis systems* referred to earlier will on request put their conclusions in graphic form.

Certain supporting systems may act in a predictive mode, based on courses of action proposed by the situation analysis elements. When situations and options become too complex, the commander's personal "interpreter" system may be assigned the task of eliminating unneeded detail.

Remotely Operated Long-Range Systems

Remotely monitored and operated surveillance systems (military or civil) cannot be considered novel, as the earth satellite had become an important class of long-range sensors before the 1980s. The capabilities of unmanned attack systems, such as standoff attack missiles and anti-air and anti-ship defensive missiles, have continually increased. The specialized systems required for launch control and coordination represent some of the most technologically sophisticated real time command and control facilities in the 1980s.

Other remotely operated sensor systems, often based on small aerodynamic vehicles, have enjoyed less success, in large part due to limited sortie times and high attrition often due to poor landing ability. Examination of the environment for manned aircraft in the future portends greatly enlarged roles for unmanned systems. Already, first line aircraft such as the F-16 have greater maneuvering ability than a human pilot can withstand, though available technology offers even more. Arguments for manned vehicles are based to a considerable degree on the concept of crews as skilled visual observers, but developments such as stealth (low observables), when pressed further, might well restrict a human operator's visibility. As onboard electronic sensors and computers increase in capability, the time may be drawing near when major aircraft systems operate either manned or—for high risk missions—unmanned. While past military unmanned aircraft often have sprung from the aircraft modelers' community, the more capable systems of the future will be derived from the best of manned systems, with performance well beyond that of present manned systems, and they will take full advantage of the new design opportunities afforded by replacement of the most vulnerable part of the system.

"Flexible" High Resolution Sensors

Conventional sensors derive high resolution from large apertures, as in radars whose antennas are half the area of a football field. Not only are these huge sensors essentially immobile, they are highly vulnerable. Furthermore, such giant apertures are virtually impractical in space based applications. Many future sensors will be constructed of small, low-power, physically independent elements, forming in a loose sense an array, and cooperating with one another through short-distance coherent communications so that they form virtual arrays of high power and resolution. The concept of cooperating sensors will be applied to both passive (emitter locating) and active (radar) sensors. Sensors on ships, aircraft and land vehicles, when within meaningful range of one another, will be able to operate in cooperation, to provide previously unknown detection range, resolution and coverage with reduced vulnerability.

New Opportunity via Precision Time Standards

High accuracy "clocks" will provide primary technology for positioning and targeting. As currently used in special targeting systems such as GPS NavStar, high precision clocks coupled with time difference of arrival analysis permit high accuracy positioning. The principle can be applied more broadly to any emitter of sufficient bandwidth allowing accurate time determination: radars, communication links, even jammers. This would permit any station or platform, emitting only briefly or inserting the timing signals incidentally into its normal emissions, to be located with precision. Counter-countermeasures would no doubt be applied to discourage exploitation of the information by enemy forces.

Precise coordinated timing also will be central to the operation of sensors such as bistatic radars, with each of the multiple transmitters illuminating a given area of interest on a preplanned time schedule. A receiver-equipped user will be able to derive high-resolution data from an area of interest in one of two ways: directly, by collecting and manipulating radar echoes, or indirectly from other receivers transmitting coded imagery on a periodic basis.

Future Communications

Improving the Quality and Value of Information

Future military command and communication will be carried out in a more difficult environment than has been enjoyed in the past. Encryption will be enforced for all communication; even the most seemingly innocuous data may provide valuable information to an adversary. In combat situations, jamming will be expected as the rule, and will be combined with cleverly devised *deception* whenever encryption can be broken or bypassed. Such techniques as rebroadcasting of encrypted signals which have been corrupted, not necessarily involving decryption, may be used to occupy the time of, and thus attempt to saturate, processing-intensive receiving and decryption systems.

In short, it will be a much more difficult environment in which to carry on communications. Techniques for decreasing bandwidth and increasing redundancy, such as the use of encoded *vocabularies* (words, rather than characters) may be used to transmit speech recognized automatically in a far more compact format.

Advanced two-way radio communications systems will embody dynamic, automatically invoked strategies for reducing information bandwidth, in response to interference. Signal formats also may be adaptive, permitting a high intermittent information rate with a lower average throughput. In such an adaptive system, the changes in communication strategy and format also will need to be signaled. Fully integrated *diversity* communications will be implemented, transmitting priority information on multiple bands and in multiple formats, with "intelligent" analysis of each received signal and formation of a composite of improved accuracy.

Great attention will be given to minimizing data at its source. Hence, most sensors will transmit *differential* information, that which has changed since a previous sweep or scan. However, unlike current systems which may distinguish only designated targets exceeding some present threshold, future systems will be able to send differential information at more detailed levels. A detail of no apparent interest from the sensor's location may correlate strongly with information from other sensors, when processed at an information fusion center.

Communication managers will place emphasis not only on *signal* quality, but also on *information* quality. Primary emphasis again will be placed *at the source,* to minimize required bandwidth time product. Computer based processing will examine quickly command messages for ambiguity and completeness, and make corrections or (in rare cases) ask for more input.

As mentioned earlier, *contextual awareness* will play a major role in the handling of communications. Redundancy needed for a contextually aware destination will be less than that required for destinations which are little aware of events taking place at a transmitting location. Hence, not only natural language communications but also those generated by sensors will adapt transmission in accordance with the receiver's understanding of the context. For example, the first time a particular unidentified aircraft, vehicle or troop movement is reported, it will be described and given an identifier. On each subsequent report, its description will be omitted unless additional identification is available or previously transmitted information must be corrected. Because of this, routine information transmissions will be adaptively formatted, rather than of a fixed format.

The overall quality of a communication link will be assessed at its receiving end not only on the basis of signal reception but also via the patterns of change as the transmitter adapts to the receiver's increased knowledge of the information context. This will form the most important basis for the detection and exclusion of deceptive information generated by an adversary. Because *emergency messages* represent exceptions to a continual information transfer process to which context based optimization can be applied, there also will be "credibility" tests based on direction of arrival, signal strength, characteristics of the transmitted signal and data providing location estimates for friendly forces.

Directivity or Chaos

In the future, an increasing proportion of communications will be carried out using highly directional transmitting and receiving antennas, at frequencies well above those traditionally used for ground or mobile communications. Radars may time-share their electronically steerable antennas with communication services, providing greater diversity.

In the past, directional communication systems have been avoided where possible, particularly for land communication. At frequencies of a few hundred megahertz and below, adequate reception is possible with non-directional antennas. Hence, a terminal can communicate with others independent of direction; there is no need for antenna steering in the direction of a transmitter (or for strongest reception). Furthermore, strongly directional antennas at these relatively low frequencies are enormous in size.

Since these bands lack the bandwidths which will be required to support future jam-resistant high bandwidth data signals, directional communications at higher frequencies will become commonplace, with electronically steered lightweight planar antennas made practical through the combination of integration of both digital control circuits and high-frequency circuits. Directivity will deter jammers and provide additional security against deceptive signals whose sources now will be more obvious. A variety of passive and active reflectors and relays, on land, in the air and in space orbits will provide connectivity for some non-line-of-sight communications, while the lower frequencies may be reserved largely for short-range non-line-of-sight communication.

Optical fiber communication—the ultimate in wireless directivity since signals are limited to the fiber itself—might play a signal role in future command and control systems, not only for fixed locations but also for short-run temporary or even transient communication links. As has been demonstrated in the U.S. Army's fiber-optic-guided missile (FOG-M), large signal bandwidths can be transmitted through the miniscule fibers during the missile's fall through the air, and therefore, most of the expensive signal handling electronics can be located at the less vulnerable end. How, then, do we extend its application in future command and control?

Jamming—Altering a "No-win" Game

At first, jamming was applied mostly to radars and other direction finding and locating devices. It has become increasingly attractive as an effective weapon against command and control systems relying heavily on radio communications. Somewhat like nuclear armament, the attainment of some level of antijam capability will encourage the development of ever more powerful and specifically directed jamming. Where adversaries develop systems operating in different frequency bands (due to technology capabilities or perception of requirements), jamming directed against an enemy may have little effect on friends. But increasing requirements for bandwidth, brought on not only by vastly larger information flow rates but also by a need to defeat known or suspected jamming systems, must eventually lead to widespread utilization of common bands. It will be difficult to coordinate jamming over geographic areas within which one's jamming efforts may be effective in disrupting one's own communications. Even the use of strongly directive jammers, which brings with it the need to locate and identify jamming "targets," will merely delay the era in which full exploitation of available jamming may effectively disrupt most radio communications within a range of one hundred miles or more of a battle area.

As an end result, the use of spread spectrum, frequency-hopped signals forces jammers to use greater power, applied in the form of wideband noise. The newer signals are more difficult to jam, but in the process, many more signals may be jammed. When viewed from the perspective of the communication system or radar designer, the more serious jamming threats imaginable are formidable indeed, but so is the problem as seen by the developer of jamming systems.

Accordingly, in the future we may expect to see a closer relationship between the design of jamming systems and of sensors and communications systems. Some jammers will be designed to have minimal effects on signals of the types used by one's own forces, with maximal effects on waveforms and in frequency bands used by adversaries. Even one's own signal waveforms, operating in the same bands used by the adversary, may be designed to have a jamming effect on the signals of the adversary while permitting clear transmission; presuming such a technique were to prove feasible. A strong coordination will be exercised between jamming and communications. Techniques making use of time variation of signals and jamming will attempt to exploit the possibilities inherent in short, well timed bursts of information (or of jamming), recognizing that occasional loss of information can be more disruptive than a full-time jamming effort against which an adaptive signaling system could defend itself. Discontinuous jamming also will provide the opportunity to detect and react to actions by the jammed systems, and to assess the actions being taken by the adversary.

Thus, in the future, once it becomes clear that continued evolution of wider band and more powerful signaling systems and jammers is a "no-win game," attention will focus on sophisticated time dependent approaches with strong coupling between the three involved sectors: signalers, disrupters and intelligence collectors.

Getting There

Though evolution inevitably will occur, there exist many inhibitors to changes in the nature of things. One of these, often characteristic of capital-intensive activities, is the economic infeasibility of short-term replacement of equipment and facilities which have been assembled over decades, notwithstanding that those systems which repeatedly have been "patched" always suffer limitations imposed by their original and more limited concepts.

A second inhibitor, particularly to strategic command and control systems but also to many other military systems and to civil air traffic control, is that the existing systems are required to operate on a 24-hour, 365-day basis. This raises the cost of replacement, puts a premium on both time overlap and gradual phase-over from old to new and raises specters in personnel management areas.

Although in some cases the slow evolution of command, control and communications systems reflects changing perceptions of need, often the development is limited by available funding. The mere replacement of an aircraft radio, in the requisite quantities of 10,000 or more, often represents a capital outlay of several hundred million dollars. Unlike the situation with many weapon system upgrades, the upgrading of systems, which (like communication and positioning systems) *by nature interact* between elements of the military force, demands overlapping of old and new capabilities. This is particularly difficult with limited space available and increasingly integrated subsystems.

Strides Needed in Design and Manufacturing

The key to controlling acquisition costs is product volume. In turn, the key to achieving product volume in such items as receiving channels, signal processors, antenna phase steering elements, and encoder/decoders is unity of design and commonalty of use. (The same will apply to some of the new elements proposed earlier.) Achieving appropriate levels of commonalty at both radio frequency and digital interfaces will permit future command, control and communication systems to be assembled from widely used subsystems (at either printed circuit board or semiconductor chip level), rather than having to be designed "from scratch." Provided that these interfaces are laid out with an eye to future technological evolution, they could serve as the basic framework for evolution of technical command, control and communications systems for at least twenty years.

Electronics is the principal hardware basis of these systems. Although electronics manufacturing (in comparison to some other military system manufacturing) enjoys the possibility of a high level of automation, this automation has evolved from high volume computer and consumer electronics requirements. For production quantities in the *hundreds,* rather than hundreds of thousands, automation and the associated quality control techniques are largely inapplicable. The results are higher costs and poorer reliability. The "flexible automation"[1] for mechanical manufacturing may be emulated in electronics; this can bring with it tighter couplings between design and manufacture, less redesign, and a shorter product acquisition cycle.

The 'ilities, Maintenance and Training

In the headlong rush toward technological advance, the only critical operational concern which is considered may be *performance*. In fact, in all technology it is characteristic that issues such as reliability and maintainability (along with their offspring availability) are given secondary consideration until "improved" technology has demonstrated its performance and performance related benefits in actual use.

Command, control and communications systems are especially vulnerable to defects in system reliability, because they operate cooperatively, in pairs or in groups. They are also not only vulnerable to normal deterioration and manufacturing defects; but they also suffer the vagaries of nature (propagation problems) and of hostile interference.

As these systems grow more complex, it will become increasingly difficult to *localize* system problems affecting "end-to-end" system operating qualities. Against this, however, can be arrayed a more powerful analysis capability and adaptivity to compensate. Future command, control and communications systems will, of necessity, contain most *or all* of the error-finding capabilities to tightly localize and work around problems. Hence, from a maintenance viewpoint, they should be less demanding than many other military systems.

The future requirements for personnel training, for complex and sometimes one-of-a-kind systems also will be met by facilities built into the system's equipment, including scenario simulation and the ability to use both parts of a facility's analytical resources and a fraction of its personnel to "game" other resources in a time- and situation-realistic fashion. It seems inevitable

[1]The ability to produce small volumes efficiently and at high quality, using robots and related technology.

that command personnel will interact more directly with the systems than they have in the past, and through their involvement the system and its parts will be able to adapt quickly to changing command styles and organization.

Conclusion

No prospective for the future can be complete, or needs to be totally self-consistent. As cautioned at the start, we have leaned heavily on what appears feasible based on the directions of technology. Our most important predictions, we believe, are those that suggest a much closer relationship of command, control and communications systems to their human users. Not only is this inevitable if humans are to retain overall control, but future system capabilities show great promise for amplifying the contributions of their human resources given this more intimate relationship.

How soon may all of this come about? That is difficult to say. History shows that some predictions, spurred by opportunity or necessity, evolve more rapidly than their prophets would have expected, others remain elusive for technical, economic or social reasons. Meanwhile, yet another group of previously *unpredicted* opportunities continues to emerge.

Project Forecast II—A Glimpse at Tomorrow's C^3I

Lawrence A. Skantze

Project Forecast II, an Air Force effort to find the most promising military technology for the next two decades, has yielded ideas that are expected to revolutionize command, control, communications and intelligence (C^3I). From late summer 1985 through February 1986, 18 panels of technological and operational experts from the Air Force, academia and industry brainstormed more than 2,000 ideas that might spark major increases in U.S. military capability; the field was narrowed down to 70 "keepers."

The Air Force Systems Command will direct 10 percent of its fiscal 1988 science and technology budget, almost $150 million, to kick off development of those 70 Project Forecast II keepers. By the end of FY 92, that investment will jump to 40 percent of the science and technology budget.

Project Forecast II's goal has been to exploit emerging technologies for the U.S. defense. One vital military capability is C^3I—the means by which battlefield commanders manage their resources on land, in the air and in space. Commanders manage those resources in a hostile, hectic environment where they must make quick decisions as they respond to rapidly changing tactical situations.

The Air Force Systems Command's laboratory for C^3I is the Rome Air Development Center (RADC), Griffiss AFB, New York. In 1963, the original Project Forecast thrust RADC into intense research and development in computers, intelligence data handling and radar. In the same way, Forecast

**This article originally appeared in the March 1982 issue of SIGNAL*

II will thrust RADC's research and development (R&D) efforts into artificial intelligence (AI), microwave integrated circuits, optics and other emerging technologies. Those R&D efforts will have a major impact on the center's mission areas of battle information management/command and control (C^2) communications, surveillance and intelligence data processing and handling.

Battle Information Management and C^2

The future field commander will have all of the information and decision aids that he needs at his fingertips to make quick decisions and to manage his resources. High precision sensors supported by high capacity wideband communications will give the commander real time information on the activities of enemy as well as friendly forces. Augmented with intelligence information, detailed geographic data and current doctrine and directives from higher headquarters, the commander will operate more efficiently, with a survivable, enduring, flexible and mobile C^3I system.

The field commander will conduct warfare from the Battle Management Work Station, which will provide an interactive and integrated work environment (Figure 1). The work station will incorporate advanced display technology with spoken commands, comprehensive AI based decision support, high speed simulations for planning and "what if" exercises and a capability to communicate through both audio and visual media.

Initially, the work station will provide a large screen, high resolution, full color display capable of sharply representing routine correspondence, pictures and maps, positioned at the total discretion of the operator. Eventually, just as in the supercockpit, the display medium will use three-dimensional technology to provide the commander with a precise view of the battlefield from any angle.

AI technology will provide direct speech communication between man and work station. It will help the commander to generate and to select targets, options and full up plans; and, at his request, it automatically will produce and will communicate the responsive tasking orders. Automatic blending and correlation of sensor and intelligence data will give the commander real time situation assessment.

Distributed information processing will provide for the survivability, mobility and flexibility of the dispersed computer and data resources necessary to drive the work stations and the other elements of future battle management systems. Functions and data at one node automatically will migrate to other nodes whenever a failure is encountered. Functions and data automatically will disperse even further as more nodes are added to enhance performance. Such redundancy allows us to "fight to the last computer."

Actual fielding of such battle management capabilities will occur with another Project Forecast II system, the Theater Air Warfare C^3I System, which will build on the 21st Century Tactical C^2 System. It will incorporate the work station, knowledge based systems, distributed information pro-

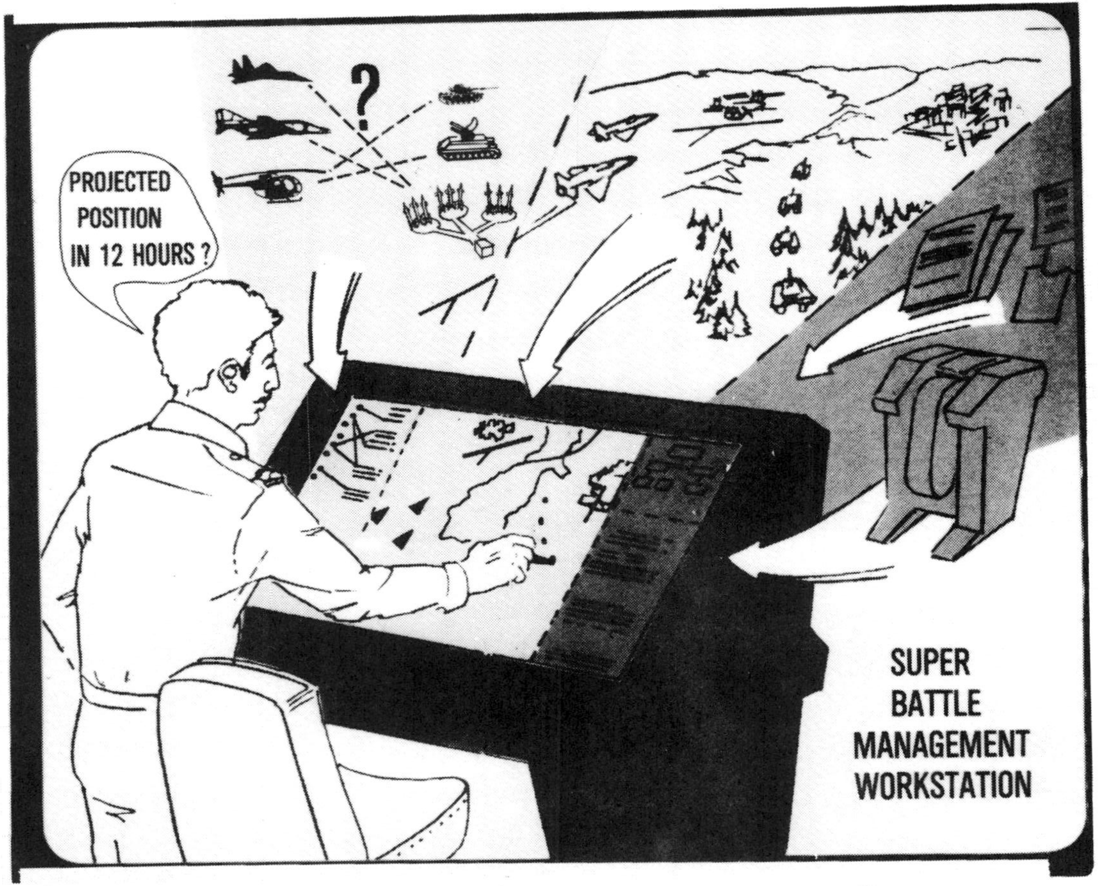

Figure 1. Commanders of the future will use the Battle Management Work Station to reduce information overload as they direct war fighting.

cessing and multimedia survivable communications to produce the most powerful, survivable, flexible and interoperable tactical C^3I system the world has ever known. It should eliminate the dreaded information overload and allow us to manage and to execute a war with a skeleton crew.

In a recent demonstration held in the Battle Information Management laboratory, a simulated war game environment showed how AI technology (automatic fusion and correlation of real time sensor data; decision aids for weapon pairing and resource allocation; AI for producing the air tasking order) can give the commander a quick assessment of battle progress and can help him to plan and to execute courses of action.

Also, RADC has a Mobile Battle Management Demonstrator for testing and demonstrating these capabilities in the field. Project Forecast II will accelerate the development of this technology and will add even more capabilities.

Communications

To maintain control of operational forces during crises and war, vital C^2 functions will have to be dispersed physically and distributed functionally so that no single failure will be catastrophic. The critical factor in achieving this goal is the ability of the supporting communications to hold these functions together in the heat of battle. Unlike communications in the commercial world, military communications must be designed from the ground up to survive all types of enemy attack.

Radio electronic combat is a well established part of Soviet war fighting doctrine and will become even more so as we approach the 21st century. In such a threat-driven environment, it is not surprising that Project Forecast II found dramatic opportunities to push technological advances in U.S. war fighting communications capability.

The next two decades will open up tremendous opportunities for U.S. communications design engineers to take advantage of the significant U.S. lead in microcircuit technology. Microprocessors are becoming as commonplace as transistors were 15 years ago. From the communications engineer's viewpoint, this will mean that communications equipment can be made abundantly smart with built-in, complex algorithms that allow the components to organize into a reactive, self-protecting system that will be virtually impossible to defeat. AI is typical of some of the up and coming "soft" technologies that may allow future networks and radios to react in the face of situations that even the designer did not anticipate.

As a result, in Project Forecast II, the day is coming when communications systems will achieve degrees of adaptability well beyond today's capabilities. Networks will act as living mechanisms with the ability to monitor extensively the performance of component parts, such as radios and switches, in order to give the user the best communications possible. Individual media assets

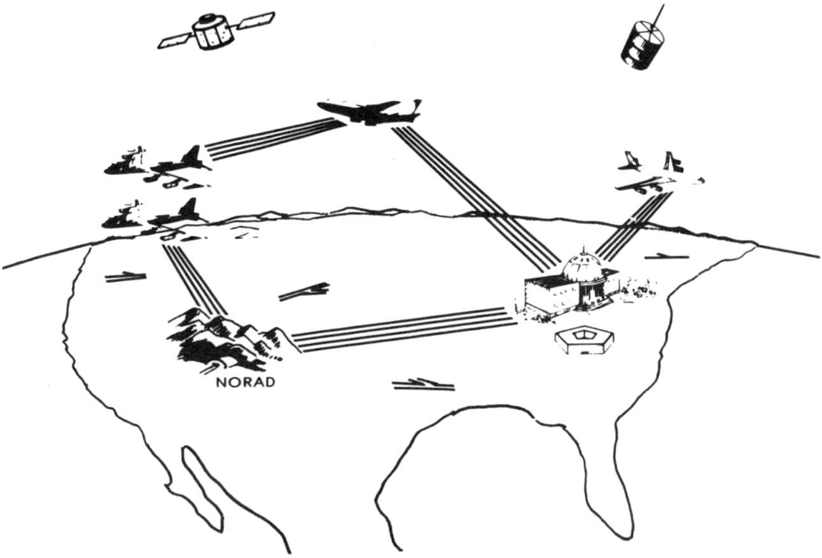

Figure 2. Multi-media node connectivity.

will be multi-mode, using adaptive modulations and coding schemes, which will optimize continuously both channel and user requirements in real time.

Radios and switches can be built to be adaptable, but management will be key. The essential technology in achieving survivable communications will be the ability to understand and to design a multimedia, multi-layer network as well as the structure and functions for a dispersed network resource management operating system (Figure 2). Project Forecast II has focused sharply on this issue.

Great inroads will be made in the area of interoperability. Radios of the future will be able to operate over various spectrums using a variety of waveforms, each chosen to suit the specific conditions. Radios will be able to sense their environment, to discern natural perturbations from enemy jamming and to protect themselves by using different modes of operations, including covert waveforms, or perhaps, stealth types of operation. Network interfaces will be designed so that two interfacing communications components automatically can resolve interface differences rather than having human operators attempt to do it manually as it is done today.

And lastly, networks will be able to go on the offensive by employing deception techniques, which will make the enemy's intelligence activities draw incorrect conclusions about the flow of operational communications.

Surveillance

Long-range, all weather radar systems will continue to provide the "eyes" of future C^3I systems. These radars will operate from the high frequencies

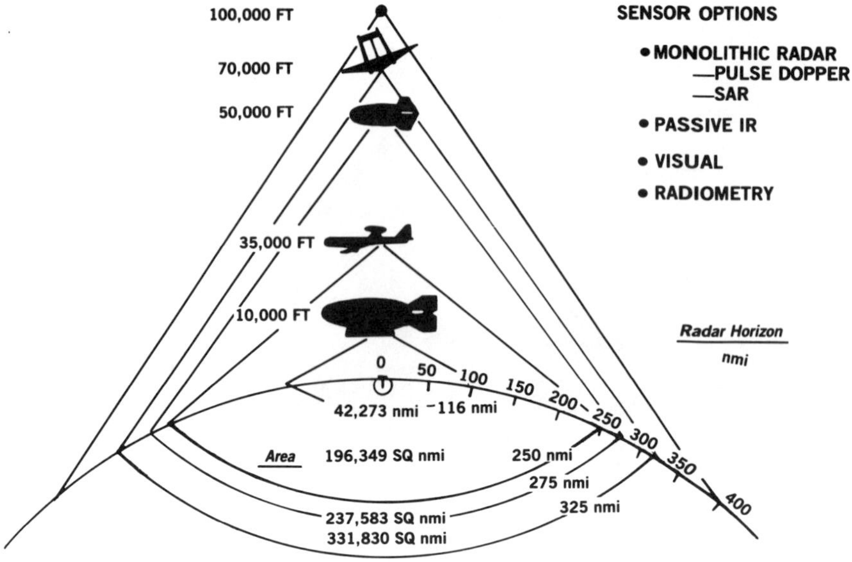

Figure 3. Airborne multispectral surveillance.

through low microwave bands to provide wide area coverage while overcoming weather and other propagation conditions that are more severe at higher frequencies. These future radar eyes will be under ever increasing pressure to survive physical and electronic attack while also satisfying the demand to detect, to track and to identify faster, more maneuverable targets of ever smaller radar cross sections. The cost of these more sophisticated radars will be kept down by using advanced signal processing, monolithic circuit, maintenance and production technologies.

Project Forecast II addresses the fundamental issue of the survivability of future C^3 sensors. Locating the sensors in deep space or at the fringes of the atmosphere provides significant security from sudden physical attack, while enabling sensor coverage of extremely large areas of the Earth's surface and space (Figure 3). Distributing what otherwise might be a monolithic sensor in small craft over large areas, while still providing the ability to operate coherently, significantly boosts survivability. The distributed sensor concept permits convenient growth in large area surveillance, detection and tracking capabilities.

Furthermore, the development of dedicated production facilities to manufacture efficiently the small, identical sensor packages should provide the opportunity to contain the cost of what might be very expensive space radar systems. Sophisticated new technology developed over the next decade will permit the efficient production of these small radars and will permit the extensive communications, distributed control and signal generation, control and processing required of such advanced radars. The long baselines anticipated in these distributed phased array systems offer advantages in secure communications with small clusters of users and in providing precision signal gathering and interpretation capabilities.

Secure, long-term surveillance of tactical battlefields can be achieved by the use of extremely high altitude, long endurance sensor platforms that use remote power and that carry lightweight, electronics packages. The long endurance platforms for future airborne surveillance systems will employ multifunction phased arrays incorporated into the structural skin of the aircraft. The arrays will be capable of reconfiguring themselves to achieve wideband operation and to continue to function when portions of the array fail or are damaged.

Advances in integrated circuits will make possible this modern array technology for radar, communications and signal gathering. These advances will permit the effective integration of microcomputers with analog microwave and optical signal generation and control devices to permit far greater levels of array control and signal processing at the skin of the aircraft. Multifunction phased arrays integrated in the aircraft skin will be tested as part of an Airborne Battle Management Laboratory, wherein the advanced sensors will be evaluated along with new battle management concepts.

The extreme bandwidth promised by today's monolithic circuit technologies should permit the use of ultra wideband surveillance radar systems. Besides being difficult to jam, these ultra-band systems should offer great potential in detecting and in identifying non-cooperatively low cross section aircraft. Key technologies to be developed are signal processing for adaptively providing coherence for the extremely wideband waveforms and wideband antennas and better understanding of the target and clutter characteristics associated with the scattering of these radar waveforms.

Technology Supporting System Development and Acquisition

Several Project Forecast II initiatives that apply to system development and acquisition promise improvements in reliability and survivability at less cost.

The current persistent problems in ensuring adequate attention in design to system reliability, maintainability, producibility and other logistics concerns can be eliminated by an approach called "unified life cycle engineering." Emerging models relating design to the various concerns affected will be coded into modules for computer aided design (CAD) packages. Advanced graphic and optimization tools in the CAD will interact with the designer to generate and to analyze design options until a design is created that satisfies all concerns. The CAD data base will be unified with all other design dependent data bases so that a design change would be acted upon immediately, for example, to update a technical order or parts list.

Automated tools also will provide great improvements in software quality and productivity. The present manpower intensive process of creating and verifying software will be replaced by a process in which software tools, some using the emerging concept of logic programming, will analyze mission software requirements, generate prototype programs, evaluate and debug the coding and manage changes. Programs far larger than present operational software will be created in less time and will be virtually error free (Figure 4).

The need for ever increasing computer power in C^3I systems is addressed in an initiative to provide low cost, high speed computing. This initiative proposes the exploitation of automated design and fabrication techniques to permit the rapid and inexpensive creation of high speed, special purpose computers. It is envisioned that the technology will help to create a computer measuring one cubic foot with the power of several present day supercomputers at less than one-tenth of the cost.

Enabling automatic reconfiguration and reducing excessive maintenance in any system is the purpose of the smart built-in test (BIT). Using AI techniques, smart BIT will discriminate among hard failures, intermittent failures, and false alarms. Since up to 85 percent of failure alerts now are false alarms and since 30 percent of supposedly confirmed failures cannot be duplicated in maintenance, dramatic reductions in maintenance are possible, which translate into higher readiness and lower support costs. Smart BIT will be used by self-reconfiguring systems to avoid unnecessary reconfiguration induced by false alarms.

The processing and memory structure of smart BIT also can be adapted for a concept called marginal checking, which is intended to predict a failure before it occurs. This would be accomplished by determining characteristic signatures preceding failures and by incorporating signature analysis algorithms into smart BIT. Useful signatures may include infrared emissions, radio frequency noise and changes in part parameters or performance. With marginal checking, weakening parts can be replaced during scheduled maintenance, precluding their failure during a mission.

Innovative C^3I Technologies

Fundamental research on innovative technologies relevant to C^3I is absolutely essential to the success of Project Forecast II. Results will be optimized by melding in-house research performed at Air Force laboratories with supporting efforts contracted to universities and industry. Representative of the innovative technologies identified by Project Forecast II for increased attention are integrated photonics, nonlinear optics, ultrastructure materials and the acoustic charge transport.

Integrated photonics. The Air Force Systems Command is planning a major initiative in photonic technology because of the potentially large payoff for future C^3I systems (Figure 5). The objective is to integrate optical systems and devices to replace electrons with photons in a variety of applications. Candidate systems for integration include optical processing/computing, communications, photonic sensors and weapons.

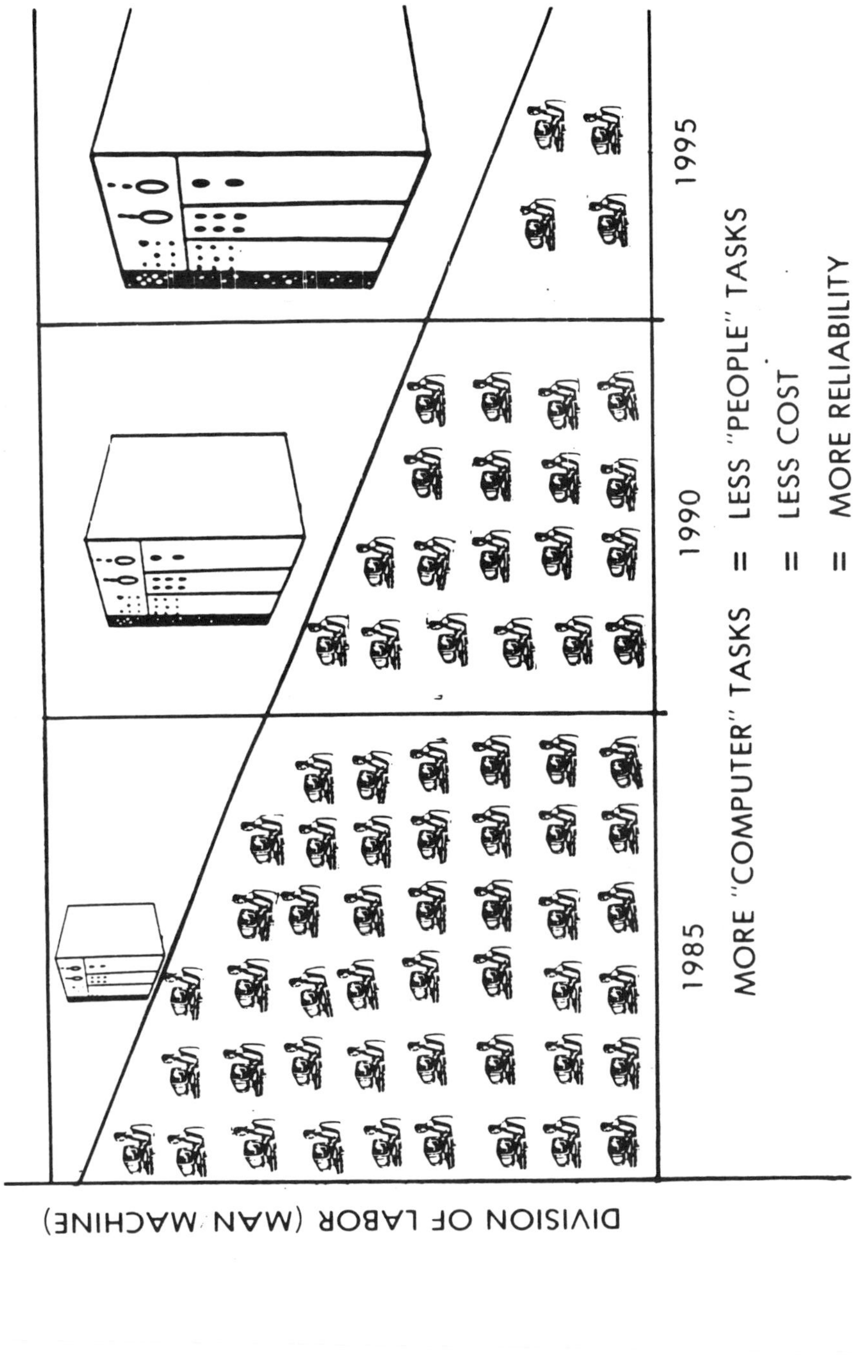

Figure 4. Ultrahigh software quality and productivity.

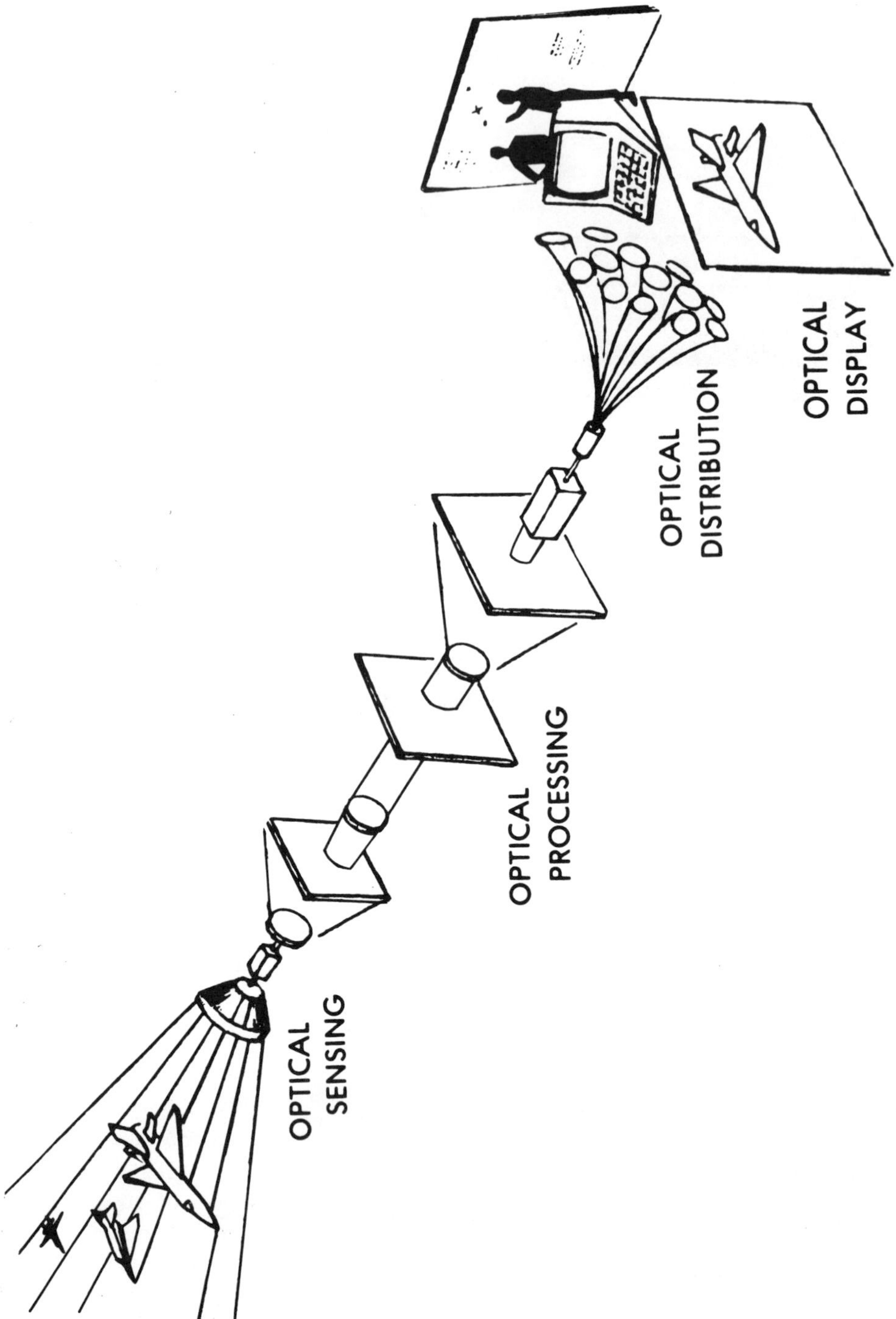

Figure 5. The concept of integrated photonics involves the detection, processing, distribution and display of information using optics rather than electronics.

Digital optical computers promise orders-of-magnitude improvements (100 to 1,000 times) in processing power over current projections of semiconductor electronic computers. Before digital optical computers themselves are developed, prerequisite optical processing systems must be developed (for example, correlators and excisors).

For radars and radios, large-scale integrated electronics would give way to smaller integrated optics systems that consume less power. Optical digital storage systems have demonstrated gigabit throughput rates at terabit storage capacities with projections to the 10-terabit range. Optical analog recording has been extended to 200 MHz information bandwidth with extensions to one GHz capacity. Introduction of holography for storage, multi-dimensional implementation of quick reference tables in signal processing and instantaneous reconfigurations of a signal processor could vastly increase processing capability and adaptability.

Photonic sensors, especially when used in conjunction with active illuminators or adaptive optics, can provide high resolution target recognition information that will be difficult to detect or jam.

Nonlinear optics. Nonlinear optics opens up an entirely new arena for future C^3I systems. The same nonlinear building blocks are used in electronic C^3 systems (for example, harmonic generation, mixers and heterodyning) and can be considered for optically implemented systems. A phenomenon, termed nonlinear optics or optical phase conjugation, has the potential to revolutionize photonic systems. The major issue that ultimately will determine the utility of nonlinear optics to C^3I applications is the field strength required to drive the nonlinear process. The achievement of optical bi-stability and four wave mixing at a low optical threshold will result in widespread use of optical nonlinear elements. Atmospheric communications, radar systems, tracking systems and weapon systems can make significant use of this phenomenon.

Ultra-structured materials. The objective of this far reaching Air Force program is not only to determine the applications of ultrastructured materials for near-term Air Force uses, but also to foster the development of materials engineering at the atomic level. Ordering on this level results directly in a host of new and, in some cases, unanticipated properties in a diverse range of materials including semiconductors, metals, ceramics, glasses and organics. The resulting technology will have a significant impact on practically all of Forecast II's programs.

Acoustic charge transport. Sponsored by the Defense Advanced Research Projects Agency (DARPA), the acoustic charge transport (ACT) device is a new high payoff technology first demonstrated at the University of Illinois in 1982. The operation of ACT devices is based on the transport of charge packets down a gallium arsenide channel due to the piezoelectric interaction of a surface acoustic wave and the gallium arsenide. This technology offers the potential to provide much greater signal processing power in a smaller volume, lower power, lower cost, yet more radiation hardened chip than can be done with other technologies for many demanding system applications.

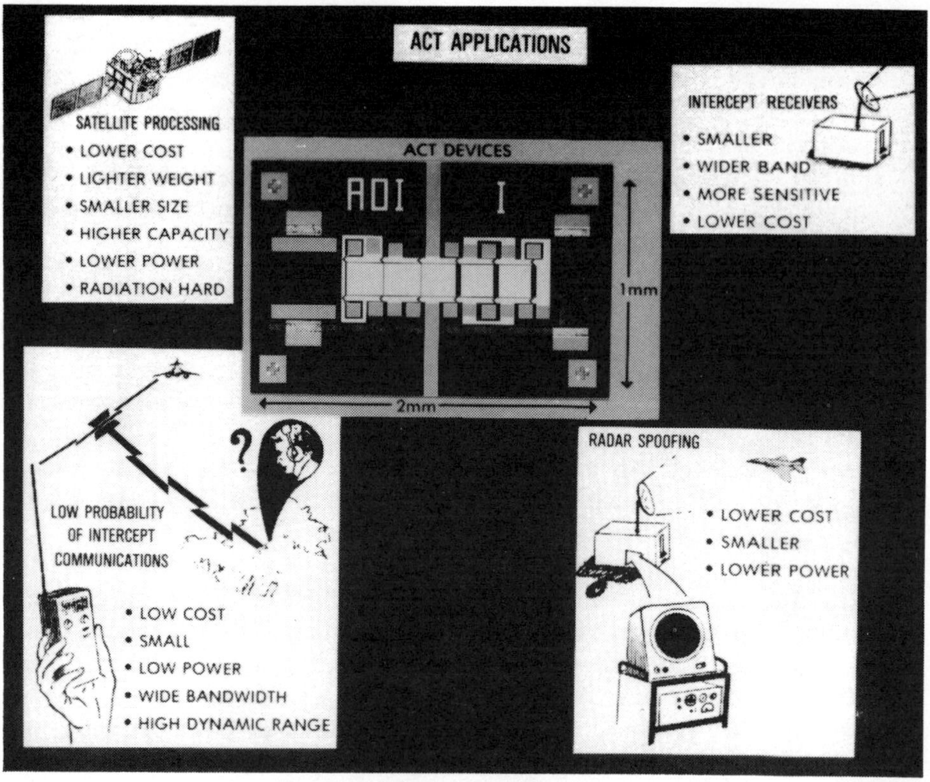

Figure 6. Acoustic charge transport (ACT) applications.

ACT devices will become part of many military systems that require these characteristics of signal processing (Figure 6).

The payoff of this technology will come from two directions: first, new system capabilities will be achieved that are not feasible using any other technology; and second, enhanced performance of current systems via cost, size and power consumption reduction will be achieved. Near-term applications of ACT technology will be in the areas of intercept receivers, radar spoofing for electronic warfare, low probability of intercept communications and satellite processing.

Conclusion

Technological progress will continue unabated. Emerging technologies such as AI, photonics and acoustic charge transport devices, along with continued advancements in software and microelectronics, will result in far greater capabilities in C^3I. Those capabilities will manifest themselves in the battle management work station; in survivable, distributed communications and information processing; and in the ability to detect, track and identify smaller, faster, highly maneuverable targets, regardless of weather or enemy countermeasures. Vastly improved and highly reliable C^3I systems, coupled with new weapon system concepts identified by Project Forecast II, can significantly affect the nature and conduct of air warfare in the next century.

C^2 in the Year 2000 (Remembering the Future)

Robert L. Edge

The tourist who was riding past the government archives building looked at the carving on the front of the building—"What Is Past Is Prologue"—and wanted to know what it meant.

"It means," said the tour guide, "that it's very difficult to remember the future if you can't remember the past."

When Dr. Boyes asked me to write an article on C^2 in the Year 2000, I was reminded, perhaps because Past is Prologue, of the WWMCCS Architecture study conducted by one of our leading U.S. contractors about twelve years ago. The purpose of the study was to try to determine what "life style" our National Command Authorities expected in their World Wide Military Command Control System (WWMCCS). The study method was remarkable, and worth spending a few paragraphs describing.

The key members of the study team constructed a "straw man" architecture. They then proceeded to brief this "straw man" to every senior leader in the Department of Defense who would sit still for an hour to listen to it. The study team very carefully recorded the salient questions and comments from those it briefed.

The team used its notes, together with the results of research and study to update the "straw man." They then proceeded to repeat the briefing circuit, once again recording salient questions and comments. After a few circuits, the team synthesized its results into a written report. The report, itself, was very interesting. If unconstrained by the real world (technological and monetary) considerations, the architecture would have sounded very much like a description of the television networks' capabilities in covering major sport-

This article was written especially for* ***Principles of Command and Control.*

ing events. Although the report didn't say it this way, the senior leaders in the DOD clearly had been conditioned by the television networks to want the following:

—"Big Picture" coverage, in near real time, of anything interesting, anywhere in the world
—"Zoom" capability to expand the detail, to whatever level of detail the viewer wishes, in areas of interest, also in near real time
—"Living color" pictures with high fidelity sound
—"Play by play" commentary, to explain what the viewer sees
—"Color commentary" by informed insiders, to explain the rules, to explain application of the rules, and to explain the significance of events
—Box scores over selectable time periods
—Relevant historical monographs interspersed at appropriate points in the coverage
—Live, face-to-face, candid interviews with the stars (on both the winning and losing teams), immediately before, during, and after major events
—Expert predictions of outcomes of individual events, and predictions of summaries
—Previews of coming events, supported by realistic simulations

Of course, no one yet knows how to do all of this. But this is not meant to be critical of our leaders for wanting it. Indeed, their aspirations for WWMCCS should not be constrained by technological and monetary considerations. Which reminds me of the time in the early '60s when one of our leading fighter pilots was asked to specify the operational requirements for the next generation of fighter aircraft for air-to-air combat. He responded that he wanted a fighter which could take off with an infinite load of ammunition, using no runway, climb to infinite altitude in zero time, cruise at any speed he chose, be able to detect and identify his enemy at any range, and have a perfect kill probability.

He was told by the technologist who was his interlocutor that such a fighter was not technically achievable; that there are certain laws of physics that can't be changed; that, besides, the available technology cannot achieve everything which is theoretically physically achievable; and that it would take our entire gross national product to try to produce even a single aircraft.

The leading fighter pilot was totally unperturbed. He said, "I'm a fighter pilot, not a technologist or an economist. If you want a technology or economic estimate, ask someone else. But I've given you the operational requirement. And if you ask me again in twenty years, the requirement will remain the same." Indeed, if the WWMCCS Architecture study were to be revisited in the year 2000, the unconstrained requirement likely would remain pretty much the same. The major differences, if any, would be prompted only by changes in imagination, stimulated by then current media tools and science fiction.

What then, about the real-world C^2 systems of the year 2000? What will they attempt to do? What will they look like? We will tackle these questions one at a time.

What Will C² Systems Try to Do in the Year 2000?

At the top level of abstraction, C² systems in the year 2000 will try to do exactly what C² systems try to do in the year 1987—and have tried to do since the beginning of recorded history: to command and control assigned forces, in the environment in which those forces must operate in order to be effective. Since it now seems to take from ten to twenty years, from concept to field implementation, to introduce new major weapons, the environment in the year 2000 can be projected from those programs which are now in development or production. Aircraft (ours and theirs) will fly faster and be more difficult to detect and to track; ships (ours and theirs) will sail faster; submarines (ours and theirs) will be quieter; tanks (ours and theirs) will drive faster. Weapons (ours and theirs) will be more lethal and each round will be more costly. The price of lack of control will be more costly, both in terms of unintended provocation at the start, and in terms of inventory depletion later. More detailed, more quantitative discussion of the forces and weapons of the year 2000 are properly the topic of another article; the key point for this article is that the need for effective command and control will not decrease—indeed, the stresses on command and control will increase. "Effective" systems today may not be effective systems in the year 2000.

Before proceeding further in discussion of "What will C² systems try to do?" we must distinguish two major categories of C² systems. The first type deals with events which are, in some important sense, determinate. For example, an Air Defense Ground Environment deals with physical things which are forced to obey well understood laws of physics. Aircraft can travel no faster than a known finite limit; they can change direction only at a known, finite rate; effective bombers or cruise missiles must be at least as large as some known size, and their radar, infrared, optical and sonic signatures must be at least as intense as some known factors. The ground environment must be "closed loop," capable of completing a control loop in near real time; hence it is usually dedicated to a single purpose. This type of system is more properly thought of as a "weapon control system" than as a "command and control system;" unfortunately, our lexicon does not often enough make this distinction, and confusion in thinking about command and control sometimes results from this lack.

The second type of system is at least one level higher in terms of abstraction; it may extend throughout the spectrum from initial pre-engagement planning, through full resource management and accounting; it may attempt to recycle planning, execution and direction, using the identical system for the full spectrum of combat and management support. These types of systems, by the nature of their attempted mission, usually cannot operate in near real time. They are not "closed loop" in the usual sense of that term in control system theory; in fact, attempting to use them in a closed loop usually results in great instability—the control signals quickly get out of synchronization with the feedback loop—and the system becomes ineffective. These systems do not necessarily submit to the laws of physics; the law of man usually must be considered by them, with indeterminate results.

The key differences, then, are in determinism, in real timeness, in closed loopness, and in narrowness of function. Today these key differences are accounted for, doctrinally, by the policy of "centralized direction and decentralized execution." This policy means that, within the control envelope of doctrine, rules of engagement and standard operating procedures, as much decision-making is delegated as can prudently be delegated. In an extreme example of decentralized execution, no central command center—not even the platoon commander—attempts to peer over the sights of the rifleman to be sure he is on the right target before allowing the rifleman to squeeze his trigger. In my opinion, the changes in force structure and weapons will increase the thirst for central control; but the time line requirements will generally force greater emphasis than today on decentralized execution. It is quite natural for a confident commander to want to centralize all decision making to his level, since he must believe that he is more competent than his subordinates; but the same commander will want no decision-making centralized above his level, since his superiors cannot possibly understand the situation as well as he does. One of the hallmarks of a good commander is the self discipline to delegate decision-making to the lowest level where sufficient information and competence are available.

What Will The C^2 Systems of the Year 2000 Look Like?

This question reminds me of Heisenberg's Uncertainty Principle in quantum mechanics. In simple language, the principle says that if there are two related variables, then the more you know about one of them the less you can know about the other; for instance, if the position of some particle is related to time, the more accurately you determine the particle's position, the greater is your uncertainty about when it was there—and vice versa. Heisenberg's principle deals with past and current observations. When we are asked to remember the future, something most of us haven't yet learned to do, then there must be even some greater principle of uncertainty which comes into play.

So in trying to remember the year 2000, I will avoid trying to be too specific about *what* for the specific year 2000; give me a tolerance of minus ten or plus twenty years if I say specifically *what;* give me a wide tolerance in *what* if you tie me down to a specific year.

Part of the uncertainty arises from the technological uncertainty—there is some theoretical technological upper limit, moving with time as particular technologies ripen. Another part of the uncertainty arises from programmatic uncertainty—given technical feasibility, what we will be willing and able to do.

There are others who are better qualified than I to remember the future of technology; all I will say about the technology is that the best qualified minds will *under*estimate what the technology will be *able* to do in the year 2000; and those same minds will *over*estimate (by as much as 20 years) the degree to which we actually *will exploit* available technology for C^2 by the year 2000.

I will try to focus on the programmatic future of C^2. Even in programmatics, however, we are reduced to guessing. We have already stated that a theoretical limit is imposed by the technology, but that we can't forecast with any accuracy when various technologies will ripen. There is an even more immediate and tighter limit to programmatics, however; that is the political limit. At a macro level, the amount of money spent on C^2 probably will never exceed some relatively small fraction of the total defense budget—say 10%. Short of general war, it is not likely that the total defense budget will exceed some percentage of the total national budget; and the total national budget will be limited by the Gross National Product. So we really need a projection of the GNP and a projection of the perceived need for defense, for the entire period between now and the year 2000, in order to be able to estimate how many dollars are theoretically available for C^2 during the period.

I can't emphasize too much the importance of the political perception. I can remember in 1979, giving a speech to a gathering of naval officers in a friendly South Pacific nation; they were "poor mouthing" their country's ability to acquire "modern" C^2 tools: their total defense budget at that time was about $2 billion per year, and they had thousands of miles of coastline to defend. I asked them to look at the small nation of Israel; to compare their GNP with that of Israel (theirs was much larger) to compare populations (theirs was much larger); and to compare their defense budgets (Israel's was larger). I then challenged them to explain the striking difference in priority which the two nations seemed to attach to national defense. It didn't take long for us to pinpoint the key: Israel saw its defense forces as vital to its national survival against a real and present threat; the South Pacific nation either did not see a real and present threat or was reluctant, for political reasons, to identify it. We have this difference in perception, today, within the U.S. Some segment of the population believes that the primary purpose of the national government is to provide for the national defense, and that there is a real and present threat; this segment supports a steady, strong national defense budget. Other segments of the population may agree that the primary purpose of the national government is to provide for the national defense, but they see no real and present threat; these would support a national defense budget, but at a lower level. Still other segments would eliminate (unilaterally, if need be) any U.S. defense forces, as their way of encouraging peace in the world.

I cannot remember any future event or collection of events which is likely to erase any of these segments. At the most, short of general war, changes in opinion seem to occur only at the margins: Only a few who do not see a real and present threat this year are likely to believe in it ten years from now. So I see continued irresolution or ambiguity or shifting in priority about our national defense forces and the budgets provided to them. This will pose, I believe, one of the greater challenges to our military leaders and to the civilians to whom they report—for defense in general, and for C^2 in particular.

After you have made your best guess about absolute defense budgets between now and the year 2000, you need to estimate how many dollars are needed to "buy out the program"—to complete the acquisition of programs already in acquisition or programmed for acquisition. And how many dollars are needed to support the operation and maintenance of fielded C^2 systems during the period. Subtracting these "buy out" and O&M dollars from the total dollars available leaves you with some figure which might be available for new initiatives. These new initiatives, in my judgment, will occur mainly in dedicated weapon control systems (the types which are deterministic, near real time, and closed loop); these *must* be funded in order to support new weapon systems. Probably very little will remain for other kinds of initiatives; instead, these kinds of initiatives may require funding from savings in O&M. This may sound more difficult than it really is.

With some optimistic assumptions about what can be done during the next five years, much more can be achieved, at less cost, than many might now imagine. For example, we might make the following optimistic assumptions:

The Joint Chiefs of Staff (JCS) will be able to convince *all* of our civilian leaders (both elected and appointed) that a C^2 system does not make a commander. We then can dispense with any notion that military commanders can be bypassed in command and control if only we can devise the perfect C^2 system. Having reached this pragmatic position, a more practical approach to C^2 system architecture is possible. The net result will be to make more real progress, faster and at less cost than otherwise.

I was once told by a senior civilian in the Office of the Secretary of Defense that he was as well qualified as Commander in Chief, Strategic Air Command (CINCSAC) to command a nuclear exchange. His reasoning was that CINCSAC had never commanded a nuclear exchange, and so any intelligent person, using SAC's command control system, was equally qualified to do so. What he didn't care to admit was that CINCSAC is intimately involved in the planning for such an exchange; that exercises, in which CINCSAC participates, are conducted often enough to establish confidence in the viability of the plans and the operational procedures which support them, and to expose the types of questions which are vital during execution of the plan; and that the Congress confirmed CINCSAC in his position based on his military experience, the military judgment and intuition which were developed from that experience and their direct applicability to the responsibilities of the Command. Even for the highly pre-structured conduct of a nuclear exchange, the key ingredients—military judgment and intuition—cannot be programmed. For the less structured form of combat—often called "conventional warfare"—the opportunities for military judgment and intuition come more frequently and, probably, over a longer period of engagement. People can learn from their experiences; successful people always learn from experiences—theirs as well as that of others. Maybe we can teach computers to learn easily—some more advanced kind of intelligence than is implied by "rule-based" systems; if so, DOD will be very careful to insure that the computers are not exposed to tutorials by the enemy. How one can arrange this duality—easy learning from us, resistance to teaching by our enemies—in the case of systems where learning might be most advantageous, is an interesting topic for a doctoral thesis, at least.

The Department of Defense will fully appreciate the clear difference between a *control* system and a *command and control* system—not the semantic difference, but the physical difference. The essential discriminator is, of course, whether the scenarios for use of the C^2 system are highly deterministic or highly non-deterministic. The extent to which realistic scenarios can be specified, *a priori,* is dependent on the degree of determinism of the mission which the system is expected to perform. Unless the mission is very deterministic, the scenarios must be very general in order to cover the full area of realistic probabilities. And, given a generalized scenario, specific application program software must be generalized, as well. This means that we can successfully acquire *control* systems from industry, usually as adjuncts to weapons systems which the control systems are meant to control. It also means that we should not try to acquire *command* and control systems from industry—instead, we should try to acquire more generalized tools which industry can provide.

The Department of Defense will have convinced itself that it is not possible to acquire a *command* and control system from industry. We have tried, many times, to do this; never have we succeeded. The reasons should not be mysterious, if we understand what such a system really is: A collection of standard operating procedures, people who perform the procedures, and equipment and facilities to support the people. The standard operating procedures must spring from the doctrine and force structure; these factors are not generally well understood by industry—no reason that they should be; so asking industry to develop the standard operating procedures, simply because computers will be used to assist in implementing them, has got to be a stupid thing to do. *Tools* to assist in the functions of command and control *can* be acquired from industry; some of the tools can be used in commercial off-the-shelf form; some of these must be tailored to fit a particular command. Some tools must be developed *ad hoc*. Many tools are best developed by the user, himself, using the software equivalent of "machine tools" developed by industry. And, having acquired an initial set of tools, they must be updated from time to time, to provide a continuous fit as the mission and/or the organization changes; in addition, the technology of the tools (and "machine tools" equivalents) is continually upgraded by the industry. Organizations with command and control systems must acquire the ability to keep their tool kits current—and either by in-house modification, commercial update of "machine tools," contractor augmentation of in-house work force or some combination of these.

The Department of Defense will recognize that the planning function in C^2 requires a different set of tools than is required for weapons control. Many of the tools are mundane, commercial off-the-shelf, such as word processors, electronic spread sheets, modern data base management and graphics systems. Confusing planning functions with near real time control functions leads only to architectural complexity and dysfunction; instead, the command and control system architecture will provide for highly functional (probably at a relatively high functional level) interface among planning tools and between the planning function and the control function.

The JCS will be able to convince itself and our civilian leaders that "interoperability requirements" never should stand as a naked term; the term always must be "required degree of interoperability." Just a little thought is necessary to reason that we never will be able to afford to throw away all of our old communications equipment, in a short time period, in order to replace it with new and modern equipment; if we insist on "complete transparency"—which most uninitiated think of when they say "interoperability"—then we probably deny important functions to the new equipment, or make it pay heavy penalties in cube, weight, power and cost. If we, instead, define the *necessary degrees* of interoperability, and force only that subset onto the new equipment, then we can considerably reduce the penalties without sacrificing necessary functionality. Similarly, different missions require differences in equipment; to take an extreme example, a clandestine mission may require a very small satellite antenna, with single channel burst transmission to make detection more difficult; whereas a major trunk subscriber of the Defense Satellite Communications System is driven by economy, and will require high power levels and different transmission protocols. The clandestine terminal and the major trunk terminal cannot easily be made transparent to each other, but necessary degrees of interoperability can be provided. Once we come to terms with the pragmatic approach, it will be a relatively simple and straightforward task to arrange the necessary degrees of communications interoperability among our own communications systems, among our Services, and with our Allies. But communications interoperability, in the technical sense, is not sufficient; unless the operators can arrange operational (doctrinal, procedural and language) interoperability, then any technical interoperability will be wasted.

The Joint Tactical Command, Control and Communications Agency (JTC^3A)—the DOD agency chartered to worry about technical interoperability—cannot arrange the operational interoperability; the operational commands must arrange this themselves. For example if the U.S. Army and the U.S. Air Force must interoperate, then they must *operationally* arrange at least a communications gateway between the two forces. This is often as simple as exchanging communications equipment and a few hours of training, plus the burden of carrying an "alien" communications package into battle; the professional communicators know how to do this if the operators will bring them into the operational planning cycle in time to contribute.

The Department of Defense will recognize that sensor systems, each exploiting a different phenomenology and usually with a unique mission segment, must be designed to fit the mission segment and the available technology. The forward communication (for tasking of the sensors) may be relatively uniform across sensor systems; but the rearward communication will remain relatively unique to the phenomenology and the events being reported. Hence gateways must be exploited in order to provide effective interface with other (often older) systems.

The Department of Defense will come to recognize that gateways offer optimum solutions to many communications problems. So long as the technologists continue to invent new communications systems—using higher portions of the electromagnetic spectrum, more powerful coding, or more sophisticated modulation—and so long as we are unable to replace rapidly large segments of our imbedded investment in communication—that is how long we must look to gateways to provide interoperability between the old and the new. In addition, there are older systems which were acquired without interoperability with some other specific communications systems in mind; gateways offer the only practical way to provide interoperability between or among such older systems. Some critics abhor gateways; they treat gateways merely as evidence that the Services, and separate missions within the Services, are *unwilling* to agree to common standards for their communications needs. It is, perhaps, easy to excuse generalists, who do not really understand communications in a technical sense, for believing such criticisms, especially when they are fed to them by professed experts. But it is difficult to have patience with the professed expert/critics, or to accept their sincerity. What they are really saying is that we should have only a single Service with a single Mission—that there is no difference between land warfare, naval warfare, or air warfare; or that there is no difference between submarine warfare and surface warfare. Certainly all of these forms can be related; but they drive communications requirements in strongly different directions, as any serious student quickly learns. These same professed expert/critics accuse the DOD of concern with obtaining the latest technology at the expense of interoperability. Presumably, then, DOD should standardize on something simple, like jungle drums. Ignore how easy it would be to jam these; ignore the short communications distance (or the very high number of relays necessary); ignore the relatively manpower-intensive system that would result; ignore the very slow information transfer rate (near zero) that would result; and ignore that *standardizing,* even at this basic level, *would not provide interoperability*. We still would need to know at what time to listen and what time to transmit, agree on protocols and codes and we would need to agree to do these things *before* an operation begins.

The Department of Defense will recognize that Artificial Intelligence (AI) (at least "rule based" programs) can provide useful tools in C^2, for deterministic scenarios, especially. But for unforeseen events, or for unforeseen sequences of events, the "rule book" probably will not fit; and except for the first round of fighting, the detailed rules are bound to change. DOD will look very carefully for places where Artificial Intelligence can be effectively used, but will not delude itself that Artificial Intelligence is as good as human intelligence in cases where judgment, imagination, or versatility are required.

DOD will not act as if a computer with artificial intelligence can solve problems which we do not know how to manually parse; the computer, at least for the foreseeable future, using the very best technology which is imaginable today, cannot solve a problem for which we do not already know the proper set of questions.

The Department of Defense will recognize the importance of parallel processing to C^2 and Intelligence. Much of the monitoring function is concerned more with pattern recognition, and "What's new?" at any instant of time, integrated over a defined period of time, than any other question. Parallel processing should be especially capable of supporting "What's new?" processing. Once the potential importance is recognized, then greater priority will be assigned to developing the necessary software development tools to maximally exploit parallel processing for "What's new?" processing. Similarly, advances in rule-based systems will be highly dependent upon parallel processing, and upon necessary software development tools to maximally exploit parallel processing to support inference engines.

Like interoperability, data security will get more pragmatic treatment than it is presently getting. The Department of Defense will learn to specify data security, and especially multi-level security, pragmatically. Internally consistent scenarios will be used in order to derive data security requirements. For example, data security will be looked at as a proper subset of Operations Security (OpSec). Physical security, Communications Security (ComSec), Tempest and Computer Security (CompuSec) will all be looked at together, using a single set of scenarios.

For example, after a nuclear attack Tempest will not be considered a major vulnerability, any more than one worries about freezing on the Serengeti Plain during an elephant stampede. That is not to say that data security should be minimized; on the contrary, strong pragmatic measures must be implemented in order to support effective distributed data bases in peacetime, if for no other reason. And computers do tend to concentrate non-sensitive data into sensitive collections of information; and—if the systems are user-friendly—make it relatively easy to get at the information for purposes good or ill. So practical measures will be taken to control access to the data.

The DOD will learn how to exploit distributed data bases as an important adjunct for command center survivability. With potent weapons possessing high kill probabilities, probably the single most effective survivability measure for C^2 is hidden backup. The hiding can come about literally, by concealment, or through position uncertainty (from a targeteer's point of view); that is, the intelligence cycle plus the targeting and firing cycle takes so long that an effective attack on a moving command center cannot be mounted, affordably. The chief problem with this concept is that the backup must, effectively, come out of hiding when called upon to back up its principal command center; and if its function must survive, then it must be backed up by a hidden center, etc. If these redundant centers are very large, the cost quickly becomes prohibitive; one must keep the redundant centers relatively small in order to be able to provide sufficient multiplicity for effective survivability. The smallness implies limited functionality; limited functionality in a single cell implies multiple cooperative cells, each capable of playing a number of different roles in the cooperative team effort. Combining this with the need to survive communications nodes and connecting links (not just the C^2 nodes), we must attempt to conserve communications bandwidth. One important way to do this is by using distributed data bases; but the concepts still need to be completely thought through, so that a high multiplicity system, with high effectiveness and survivability at affordable cost can be designed and implemented.

The Department of Defense will have defended successfully against well-meaning initiatives to force bilingualism upon the Military Departments. One of the key arguments which will have led to the successful defense is the criticality of command and control, and the difficulty of clear and unambiguous commands, even using a controlled subset of a single ethnic language. Recognition that lack of interoperability stems more from operational deficiency than from technical deficiency, and recognizing the vital importance of a common language to operational interoperability, will carry the day. In addition, the very high cost of reproducing all of the command and control supporting data in two languages will have proved to be compelling.

As electronic warfare becomes more complex and more pervasive, the Department of Defense and the Services will come to recognize the singular importance of the electronic battle and its C^2. Command and control of the electronic battle will be centralized to a greater extent than that of the conventional battle (and to a greater extent than at present). Decisions to monitor, jam or physically attack will be made centrally, assisted by optimizing algorithms. Joint doctrine will be developed carefully to attempt to cope when the central electronic warfare command center is inoperative or communications with it is severed. Rule-based system tools will play an important part in executing the doctrine.

Command and control of Low Intensity Conflict provides an exception to "decentralized execution." Because it is sharply focused, it can sometimes be successfully controlled centrally; the television network sport coverage has created a pervasive notion of feasibility of such central execution. Because it may be "the only game in town," the desire to control it centrally may well be overpowering. Hence, the JCS will specify some measure of capability for central control of Low Intensity Conflict; but the OJCS will prevent this requirement from seriously damaging the command and control systems, in general.

Summary and Conclusion

There will be changes in C^2 by the year 2000. The changes will be motivated by changes in the environment—force structure, etc.—and limited by technology and political support. Limitations can be minimized by greater pragmatism, and by better communication of that pragmatism to the Congress and to other civilians. The C^2 challenges are not just technical; the operational challenges—and congruent opportunities for improvement—are at least as great as the technical challenges, and can be met without significant increased financial support or technological progress. We optimists assume that the operational challenges will be met.

Command and Control in the Years 2000+

Eberhardt Rechtin

It helps in projecting command and control (C^2) ahead 15–20 years to look at the past. Command and control history is one of steady evolution interspersed with abrupt advances. Such an advance is in the making—the arrival of wide area (global), near-continuous surveillance as a critical factor in effective command and control. The effects may be as great as those from the addition of global navigation in the 1400s and long distance communication in the late 1800s. Global surveillance is already affecting deployed systems and operations. It may affect the structure of the chain of command in the early 2000s. This advance, like those before, is the result of a new technology—in this case, space based surveillance systems.

This would not be the first time that space dramatically changed military operations. Long range navigation was made possible by the use of the stars and worldwide communication by the ionosphere. Each in their time expanded the geographic interests of nations—their political spheres of influence and the range of their military operations. Not surprisingly, space navigation and communication systems have been integrated into operations quickly and easily. But, as evolutionary improvements to existing functions, they have not mandated a change in previously established principles of command and control.

Wide area surveillance, however, adds a new dimension. From the beginning of the space age thirty years ago, it was apparent that the high altitudes of space systems could give them unprecedented fields of view. Since then, space systems have been built and demonstrated for radio, radar, infrared,

This article was written especially for* ***Principles of Command and Control.*

light, X-ray and gamma ray surveillance. The main applications, in addition to scientific, have been strategic—intelligence, warning, arms control treaty verification and deterrence. The primary impact of space surveillance on C^2 has been on command and control of the nuclear single integrated operational plan (SIOP) forces; that is, on pre-SIOP deterrent posture more than C^2 dynamics. But that focus is now changing as concerns develop regarding post-SIOP endurance and battle management of strategic defense (SDI BM/C^3). Resolving those concerns will require near-continuous global surveillance.

A comparably important impact is likely in *tactical* command and control with the deployment of wide area surveillance of air weather, ocean weather, ships, planes and land targets. For all practical purposes, there was little wide area surveillance prior to the advent of space systems. The closest approximation had been time-limited local observations from surface or air platforms relayed over relatively unreliable communication links. This tactical situation was both a blessing and a curse. It was a blessing because one usually could accomplish surveillance while hiding one's main force over the enemy's horizon; it was a curse because the enemy could do likewise and attack from ambush.

Space systems, limited only by their numbers, can and will change all of that. Surveillance can be near-continuous. There will be no easy place to hide. The implications for tactical operations are just beginning to be assimilated and exploited by both the Allies and the Warsaw Pact nations. The pace is likely to increase when it is realized that present tactical surveillance systems are but primitive examples of things to come in the next decades, well within the lifetimes of present Service inventories of ships, planes and armies.

In principle, it is a short step from the present space systems to those that could image almost continuously whole armies and naval battle groups and could considerably increase the effectiveness of submarines and other low observable weapons platforms.

This is not to say that all nations will incorporate tactical wide area surveillance at the same rate. The U.S. and Soviet navies, for example, face different military situations. The U.S. need to maintain sea control requires major surface fleets, the surveillance of which is clearly important to the Soviets. And, indeed, Soviet wide area surveillance of the U.S. Navy has been underway for decades. On the other hand, the Soviet surface fleet would not yet seem to demand intensive U.S. space surveillance; a greater U.S. concern instead would seem to be surveillance of Soviet land based naval aviation, a much tougher problem for which progress has necessarily been slower. Both countries are expected to be interested in ocean weather (oceanography) for submarine and anti-submarine warfare purposes.

It does not take much foresight to predict that wide area surveillance will be commonplace in the years 2000+. There is more than enough motivation for it. Future confrontations, barring complications, are likely to be brief, intense and costly both financially and politically. Responses will have to be detailed and timely thousands of miles away. The side with the best surveillance information clearly will have a considerable advantage.

The trend is there. Initial signs of it were present in the Arab-Israeli wars in the late 1960s and 1970s, the Mayaguez incident, the Falkland Islands war, the Central American wars, Chernobyl and recent submarine accidents. The leaders of nations wanted precise, complete information *now,* anywhere on the globe. Recently, when the capabilities of U.S. surveillance systems were temporarily reduced by launch vehicle failures, considerable concern was expressed over the possible consequences to national security.

The principal difference in wide area surveillance now and in the years 2000+ is in investment; that is, in the number of satellites, in the degree of sophistication in information processing on board and on the ground, and in the distribution of results to the users. But these investments are quite likely, given the realization that the massive inventory of ships, planes and armies of the world could be held hostage to space surveillance in a relatively short time. The debate is not whether this will happen, but how soon.

Given that wide area surveillance will be commonplace early in the next century, what are the likely consequences to the principles of command and control? To answer that question, one first must project what wide area surveillance systems are likely to be like in the early 2000s.

Wide Area Surveillance in the Early 2000s

Presently available wide area surveillance is already remarkable in sensor capabilities—sensitivity, resolution, field of view, size, weight, orbital lifetime and coverage of the electromagnetic spectrum. From the commander's point of view, however, present surveillance information is too infrequent, too slow in arriving and often too voluminous (at least in present form). The first problem is solved by more satellites in orbit. The second and third problems are solved by faster and more intelligent information processing. In particular, more information processing must be done earlier in the chain from sensor to end user; that is, more on board processing, earlier comparison of related information, etc. Considering how the human sensors and brain work, a vast improvement should be possible, and considering the rate of improvement in computing capability, efficient handling of huge amounts of data is not likely to be a limitation.

A legitimate concern does exist over the survivability of surveillance systems—a concern that is inherent in any system that must be in line of sight to its target. This is not a new problem, nor are the solutions new—warning, speed, height, maneuvering, hardening, low observability, cover, deception, tactics and not least, uncertainty on the part of the enemy of the merits of attacking the surveillance system itself. Those problems no doubt will be addressed in due course.

The critical surveillance advance which will change the principles of command and control is *continuity and field of view*. The immediate effect is to extend the combat horizon drastically.

Extending the Combat Horizon

As noted earlier, the great changes in command and control in the past, and today, have been associated with great changes in the combat horizon. Global navigation permitted efficient strike at a distant target, provided that the target did not change much while traveling to it. Global communication permitted better control of the forces in transit and at the target, provided that the forces could assess the local situatión. Long range navigation resulted in navies operating autonomously, empowered with great discretionary authority, cut off not only from government control but also from that old nemesis—shore control. Though long range communication subsequently reduced that autonomy, the distant commander on the spot with organic surveillance assets was still seen to be in a better position to command than anyone else. In any case, the only weapons that could be brought to bear effectively were those of that distant commander. That situation prevailed well into World War II.

Long range aircraft (land and sea based) extended the combat horizon appreciably. But, in general, the principles of command and control remained much the same. Aircraft fleets were like ship fleets: once over the horizon they were relatively autonomous. They contained their own surveillance and their own weapons. Report back was limited by concern over being detected en route. As with navies, aircraft fleets, once engaged with the enemy, were in command of their own battle.

However, World War II demonstrated one of the earliest types of effective wide area surveillance—communications intelligence—the interception of communications and the cracking of the codes intended to make them secure. The same ionosphere that made long distance communication possible made remote interception possible. The same technology that made codes possible made code cracking possible. The results were dramatic—insight into enemy operations deep within enemy territory, ambushing of ships at sea, defeat of submarine wolf packs, effective deceptions and the victory of Alamein. Not the least of the results was much more effective centralized command and control. More centralized control came about for a variety of reasons. Communications intelligence collection by frontline combat groups was limited at best. Analysis of the collected communication typically required expertise available only at central headquarters, and use of the results was limited by concern over revealing sources and methods.

Integrating this new surveillance into C^2 was not always successful. Some commanders openly distrusted it (*ULTRA Goes to War,* Ronald Lewin, McGraw Hill, N.Y. 1978). But without question, it dramatically increased the nature and effectiveness of the command and control of submarine and antisubmarine warfare. This mating of wide area surveillance, a stealthy weapon platform and near real time remote C^2 may be seen by future historians as the real beginning of this next command and control era.

Conversely, the lack of continuous, wide area photographic surveillance by the Germans helped make possible the great Allied deception of the phantom Allied army prior to the invasion of Normandy. The lack of timely, wide area surveillance by the Americans made possible Pearl Harbor, and its lack by the Japanese, Midway (*And I Was There,* Edwin T. Layton, Wm. Morrow & Co., Inc., N.Y. 1985).

World War II also saw the first of the long range ballistic missiles, a weapon that now can reach anywhere in the world in less than an hour—a time comparable to the reputed 15 minute time constant of tactical battles. But lack of precise and timely targeting information has kept them from effective tactical use so far. Using continuous, remote wide area surveillance to provide updates and on board sensors to provide final targeting could change that situation considerably.

The 1950s saw the development of data distribution and display systems, such as the Naval Tactical Data System (NTDS) which gave commanders a continuous, wide area view of the local combat zone. Later developments increased the radius of his field of view to more than 600 miles. All that was needed to display still more was wide area surveillance. The potential for that was demonstrated in the 1960s and 1970s with satellites.

The advent of space based, wide area surveillance by the early 2000s thus adds the final and crucial element to a world in which:

- There are no longer any horizons to provide concealment.
- There are no sanctuaries out of reach of global weapons.
- The battle zone (the zone containing all combatants) is global.
- The battle commander may be anywhere in the world.
- Timely and accurate information will be critical, leading to what is often called "the information war" as the combatants struggle to obtain essential information and deny it to their enemies.

Consequences to the Principles of Command and Control

Probably the most important consequence of wide area surveillance to C^2 will be a further diminution, if not a sharp drop, in the autonomous authority of the commander on the spot. Others, elsewhere, will know more, be less vulnerable to capture or defeat, be able to call on more resources (communications, surveillance, analysis, weapons, counterattacks) and be in a better position to affect the outcome of the battle.

In a very real sense, there will not be a single "spot" to be *on* when the battle zone is global. Being on a single "spot" is too dangerous—fixed targets can hardly be called survivable. The top commander, therefore, must be both everywhere (global) and nowhere (mobile and not identifiable to an enemy). Concentration of weapon platforms will have to be rare. Instead, there will be concentrations of weapons on target, launched from many points. Ideally, there would be no "nodes" in one's force structure; in practice, the force structure will have to be configured such that single nodes are not critical. In communications, this means alternate routes and dual or triple homing. In a chain of command, this means skip echelon.

Nor will there be a clear front line, a forward edge of the battle area (FEBA) or rear echelons.

As for C^2, it is worth recalling that the present chain of command structure was invented for a world where there was a relatively protected central government, radially expanding logistics and communications, and combat units at the outer periphery.

The new world can be quite different. We might expect a new C^2 structure to be different as well. A self-managing network with fewer levels perhaps.

Though for somewhat different reasons than survival in the new global battle zone, some C^2 changes began occurring in the Vietnam War—a war of international political scope, rapid communications and good battle zone surveillance. Washington increasingly became involved in real time battle management. The formal chain of command was at times severely stressed. The last war in which the commander on the spot was autonomous—or believed he was and should be—probably was MacArthur's Korean War.

Predictably, there will be considerable resistance to modifying a command structure that has been around for so long that it seems self-evident. It may take the imperatives of war to force structural change.

A far less controversial development will be the defensive counter to wide area surveillance. Hiding techniques—cover, concealment, deception, "stealth," low observables, low probability of intercept, secure communications, distributed inventory, undersea and underground operations—will be further refined and developed out of necessity. Not the least to require such protection will be the commander in chief. Not only must his whereabouts be unknown, but also he must be able to command from wherever he is. Headquarters, in other words, will not be a place, it will be wherever the commander happens to be. And wherever that is, there must be access to all required information and to all required command channels. Technically, that means a more or less uniform communication network of many access points, each of which provides access, in turn, to the command structure, surveillance systems, weapon systems and political structure.

Huge headquarters staffs can no longer be physically collocated with the commander; every movement of such staffs will be too visible. The staff will have to be distributed, with communications tying it together. Somehow, everyone will have to be on the move most of the time. Alternatively, no single individual or group must be an attractive target. It is by no means clear how to accomplish this for such headquarters support as intelligence analysis, logistics coordination and communications control with their associated information processing facilities.

The resulting command configuration, however, does have a precedent, though at a much lower level. It is the configuration of a guerilla or underground organization—designed, not coincidentally, to hide from all-seeing surveillance.

Weapons and their platforms will have to be concealed at all times; three will be no rear areas where they can be stockpiled in the open. The problems and opportunities to accomplish this are evident in today's submarines, stealth aircraft, low radar cross-section missiles, mines and mobile strategic missiles. These systems present special problems to command and control. How can they receive commands and respond without giving themselves away? How can they operate in conjunction with other forces without the knowledge of those forces—and hence possible destruction by them?

The earliest scale model of a C^2 structure built for this new world of surveillance and concealment is likely to be that for modern attack submarines with their low probability of intercept (LPI) communications and consequent access to remote surveillance assets. The most difficult C^2 structures to construct probably will be those for C^2 of low observable aircraft and strategic defense. Accomplishing survivable C^2 in a world of extraordinary surveillance will be no easy task.

Conclusion

The advent of global surveillance in the early 2000s will mean far more to the principles of command and control than simply the addition of a new capability to an existing force structure. By completing the macrosystem of global communications, global weapons and global information processing systems, it changes the foundations upon which many of the principles of today's C^2 were built.

About the Authors

Dr. Stephen J. Andriole is a Professor of Information Technology at George Mason University's Department of Information Systems and Systems Engineering. He is also the President of International Information Systems, Inc. and former Director of the Defense Adanced Research Projects Agency's (DARPA) Cybernetics Technology Office. Dr. Andiole has written and edited numerous books including *Interactive Computer-Based Systems Design and Development*, *Applications in Artificial Intelligence* and *High Technology Initiatives in C^3I*. Dr. Andriole received his Ph.D. from the University of Maryland in 1974.

Dr. Walter R. Beam is Visiting Professor of Systems Engineering at George Mason University in Fairfax, Virginia. He has been involved in electron device and computer design and applications since 1952, has been a researcher at both RCA's and IBM's central research labs and an executive at IBM and at Sperry. He served as Deputy for Advanced Technology (1974–81), Office of the Assistant Secretary (R&D), U.S. Air Force. From 1977–81, he had oversight responsibility for Air Force command, control and communications systems programs. He was a member of the Air Force Scientific Advisory Board from 1981 through 1986.

Dr. Roger A. Beaumont is a history professor at Texas A&M University and the North American editor of *Defense Analysis*. He taught history at the University of Wisconsin, Marquette University and Kansas State University and has lectured at the Industrial College of the Armed Forces, the U.S.D.A. Graduate School, the Institute for Defense Analysis and the U.S. Army Institute for Personnel and Resource Management. Author of more than 50 articles and books including *Military Elites* and *Sword of the Raj: The British Army in India, 1747–1947,* and co-editor of *War in the Next Decade,* he has been published in *The Naval War College Review, Parameters, Aerospace Historian* and *Military Review,* among others.

Harlan H. Black is with the U.S. Army Communications Electronics Command (CECOM), COM/ADP Center, where he conducts research on man-machine interfaces. As a member of the CECOM/COM/ADP's Computer Research Division, he has conducted research in tactical symbology, adaptive interfaces and C^3 graphics. As a former member of the CECOM Ada Division he conducted research on Ada development and maintenance and developed the first audio-visual demonstration of the Ada Language System.

Dr. Bruce G. Blair is a senior fellow in the Foreign Policy Studies Program at the Brookings Institution. A specialist in nuclear operations policy, he was formerly a Minuteman launch control officer and a project director for a congressional Office of Technology Assessment study of nuclear command performance. He is the co-editor for an American Academy of Arts and Sciences report, *Crisis Stability and Nuclear War* (Oxford University Press, 1987). His 1985 Brookings book, *Strategic Command and Control: Redefining the Nuclear Threat,* won the 1986 Edgar S. Furniss National Security Book Award. Dr. Blair earned a Ph.D. in operations research at Yale University.

Anthony G. Bohannan, a former senior officer in the British Army, was Assistant Marketing Director of Marconi Secure Radio Systems at the time his article was written. At Marconi, he had special responsibilities for NATO, and was Manager of the Communications Systems Group. Prior to this, he was Director of Defense Operational Requirements Staff in U.K. MOD.

Dr. Jon L. Boyes, Vice Admiral, USN (Ret.), is the International President of the Armed Forces Communications and Electronics Association (AFCEA). Retiring from the Navy after a distinguished thirty-four year career, Admiral Boyes has experience in destroyers and submarines and in the technical and operational command, control and communications fields. His educational background includes graduate studies in communications and computer science, a masters degree in international law and political science and a Ph.D. in international affairs from the University of Maryland.

Commander William J. Brennan, USN, was Director of the Platform Design Division of the Navy Space and Warfare Command at the time his article was written.

Dr. Dennis M. Buede, has been an independent consultant since 1982, applying the modeling techniques of decision analysis to decision making problems. He has an extensive background in decision analysis, probabilistic analyses, systems analysis and operations research. Dr. Buede has participated in the design and/or development of several decision/expert support systems. These DSS/ESS have been in the domains of project prioritization, planning, program management, situation assessment and training management.

Colonel Alan D. Campen, USAF (Ret.), is Vice President of C^3 and Information Systems at The BDM Corportation. He is a former Director of C^2 Policy in the Office of the Secretary of Defense. Key assignments in a 25-year Air Force career included Commander, DCA-Europe; Chief of Staff, AFCS; Commander UK Communications Region; and DCS C-E, Third Air Force. He served as President of the London Chapter and Vice President of the European Region of AFCEA, and is currently on the *SIGNAL* Magazine Advisory Board.

Commander Keith P. Collard, USN (Ret.), is the head of the Command and Control Department of Systems Exploration, Inc. in San Diego, California.

His Navy operational background has included tours as CIC officer on an aircraft carrier and the C^2 officer on the staff of Commander Carrier Group 5/Commander Task Force 77 during the height of the Vietnam war. While on active duty he was designated a proven subspecialist in C^2. For the past 14 years he has been directly involved in the systems engineering and integration of Navy afloat C^3I and combat systems.

Admiral William J. Crowe, Jr., USN, is the eleventh Chairman of the Joint Chiefs of Staff. Admiral Crowe graduated from the U.S. Naval Academy at Annapolis in 1946 and served in submarines. He received a master's degree in education from Stanford University and a Ph.D. in politics from Princeton University. Prior to his appointment as the nation's senior military officer, he served as the Commander in chief of U.S. Forces in the Pacific. Other significant assignments have included tours as the NATO Commander in Southern Europe, the Deputy Chief of Naval Operations for Plans, Policy and Operations and the Commander of the Middle East Force in the Persian Gulf.

Dr. J. Bryan Danese is a senior staff engineer in SDI studies at Martin Marietta Denver Aerospace (MMDA) in Denver, Colorado. His work at MMDA has included target and background phenomenology, SDI sensor, discrimination and BM/C^3 analyses. He is a former member of the research staff at Riverside Research Institute in Arlington, Virginia, where he concentrated on BM/C^3 for SDI. As a systems engineer at the MITRE Corporation, he covered BM/C^3 for BMD, strategic C^3 and survivability, and strategic and tactical sensors. He is a member of the APS and SPIE and holds a Ph.D. in physics from the University of Florida (1973).

Dr. John T. Dockery is a senior analyst with the Organization of the Joint Chiefs of Staff specializing in tactical command and control problems. He is a physicist by education, and has extensive experience with operations research for the Army. At the SHAPE Technical Centre he led a group of international scientists on research into C^2 analysis methods. Dr. Dockery has written numerous papers on C^2 analysis, and was an early user of fuzzy sets. He has been programming computers since 1959, and is deeply involved in C^2 simulation.

General Russell E. Dougherty, USAF (Ret.), is an attorney of counsel with McGuire, Woods, Battle and Boothe. At the time of his retirement from the U.S. Air Force in 1977, he was the Commander in Chief of the Strategic Air Command and the director of U.S. Strategic Target Planning. Previously, General Dougherty served as the Chief of Staff of NATO's Allied Command in Europe. For six years he was the Executive Director of the Air Force Association and Publisher of *Air Force* Magazine. He is a graduate of the Law School of the University of Louisville and the National War College.

Major (P) Robert Driscoll, USA, is the Assistant Political Advisor and Speech Writer at Headquarters, USAREUR. He is licensed as a professional engineer and has a Ph.D. in international studies from the American University in Washington, D.C. He earned a master's degree from the University of Konstanz as an Olmstead Scholar. Major Driscoll is a graduate of CGSC, and was an instructor in the Department of Social Sciences at the U.S. Military Academy. In his last assignment, he served as the Executive Officer of the 79th Engineer Battalion in Karlsruhe.

Major General Robert L. Edge, USAF (Ret.), is an active consultant with the U.S. defense industry. He retired from the position of Assistant Chief of Staff, Communications and Computer Resources, Headquarters, U.S. Air Force, in 1977. Prior, he had been Deputy Chief of Staff, HQ AFCS, and then Director of Command, Control and Communications, HQ USAF. In all three of these positions, he had responsibilities for developing, defining and/or supporting Air Force requirements for what is now called C^3. He is also a command pilot with over 6,000 flying hours in various operational commands.

Ronald W. Foss is Principal Systems Engineer for the Computer Systems Division of Unisys Defense Operations in St. Paul, Minnesota. He has participated extensively in the definition of advanced architectures for future tactical C^2 systems, and is currently responsible for programs to prototype elements of advanced C^2 systems. He has 15 years of Air Force and industry experience in the definition, development and support of C^3I and avionics systems including 407L, 412L, WWMCCS, JTIDS and GPS programs. He holds B.E.E and M.S. degrees from the University of Minnesota.

Dr. Jude E. Franklin is Vice President of Research and Development at Planning Research Corporation. He directs research in advanced computer science and artificial intelligence. Prior to this, Dr. Franklin started and managed the Navy Center for Applied Research in Artificial Intelligence at the Naval Research Laboratory. He was Vice President and Division Director of Applied Engineering at MAR Inc., where he supervised research and development in computer science, communications, underwater acoustics signal processing and hydrodynamics. Dr. Franklin has managed several programs that delivered ASW systems for Naval ships and submarines.

Peter G. Freck is an Associate Technical Director of The MITRE Corporation's Battlefield Systems Division and is responsible for the Division's support to the U.S. Army and the Office of the Assistant Secretary of Defense (C^3I). A graduate of the aeronautical engineering program at Princeton University and of the Harvard Business School, his experience spans thirty years with extensive involvement in studies regarding operational, technical and policy issues. Prior to joining MITRE, Mr. Freck was Deputy Director of the Systems Evaluation Division of the Institute for Defense Analyses.

General Robert T. Herres, USAF, is the Vice Chairman of the Joint Chiefs of Staff. He began his flying career in fighter-interceptors and served as Chief of the Flight Crew Division with the Manned Orbiting Laboratory Program. He has commanded a SAC bomb wing, an air refueling wing, the Air Force

Communications Command and the Eighth Air Force. General Herres subsequently served as Director for Command, Control and Communications Systems in the Organization of the Joint Chiefs of Staff and most recently as Commander in Chief of the North American Aerospace Defense Command and the U.S. Space Command.

Dr. Gerald W. Hopple is a Senior Scientist at International Information Systems in Marshall, Virginia. He is the author of more than ten books and a number of published articles on psychology and computing, the design and development of decision support systems, methodologies for defense problem solving, artificial intelligence, and command and control decision aiding. Dr. Hopple is currently Co-Editor of the Series on Decision Support Systems, QED Information Sciences. He received his Ph.D. from the University of Maryland (College Park) in 1975.

Major General John Paul Hyde, USAF, is Deputy Director for Defense-Wide C^3 Systems, Organization of the Joint Chiefs of Staff. His last four preceding assignments were: Assistant Deputy Directory for Plans at Headquarters, Defense Communications Agency; Commander of AFCC's Tactical Communications Division; Commander of the European Communications Division; and Deputy Chief of Staff for Communications and Computer Systems at Headquarters, NORAD and Space Command, and Commander of the Space Communications Division. He is a graduate of the University of Cincinnati (B.S.E.E.), the University of Pittsburgh (M.S.E.E., Ph.D.), the Air Command and Staff College and the Air War College.

David R. Israel is the Director of the Special Projects Office at the Defense Communications Agency. Prior to this, he was Deputy Director for Planning and Systems Integration at DCA. In February 1979 as the WWMCCS Systems Engineer, Mr. Israel was a Special Assistant to the Deputy Under Secretary of Defense for Research and Engineering (C^3I). Previous positions include Deputy Associate Administrator for Engineering and Development with the Federal Aviation Administration and Deputy Director for Engineering of the Defense Communications Planning Group of the Department of Defense. Mr. Israel did his undergraduate work at the Massachusetts Institute of Technology.

Lieutenant Colonel Charles H. Jundt, USAF (Ret.), has been with The BDM Corporation since his retirement from the military in 1977. Presently, he is Manager for Military Capabilities Analysis, applying an historical analysis of military force development processes to project future force composition. He has specialized in Warsaw Pact and NATO C^3 and the degradation to those C^3 capabilities caused by C^3 countermeasures. Colonel Jundt has performed threat analyses of NATO and PACOM nuclear forces, and provided C^3 countermeasures impact studies for the Army Command and Control System, the Rapid Deployment Joint Task Force (Now USCENTCOM), U.S. Strategic Forces and the AWACS.

Commander Charles E. Kesson II, USN, is a staff officer in the Contingency Support Division, C^3 Systems Directorate Organization of the Joint Chiefs of Staff.

Donald W. Kroening is currently serving as Chief of the Command and Control Analysis Division, Studies and Analysis Directorate, TRADOC Analysis Command at Fort Leavenworth, Kansas. He is the former Deputy Director of the C^3I Directorate, Combined Arms Combat Developments Activity (CACDA). Mr. Kroening is a Major in the Kansas Army National Guard, currently serving on the staff of the 35th Infantry Division (Mechanized). He is also a member of the Adjunct Faculty at Kansas City Kansas Community College, where he teaches introductory computer science courses. Mr. Kroening holds advanced degrees from Kansas State University (Computer Science) and from the University of Missouri (Business Administration).

Donald C. Latham is the Assistant Secretary of Defense (Command, Control, Communications and Intelligence) in the Office of the Secretary of Defense. Previously, he was Division Vice President, Engineering, RCA Government Systems Division. Mr. Latham has also served Martin Marietta Aerospace as Manager, Advanced Programs; Manager, Air and Ballistic Missile Defense Systems; and Director, Research and Engineering, at the Orlando, Florida facility. He received a B.S. degree from the Citadel, and M.S. and advanced electrical engineering degrees from the University of Arizona.

General William J. Livsey, USA, is the Commander in Chief, United Nations Command/Combined Forces Command, and Commanding General, United States Forces Korea/Eighth U.S. Army. He was commissioned as a second lieutenant in the Infantry in June 1952. During his 35 years of active duty, General Livsey's overseas assignments include tours in Vietnam, Germany, and two previous tours in the Republic of Korea. His most recent assignments include Commanding General of the U.S. Army Infantry Center, Fort Benning, Georgia; Commanding General, Eighth Infantry Division and Commanding General VII Corps, Germany; and Deputy Commanding General, U.S. Army Forces Command and Commanding General, Third U.S. Army.

Lieutenant General Clarence E. McKnight, Jr., USA, is the Director for C^3 Systems, Organization of the Joint Chiefs of Staff. His communications experience includes tactical, strategic, systems engineering and R&D assignments. He has had four tours in Europe, plus tours in India, Panama, El Salvador, Korea and Vietnam. He has served as Commandant of the signal School and, prior to his present assignment, was Commanding General, U.S. Army Communications Command, a global responsbility with 1400 installations, 14 countries and 33,000 personnel.

Commander James H. Offutt, USN, is the Assistant Director for Systems Analysis at the Office of the Secretary of Defense, Strategic Defense Initiative Organization. An NROTC midshipman, he was commissioned in 1966 and received his pilot wings in May 1967. His aviation tours have included VS-21 onboard the USS *Kearsarge,* VA-196 onboard the USS *Enterprise,* VS-41 as a flight instructor for the start of the S-3A training and VS-37 onboard the USS *Constellation.* In 1978, Commander Offutt served as the Executive Assistant to the Director, Total Force Programming, on the Chief of Naval Operations' Staff. Subsequently, he served as Assistant Head of Training Program Development. In 1981, Commander Offutt joined the Office of Naval Research (ONR), Western Regional Office. He became Commanding Officer in February 1982 and, later, Officer in Charge, ONRDET, Pasadena, California.

General Glenn K. Otis, USA, has held a wide variety of command and staff assignments. He has commanded in mechanized infantry, armored and cavalry units from platoon through division. He has had four tours in Germany, two in Korea and one in Vietnam. He has also served as an instructor at West Point; the Army's Deputy Chief of Staff for Operations and Plans; Deputy Commander of both the Army's Armor Center and Combat Development Activity; the Commander of the United States Army Training and Doctrine Command. In April 1983, General Otis assumed command of the United States Army Europe, and Central Army Group (NATO).

Dr. James D. Palmer is BDM International Professor of Information Technology at George Mason University. His current research interests include investigation of macro-enhancement technologies to improve software productivity, application of expert systems to various C^3 systems such as open battle management architectures, and graphics for man-machine interface technology. He has managed programs in industry in the design and development of systems employing C^3 centers; has served in the U.S. Department of Transportation as Administrator, Research and Special Programs Administration; and has held many research, faculty and administrative positions in higher education.

Dr. George H. Quester is Chairman, Department of Government and Politics at the University of Maryland. Professor Quester's areas of specialization include military strategy and arms control, American foreign policy and international politics. Prior to coming to the University of Maryland, he taught at Cornell University, the National War College, UCLA and Harvard University. He is the author of *The Future of Nuclear Deterrence, American Foreign Policy: The Lost Consensus, Offense and Defense in the International System* and *The Politics of Nuclear Proliferation.* Professor Quester received his Ph. D. from Harvard University.

Dr. Eberhardt Rechtin is President and Chief Executive Officer of The Aerospace Corporation. He previously held the positions of Assistant Secretary of Defense (Telecommunications) and the Director of the Advanced Research Projects Agency. He also has served as the Chairman of the Naval Studies Board of the National Research Council and as Chairman of the Chief of Naval Operations' Industrial Advisory Committee on Telecommunications.

W. Robert Reeves was Senior Analyst and Branch Manager for Special Projects at Delex Systems Incorporated at the time his article was written.

Captain William J. Ruhe, USN (Ret.), is Editor of the *Submarine Review* and is a consultant on naval matters. Although a generalist in naval warfare with a particular focus on submarine and missile warfare, he has recognized the major role played by electronic warfare systems in changing the character of warfare at sea. As one of the initial (prior to World War II) radar specialists of the U.S. Navy, he has held electronics staff and planning jobs throughout his naval career. After retirement from the Navy, he was General Dynamics' Corporate Director for Marine Programs, again relating EW systems to the many naval programs pursued by General Dynamics.

Dr. Andrew P. Sage is First American Bank Professor of Information Technology and Dean of the School of Information Technology and Engineering at George Mason University. He has been a faculty member at the University of Arizona, the University of Florida and Southern Methodist University (SMU). At SMU, he was the Head of the Electrical Engineering Department and Director of the Information and Control Sciences Center. From 1974 through 1984, he was the Lawrence R. Quarles Professor of Systems Engineering at the University of Virginia. Dr. Sage is the editor of *IEEE Transactions on Systems, Man and Cybernetics, AUTOMATICA,* and *Large Scale Systems.*

Dr. Randall P. Shumaker is the Director of the Navy Center for Applied Research in Artificial Intelligence at the Naval Research Laboratory. His previous assignments include management of aircraft technology programs at the Naval Air Systems Command and research group supervision at the Naval Air Engineering Center. Dr. Shumaker has a broad background in computer science and digital electronics and has taught and published in the fields of digital control and artificial intelligence technology for the past ten years.

General Lawrence A. Skantze, USAF, is Commander of the Air Force Systems Command at Andrews Air Force Base, Maryland. He directs the research, development, test and acquisition of aerospace systems for Air Force operational and support commands. General Skantze earned a B.S. degree in engineering from the U.S. Naval Academy in 1952, and a master's degree in nuclear engineering from the Air Force Institute of Technology in 1959. A command pilot, General Skantze has been the Commander of Aeronautical Systems Division, Deputy Chief of Staff for Research, Development and Acquisition at Air Force headquarters, and Air Force Vice Chief of Staff.

Frank M. Snyder is on the Operations faculty of the U.S. Naval War College, where he is Head of the Planning and Decision Making Division and also teaches an elective course on command and control. As a U.S. Naval officer, he served at sea in nine ships and for three years was the Communications Officer of the U.S. 6th Fleet in the Mediterranean. Ashore, he served for ten years in Naval communications planning and operations assignments including an assignment as the Navy's Command, Control and Communications Architect. He has a B.S. degree from the U.S. Naval Academy and an M.A. degree from Stanford University.

John R. Thompson is with the Artificial Intelligence and Decision Aids Division of Science Applications International Corporation in Albuquerque, New Mexico. He has been the principal investigator on several Army studies of the cognitive processes in military intelligence and tactical planning. Results of his research have gone into revising field manuals and training programs for intelligence analysts and automated C^2 planning aids. He has been involved in the application of AI technology to battlefield sensor data fusion and real time advisory systems. He is currently involved in the design of expert systems to support survivability/vulnerability prediction and the development of intelligent interfaces for tactical message handling systems.

Edward L. Waltz manages the Identification Systems Section of the Air Traffic Control/Identification Friend or Foe Department of the Bendix Communications Division. He is responsible for data fusion developments for tactical fighter aircraft and shipboard tactical surveillance systems. He has managed conceptual development studies and large scale fusion simulation efforts directed toward the Advanced Tactical Fighter as well as full scale hardward development of the Navy's AN/UPX-30(V) shipboard multisensor identification system. Currently, he is directing the development of advanced data fusion applications using artificial intelligence techniques in integrated avionics architectures.

Captain R.E. Ward, USN (Ret.), is an independent consultant specializing in Navy C^3. Formerly, he was a Vice President of MAR, Inc., where he was involved in a wide range of requirements analysis and operational concept development tasks. He spent many years in destroyers serving in the full range of operational and command assignments. Shore tours include Naval communications stations, electronics support and engineering activities and assignment to the Office of the Chief of Naval Operations (OPNAV) and to the J6 Division of CINCPAC. He is a graduate of the Naval War College and George Washington and Tufts Universities.

Colonel Joe B. Warren, Jr., USAF, is Chief of the Contingency Support Division, Directorate for Command, Control and Communications Systems (J-6), Office of the Joint Chiefs of Staff (JCS). Commissioned through the Reserve Officer Training Corps (ROTC) in 1965, Colonel Warren's career has included numerous assignments to USAF tactical communications units and Headquarters, U.S. Air Force.

David J. Welch is the Director of the Command Systems Division of R&D Associates in Tacoma, Washington. He has held technical staff positions and program management positions at McDonnell Douglas, the Aerospace Corporation and RDA during his 23-year career. In these positions, Mr. Welch was involved in a wide range of systems analysis and design studies. In 1978, he was assigned to RDA's Munich Office where he participated in studies of U.S. and NATO command systems. He was subsequently assigned to RDA's Tacoma Office as a Deputy Program Manager on the Distribution Command and Control System. He became the cognizant Division Manager in 1985.

C. Richard Whelan is President of Automatary, Inc., a small consulting firm specializing in market research and business planning in the aerospace and communications fields. He directs market research studies for Frost and Sullivan, Inc. in military markets including SDI and space defense, C^3I, artificial intelligence, optical computing, reconnaissance and surveillance, military satellite communications, TEMPEST equipment and federal government telecommunications. He also conducts market consulting assignments for industry clients. He is the author of the *Guide to Military Space Programs* and co-author of the *Guide to the Strategic Defense Initiative*.

General H.F. Zeiner-Gundersen, Norwegian Army (Ret.), is a part-time general consultant. He is a graduate of the Norwegian Military Academy, Norwegian Prewar Artillery Staff College, British Army Staff College (1948) and NATO Defence College (1959). General Zeiner-Gundersen served with the Norwegian Field Artillery through World War II. Thereafter, he served in various command and staff appointments including Commander, Norwegian Army Staff College; Chief of Staff to the Commander in Chief, Norwegian Army; Commander Trondelag District; Chief of Staff to Chief of Defence Norway; Commander, North Norway; Chief of Defence, Norway; and Chairman NATO Military Committee. General Zeiner-Gundersen retired from the services in 1980 after a 44-year military career.